MATHEMATICS FOR BIOSCIENCES

From Theory to Worked Examples and Applications

MATHEMATICS FOR BIOSCIENCES

From Theory to Worked Examples and Applications

Elspeth F. Garman • Nicola Laurieri

University of Oxford, UK

World Scientific

NEW JERSEY • LONDON • SINGAPORE • BEIJING • SHANGHAI • TAIPEI • CHENNAI

Published by

World Scientific Publishing Europe Ltd.
57 Shelton Street, Covent Garden, London WC2H 9HE
Head office: 5 Toh Tuck Link, Singapore 596224
USA office: 27 Warren Street, Suite 401-402, Hackensack, NJ 07601

Library of Congress Cataloging-in-Publication Data
Names: Garman, Elspeth F. author | Laurieri, Nicola author.
Title: Mathematics for biosciences : from theory to worked examples and applications / Elspeth F. Garman, Nicola Laurieri, University of Oxford, UK.
Description: New Jersey : World Scientific, 2025. | Includes index.
Identifiers: LCCN 2024023916 | ISBN 9781800616059 ebook for individuals | ISBN 9781800616042 ebook for institutions | ISBN 9781800616080 paperback | ISBN 9781800616035 hardcover
Subjects: LCSH: Biomathematics | Life sciences--Mathematics
Classification: LCC QH323.5 .G375 2025 | DDC 570.1/51--dc23/eng/20240827
LC record available at https://lccn.loc.gov/2024023916

British Library Cataloguing-in-Publication Data
A catalogue record for this book is available from the British Library.

For any available supplementary material, please visit
https://www.worldscientific.com/worldscibooks/10.1142/Q0473#t=suppl

Desk Editors: Aanand Jayaraman/Rosie Williamson/Shi Ying Koe

Typeset by Stallion Press
Email: enquiries@stallionpress.com

Foreword

Question: What do you do when you see scores of students struggling in their upper-level Biochemistry courses because of inadequate preparation in Mathematics? These are bright, capable students whose A-Levels in Biology or Chemistry were first rate but whose study of Mathematics did not include any calculus, and for many their Mathematics education had all but stopped after the completion of their GCSE.

Answer: If you are Elspeth Garman, you spend several months thinking about all the Mathematics that an undergraduate will need for a degree in Biochemistry, you design a syllabus that will develop this material in a single term, and you call the course "Friendly Maths for Biochemists."

I first met Elspeth when we found ourselves in the same small group during a parents' meeting at the school in Oxford where our daughters were both enrolled. I was on sabbatical at the mathematical Institute that year, and this parents' meeting was an early chance for me to do something that was not job-related. Over the rest of that year our two families became close. Since then, when my work has brought me back to Oxford I have usually been able to spend some time with Elspeth and her family. In this way, I have been able to watch the evolution of "Friendly Maths" as she has refined it over the years. Because I have been teaching this same material to undergraduates in Sciences and Engineering for my entire career, we've had lots to talk about.

About ten years after we first met, Elspeth and I were both invited to talk at the summer meeting of the British Biochemical Society in

Cardiff: me, to give a presentation on using a graphing calculator to visualise solutions to differential equations; Elspeth, to give a report on the "Friendly Maths" course.

A few years later Elspeth invited me to teach this course while I was on sabbatical leave from my own institution. Over my entire career of teaching Mathematics to scientists, I had never done anything on such an ambitious scale as this: in a single course to move from first principles, through differentiation and integration, and finish with solving first-order differential equations. I confess that I had my doubts. But over the eight weeks of Michelmas Term in 2006, using Elspeth's syllabus, my ninety-five first year Biochemistry students and I did just that!

This book is an expanded version of that course.

If you are planning to use this book for a course like this, or just to study it on your own, I suggest that you treat it like a manual for learning another language: Spanish, say, or Arabic. Like natural languages, Mathematics too has its own vocabulary, grammar, and notation. But unlike these languages, the objects that Mathematics treats are not part of the natural world. Rather, they reside only in the human imagination: straight line, perfect circle, or exact value of π.

Perhaps that's why this language of the imagination turns out to be also the language of science. The mathematical formulation of an hypothesis orients us towards features in the natural world that will support it. And, conversely, it is usually a mathematical argument related to the hypothesis that leads to its rejection.

Here are five tips for using this book to gain the mathematical fluency that you seek:

Be mindful of your questions and openly embrace them: Your question is a message from your brain that it is trying to fit this new idea into your growing understanding. Sometimes, to make things fit you need to look at the new idea in a different way; other times, you need to make some adjustment to what you thought you already understood. A better formulation of the new thing or a revised understanding of the old: either way, you have grown.

Let your body learn this language along with your mind: Read this book with a pencil and paper at your side. Write out the formulae; perform the computations by hand; draw the graphs. The muscle memory that you accumulate in this way will help make these abstractions more real for you.

Make friends with a graphing calculator or computer algebra system (CAS): The computing environment in your university may already have a system like Maple or Mathematica, but these are very powerful packages and so have rather steep learning curves. If you have your own computer, you can use a website like Geogebra or Wolfram Alpha. Caveat: Such websites are a great tool for checking your work and seeing patterns that you might otherwise have missed. However, for the kind of deep learning you are working towards, it is no substitute for the writing and drawing that you do by hand. Seeing "concavity" on a screen, for example, involves only the eyes, but producing "concavity" with a pencil on a piece of paper involves lots of thought as well as the muscles in your arms and hands. You can feel it!

Engage in conversation with others who are using this language, fellow learners and fluent speakers alike: In most language courses, there's an opportunity for small groups of students to meet on a regular basis with a fluent speaker of the language. This gives the learners experience in using the language in a variety of contexts and making it their own. The same holds for learning the Mathematics in this book. The more opportunity you have to use Mathematics in different contexts, the more real it will be for you and the more fluent you will become. If your university offers a lab or discussion section to accompany a course like this, that's where you can begin to get this kind of experience.

Tolerate confusion and frustration from time to time: This is surely part of the experience in learning a foreign language. It is a sign of the complexity of what you are undertaking, not of any permanent incapacity on your part. When I find myself in this state, I try to remember that new ways of thinking are almost always hard at first.

The authors of this book are both highly fluent in this language and have spent their entire careers using it to explore and solve problems in crystallography, biochemical and biotechnological engineering, pharmacology, and other fields. Their approach is the same as that used in the "Friendly Maths" course.

The Mathematics is built up clearly and rigorously from first principles, never straying into needless mathematical detail, and always with an eye towards its application to problems in the Biosciences. I do not know of another book that presents this amount of material as directly and well suited to its audience as this one.

Prof. John Fink
Kalamazoo College
Michigan (USA)

About John Fink

After graduating with a Bachelor of Arts from the University of Iowa, J.F. began his graduate work in Mathematics as a Fulbright Fellow at the University of Tübingen in Germany.

He continued his studies at the University of Michigan with support from the National Science Foundation. He was awarded a Ph.D. from that institution in 1976.

He then joined the faculty at Kalamazoo College, in Michigan, where he spent the next four decades of his career teaching Mathematics to undergraduate students of Science and Engineering. Although he was trained as a pure mathematician, his teaching of Mathematics to these students has always been grounded in its application to their chosen fields of study.

While at Kalamazoo he has also held brief visiting positions at other universities: University of Iowa and Washington State University in the USA, Universidad San Francisco de Quito in Ecuador as Fulbright Lecturer, and, most recently, Oxford University in the UK, as Plumer Fellow at St Anne's College.

In 2005, Professor Fink was named Rosemary K. Brown Professor of Mathematics at Kalamazoo College, a position that he held until his retirement in 2016.

Preface

The book of nature is written in the language of Mathematics.
— *Galileo Galilei (1564–1642)*
astronomer, physicist, engineer, polymath

This textbook is aimed at first-year undergraduate students reading for degrees in a range of Bioscience disciplines, particularly Biochemistry, Biology, Natural Sciences, Chemistry, Medicine, and Biomedical Sciences and who need a complete course of mathematical Calculus and its applications.

Knowledge of Mathematics is essential for all scientific disciplines, and its applications within Biosciences are numerous and wide-ranging. Mathematics allows a bioscientist to understand a set of key concepts, form models for physical scenarios and efficiently solve problems arising in Natural, Life, or Medical Sciences.

Our aim in this book is to bridge the gap between the 'haves' and 'have nots' in terms of those with/without mathematical training. Its purpose is to equip those students who did not study Mathematics to Advanced Level (A2) (or beyond the age of 16) with the mathematical skills and tools required for the successful understanding of their chosen discipline within the Biosciences. For those with a more extensive mathematical background, this book will be a guide to revision of basic concepts and presents relevant applications of them.

In covering the material, the authors bring both biochemical experience and mathematical rigour.

The course on which this book is based has been developed and used for teaching first-year Biochemistry undergraduates at the University of Oxford (UK) for over 25 years. It includes suggestions, edits, and recommendations contributed by many students and also by the Maths Class Tutors who delivered the weekly companion problems classes. The course has helped ensure the analytical competence and ability, but most importantly, the confidence of Bioscience students to tackle problems in quantitative ways.

This textbook presents a complete course of Mathematics from basic Algebra to advanced Calculus. It includes the application of these concepts to problems in the Biosciences and is not simply a compendium of the necessary mathematical tools. Where appropriate, the material is developed beyond the level needed and required for basic familiarity: we aim to extend the mathematical skills of the student to a working knowledge of some of the more advanced concepts for that particular topic. For example, basic theory is expanded to include some challenging examples such as complex arithmetic for Argand diagrams and their relevance to the phase problem in macromolecular crystallography and structural biology, and differential Calculus appropriate for modelling the SARS-2 Corona virus pandemic. With the exception of a few very basic concepts (such as averages and mean values), statistics are not covered.

The textbook starts from very basic maths knowledge and progresses to advanced topics in Calculus. It is organised into nine chapters.

Our approach is based on a pedagogical philosophy that has evolved over decades of successful teaching of Mathematics to undergraduate students studying Biosciences:

- Every chapter is introduced through a preamble which briefly describes the topic and its most significant applications in Biosciences for the mathematical tools explained within the chapter.
- Then, the Mathematics necessary for the mathematical interpretation of the bioscientific problems is developed and extended.
- Lastly, fundamental problems in Biosciences are presented and solved in the context of the original preamble.

To further develop the student experience and confidence levels, where appropriate the Mathematics developed in one context (*e.g.* bacterial growth) is applied to other scenarios that can be analysed using these same tools (*e.g.* radioactive decay, Newton's Law of Cooling, population growth, or drug absorption rates).

The authors' intention is that students will find that the material is clearly presented and explained, with a well-defined chapter structure. Bulleted lists are used to make the sections readable, usable, and exhaustive but still concise.

Key formulae and equations are highlighted by boxes like this

formula OR equation

with fundamental equations being numbered in sequential order for future referencing to ensure easy readability of the text.

Small boxes of the type formula are used to emphasise particular formulae within the paragraphs.

Larger boxes are used to state fundamental theorems or mathematical proofs:

Theorem or **Proof:** TEXT

The chapters are divided into Sections and include immediate worked examples to check the student's understanding after being introduced to each new concept.

Example: TEXT.

Solution: TEXT.

A straightforward worked application in one field of Biosciences is then shown in boxes appropriate to the particular area, as shown here:

* **Example in Chemistry:** TEXT.

* **Example in Biochemistry:** TEXT.

* **Example in Biology:** TEXT.

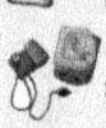** **Example in Medicine:** TEXT.

Solution: TEXT.

*These figures were generated using ChemDraw, PerkinElmer, RRID:SCR_016768.

**This figure was partly generated using Servier Medical Art, provided by Servier, licensed under a Creative Commons Attribution 3.0 unported license.

All the chapters end with a set of exercises to provide practice at problem solving. These exercises are ordered according the appearance of the topics in the corresponding chapter. Solutions plus some hints are also provided at the end of each exercise section.

Electronic resources and Internet of Things (IoT) tools, such as Geogebra or some other graphing utility (Maple, Mathematica, WolframAlpha, Symbolab, Desmos, etc), are very useful to support student learning and check the results of problems. However, it should be stressed that they are no substitute for the intellectual challenge involved in solving the problems first 'by hand'. Additionally, it is a mistake to think that it is beneficial to look at 'worked solutions' of problems before pitting your brain against them first and testing if you can do them by yourself.

Finally, the main aim of our book is that after working through it, Bioscience students will enjoy a feeling of confidence and achievement when they are challenged with mathematically-based problems and can successfully tackle them.

Prof. Elspeth F. Garman
Department of Biochemistry
University of Oxford

Dr. Nicola Laurieri
Department of Pharmacology
University of Oxford

About the Authors

Elspeth F. Garman

E.F.G. started her professional life aged 18 as a volunteer Science and Maths teacher in Manzini, Swaziland, Southern Africa, in a large girls Secondary school. She then studied for a degree in Physics (experimental option) at Durham University, UK, including a summer studentship in high energy nuclear physics at CERN, Geneva.

She carried out research for a D.Phil (Ph.D) in experimental nuclear structure physics at Oxford University. After 7 years as a Research Officer in the Nuclear Physics Department and an Oxford Physics Tutor, she changed fields completely to Structural Biology (protein crystallography) to be technical manager of a new X-ray facility in the Molecular Biophysics Laboratory, and eventually started her own research group.

She was Director of an interdisciplinary Doctoral Training Centre (Life Sciences Interface and then Systems Biology Programmes) from 2009 to 2014, supervising over 85 graduate students for their first year of study as well as 10 in her own group through to completion. She has lectured a first-year undergraduate Biochemistry course on 'Friendly

Maths for Biochemists' (then changed to 'Principles of Mathematics for Biochemistry' and now called 'Quantitative Biochemistry') for 26 years at Oxford, and this textbook is based on the course material she developed for it.

Her main research interest is in improving methods for structural biology, particularly in unambiguously identifying metals in proteins using particle induced X-ray emission (PIXE, a technique imported from nuclear physics), optimising crystal cryocooling, and understanding radiation damage effects during macromolecular crystallography X-ray diffraction experiments as well as finding methods to mitigate them. Her group has also written and released several software programs for use by the structural biology community. She has published more than 190 peer reviewed research papers and taught on over 120 advanced crystallography schools and workshops internationally.

She has been involved in regular outreach activities and also some radio and TV work. She was proud to be interviewed in 2014 on BBC Radio 4's 'Life Scientific' by Jim Al-Khalili and to explain the crystallography work of Dorothy Hodgkin on Michael Portillo's Great British Railway Journeys TV programme. She is currently an Emerita Professor of Molecular Biophysics at University of Oxford, attached to the Biochemistry Department and Brasenose College.

Nicola Laurieri

N.L., who is the son of two high-school Maths professors, completed his high-school studies focusing on classics and humanities in Italy and then moved abroad to study Biochemical and Biotechnological Engineering (BSc and MSc) at the National Institute of Applied Sciences (INSA) in Lyon (France), where he was also a one-year academic exchange visiting student at the Department of Biochemistry, University of Oxford (United Kingdom) at St. Hugh's College.

After graduating from INSA, he then studied human arylamine N-acetyltransferase enzymes, focusing on inhibitors, for a doctorate funded by Cancer Research UK in the Pharmacology Department at the University of Oxford, and at St. John's College.

Since the award of his D.Phil (Ph.D) in 2012, he has been an Academic Visitor in the Department of Pharmacology at the University of Oxford.

N.L. then returned to his native country, where he has conducted post-doctoral research in the Department of Emergency and Organ Transplantation at the Medical School of the University of Bari "Aldo Moro" (Italy) and in the Department of Mathematics, Mechanics and Management of the Polytechnic University of Bari.

He has contributed to various research projects, publications, reviews, and academic book chapters in diverse fields with different universities worldwide. In 2018, in collaboration with Prof. Edith Sim from the University of Oxford, he published the monograph Arylamine N-acetyltransferases in Health and Disease, which is a comprehensive overview of this family of enzymes as the first example of pharmacogenetics with very important consequences in tuberculosis and cancer, and takes a wide-ranging approach spanning genetics, chemistry, and structural biology, from humans to bacteria and fungi.

He has had the opportunity to tutor university students in different disciplines, including Biochemistry, Chemistry, and Mathematics, and in addition has some experience of teaching Mathematics in high schools.

He currently works as a scientific researcher at Item Oxygen S.r.l. (Italy), a highly innovative company where high-technological and robotic biomedical devices are currently under development in cooperation with different eminent Italian Universities and with funding support awarded from the regional, national, and European agencies.

Planning experimental tests, defining research methodologies and data analysis, managing and implementing research programmes, and tutoring finalist students from university, all require mathematical tools and strategies.

Acknowledgements

The birth of this textbook would not have been possible without the help of many people who we gratefully acknowledge.

We would particularly like to thank Professor John Fink for meticulously reading the whole manuscript, for his constructive suggestions, his edits, his encouragement, and lastly for writing the Foreword.

Following a Teaching Prize and Project Grant being awarded to E.F.G. in 2008, Olga Kuznetsova was funded as a summer intern to invent suitable companion exercises, and many of these are now included here.

The Maths course on which this book is based has evolved and greatly benefited over 25 years due to feedback from the more than 2,800 first-year Oxford undergraduate students (mainly studying Biochemistry but also variously Earth Sciences and Biomedical Sciences) who have attended it. Especially valuable comments and suggestions were made by over 120 Problems Class Postgraduate and Postdoctoral Teaching Assistants (TAs) who tutored the students weekly during the course.

Selection of these TAs was competitive and they not only agreed to have their classes observed by E.F.G. with subsequent feedback and mentoring to improve teaching practice, but also met together three times (unpaid) during the course to share their experiences and eat cake. These sessions were particularly helpful to E.F.G., contributing greatly to improvements in the course. She would particularly like to thank the Problems Class Coordinators for their help following her secondment to the Doctoral Training Centre and when she was very short of time: Helen Carstairs (2011), Greg Ross (2012), Aiman Entwistle (2013, 2014), Mcebisi Ntleki (2015–2018), and also Joseph Caesar, Nicolas Delalez, and Jack Miller who all faithfully taught on the course for more than five consecutive years.

The authors would sincerely like to thank Prof. Edith Sim, Department of Pharmacology, University of Oxford, who originally suggested that the authors work together on this project and introduced them to each other.

The authors are also grateful to Laurent Chaminade, the World Scientific Publishing Editor who first commissioned us, for his patience and faith in us, Rosie Williamson for her editorial help, and to other W.S.P. staff for their guidance.

They also extend their personal thanks to their various family members, colleagues, and friends who have monitored the progress of this project over the last few years and who will be most relieved at its final conclusion!

N.L. is extremely pleased to have completed the writing and editing of this textbook despite some serious and challenging health issues. He is immensely thankful to Prof. Elspeth Garman for collaborating on it and sharing her invaluable knowledge of teaching from her long career at the University of Oxford. In addition, he is wholeheartedly grateful to Prof. Edith Sim to whom he dedicates this book as his principal scientific mentor in Biosciences and kind supportive guide. He also thanks the past and present Heads of the Department of Pharmacology (University of Oxford) for allowing him to retain the status of Academic Visitor since his Ph.D graduation.

E.F.G. is in awe of the self-discipline and persistence of N.L., without whom this project would never have been completed: many thanks Nicola!

Contents

Useful Notes

Abbreviations

i.e.: *id est*, that is
e.g.: *exempli gratia*, for instance

Mathematical Symbols

Symbol	Meaning	Symbol	Meaning
$=$	is equal to	∞	infinity
$\simeq$ or $\approx$	is approximately equal to	$\|a\|$	modulus of a
$\neq$	is not equal to	$n!$	factorial n
$\equiv$	is identical to	$\to$	tends to or approaches to
$\cong$	is congruent to	$\lim\limits_{x \to a}$	limiting value as $x \to a$
$>$	is greater than	$'$	prime derivative
$\geq$	is greater than or equal to	$''$	second derivative
$<$	is less than	$'''$	third derivative
$\leq$	is less than or equal to	$\int$	integral
$\in$	is an element of	$\Re$	the real part of
$\subset$	is a subset of	$\Im$	the imaginary part of
$\cup$	is united to	π	$3.14159\ldots$
$\varnothing$	empty set or no real solutions	e	Euler's number $(2.71828\ldots)$

Ancient Greek Alphabet

Ancient Greek small letters are used in Mathematics as symbols for constants and variables representing certain quantities, while the capital letters represent distinct mathematical operator.

Lower-case letters:

α	alpha	ι	iota	ρ	rho
β	beta	κ	kappa	σ	sigma
γ	gamma	λ	lambda	τ	tau
δ	delta	μ	mu, micro	υ	upsilon
ϵ	epsilon	ν	nu	ϕ	phi
ζ	zeta	ξ	xi	χ	chi
η	eta	o	omicron	ψ	psi
θ	theta	π	pi	ω	omega

Upper-case letters:

Δ	Delta, difference/change	Σ	Sigma
Θ	Theta, temperature	$\sum$	summation
Π	Pi, set of irrational numbers	Ψ	Psi, wave function
$\prod$	product		

Chapter 1

Revision of Basic Tools

Preamble

Becoming comfortable with using mathematical methods is rather like building the wall of a house: without solid foundations, the wall is insecure. This chapter aims to provide revision of the basic tools you will need later on, as you may not have used some of them for a while.

1.1 Sets of Numbers

Each of the following sets of numbers contains an infinite number of elements!

$\mathbb{N}$: natural numbers or positive integers.

Examples: $1, 2, 3, 4, \ldots$.

$\mathbb{Z}$: positive and negative integers.

Examples: $-11, -2, 0, 6, 87, \ldots$.

$\mathbb{Q}$: rational numbers, which can be expressed as an integer numerator over an integer denominator.

Examples: $-\frac{13}{5}, -4.9, -0.\overline{7}, \frac{1}{3}, 5.3\overline{2}, \frac{19}{2}, 14.53, \ldots$

In particular,

- $-4.9 = -\frac{49}{10}$
- $5.3\overline{2} = 5.322\ldots = \frac{532-53}{90} = \frac{479}{90}$
- $-0.\overline{7} = -0.777\ldots = -\frac{7}{9}$
- $14.53 = \frac{1453}{100}$

Π: irrational numbers, which are numbers that cannot be written as a simple fraction and have an infinite length.

Examples: $-\sqrt{3}, \sqrt{2}, \pi, e, \log_2(5), \sin(60°), \ldots$

In particular,

- $-\sqrt{3} = -1.73205\ldots$
- $\sqrt{2} = 1,41421\ldots$
- $\pi = 3.14159\ldots$
- $e = 2.71828\ldots$
- $\log_2(5) = 2.32192\ldots$
- $\sin(60°) = 0,86602\ldots$

$\mathbb{R}$: real numbers; it represents all rational and irrational numbers.
$\mathbb{C}$: complex numbers, which have a real part ($\mathfrak{R}$) and an imaginary part ($\mathfrak{I}$).

Examples: $-1+2i, 2-i\sqrt{3}, \ldots$

Important note: i is the imaginary unit that has the unique property that $i^2 = -1$.

For instance in the complex number $1-i\sqrt{2}$, the real part is $\mathfrak{R} = 1$, while the imaginary part is $\mathfrak{I} = -i\sqrt{2}$.

$\mathbb{N}$ is a subset of $\mathbb{Z}$, $\mathbb{Z}$ is a subset of $\mathbb{Q}$, $\mathbb{Q}$ and Π are two separate subsets of $\mathbb{R}$, and $\mathbb{R}$ is a subset of $\mathbb{C}$ (Figure 1.1).

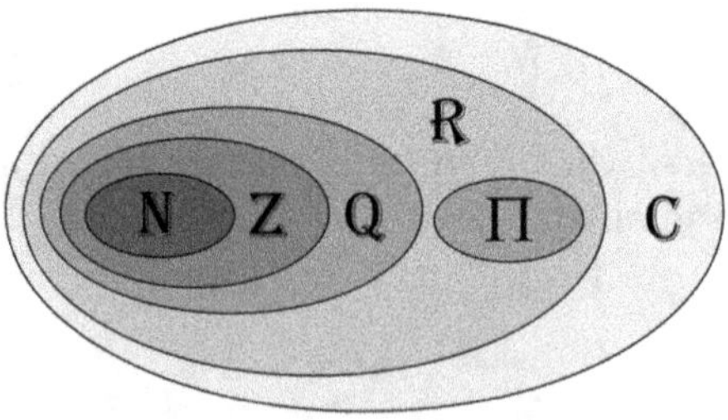

Figure 1.1: Sets of numbers

1.2 Manipulation of Fractions

Fractions express divisions or ratios and are written in the form $\frac{N}{D}$, where the alpha-numerical expression N is the **numerator** and the alpha-numerical expression D is the **denominator**.

As a result of the division of integers, each fraction (F) is called a **rational number** and belongs to the set $\mathbb{Q}$:

$$\frac{N}{D} = F \quad \text{is equivalent to} \quad N = F \times D$$

Important note: Let a be a real number, with $a \neq 0$:

- $\frac{0}{a} = \mathbf{0}$, since $0 = \mathbf{0} \times a$.
- $\frac{a}{0}$ has no meaning, as this implies $a = 0 \times b$, but no number b which satisfies this equation can exist unless $a = 0$. Thus, the denominator must always be $D \neq 0$.
- $\frac{0}{0}$ is **indeterminate**, since all numbers will give 0 when multiplied by 0.

1.2.1 Simplifying fractions

Fractions can be simplified in order to handle quantities in the calculations more easily: for instance, $\frac{13}{39}$ is the same as $\frac{1}{3}$ but less convenient to manipulate.

To reduce fractions to simpler forms, it is necessary to factorise (or factor) both the numerator and the denominator and then find a common divisor.

Example: Put $\frac{84}{630}$ into its simplest form.

Solution: If we factorise $84 = 2 \times 2 \times 3 \times 7$ and $630 = 2 \times 3 \times 3 \times 5 \times 7$, we notice that 2, 3, and 7 are common factors. So, they can all be taken out of both the numerator and denominator, leaving as the final result

$$\frac{2 \times 2 \times 3 \times 7}{2 \times 3 \times 3 \times 5 \times 7} = \frac{\cancel{2} \times 2 \times \cancel{3} \times \cancel{7}}{\cancel{2} \times \cancel{3} \times 3 \times 5 \times \cancel{7}} = \frac{2}{15}$$

Example: Simplify the fraction $\dfrac{5ab}{15ac - 10ab^2}$.

Solution: We notice that all the terms have 5 and a as common factors, so we have $5a$ as a common factor in both the numerator and denominator (see Section 2.1.2). We can then cancel $5a$ out and get a fraction which cannot be further simplified:

$$\frac{5ab}{15ac - 10ab^2} = \frac{5a \times (b)}{5a \times (3c - 2b^2)} = \frac{\cancel{5a} \times b}{\cancel{5a} \times (3c - 2b^2)} = \frac{b}{3c - 2b^2}$$

Warning! Only factors can be simplified! $\dfrac{x-y}{y+z} \neq \dfrac{x-\cancel{y}}{\cancel{y}+z} \neq \dfrac{x}{z}$.

Example in Chemistry: Eight moles of sodium chloride (NaCl) are dissolved in 192 moles of water. What is the molar fraction of the solute?

Solution: The *mole fraction* is one of the ways of expressing the concentration of a solute A in a solvent B. The mole fraction (x_A) is the ratio of the number of moles (n_A) of the solute A to the total number of moles of all the substances in the solution (in our binary solution, we have only the substances $A =$ NaCl and $B =$ water), and thus x_a does not have measurement units:

$$x_A = \frac{n_A}{n_A + n_B} = \frac{8}{8+192} = \frac{8}{200} = \frac{\cancel{2} \times \cancel{2} \times \cancel{2}}{\cancel{2} \times \cancel{2} \times \cancel{2} \times 5 \times 5} = \frac{1}{25} = 0.04$$

1.2.2 Operations with fractions

- **Adding and subtracting fractions**

To add and subtract fractions, we find a common denominator (P) and re-express our fractions over this common denominator by multiplying N and D by $\dfrac{P}{D}$. Then, we can add or subtract the numerator.

A common denominator is a number that is a multiple of both denominators. If we choose the lowest common denominator, this simplifies adding, subtracting, and comparing fractions.

Example: Calculate $\frac{3}{10} + \frac{1}{4}$.

Solution: For $\frac{3}{10}$ and $\frac{1}{4}$, the lowest common denominator is $P = 20$, so to add them we express $\frac{3}{10}$ as $\frac{6}{20}$ (multiplying top and bottom by $\frac{P}{D} = 2$) and $\frac{1}{4}$ as $\frac{5}{20}$ (multiplying top and bottom by $\frac{P}{D} = 5$). Then, $\frac{6}{20} + \frac{5}{20} = \frac{6+5}{20} = \frac{11}{20}$.

Warning! If 40 is chosen as common denominator, we express $\frac{3}{10}$ as $\frac{12}{40}$ (multiplying top and bottom by 4) and $\frac{1}{4}$ as $\frac{10}{40}$ (multiplying top and bottom by 10). Then, $\frac{12}{40} + \frac{10}{40} = \frac{12+10}{40} = \frac{22}{40} = \frac{\cancel{2} \times 11}{\cancel{2} \times 20} = \frac{11}{20}$, which gives the same result as above but with more steps.

- **Multiplying fractions**

When two fractions are multiplied together, their numerators are multiplied together and their denominators are multiplied together.

Example: Calculate $\frac{3}{8} \times \frac{5}{6}$.

Solution: $\frac{3}{8} \times \frac{5}{6} = \frac{3 \times 5}{8 \times 6} = \frac{15}{48} = \frac{\cancel{3} \times 5}{\cancel{3} \times 16}$.

- **Dividing fractions**

To divide by a fraction, we multiply by the reciprocal (inverted or 'turned upside down') fraction.

Example: Calculate $\frac{2}{5} \div \frac{3}{7}$.

Solution: $\frac{2}{5} \div \frac{3}{7} = \frac{2}{5} \times \frac{7}{3} = \frac{14}{15}$.

Example in Chemistry: To what volume should $40\,\mathrm{mL}$ of a solution containing $5\,\mathrm{M}$ potassium hydroxyde (KOH) be diluted to obtain $0.2\,\mathrm{M}$ KOH?

Solution: *Molarity* (or molar concentration) is one of the ways of expressing the concentration of a chemical species, generally called the *solute*, in a solution, in terms of amount of substance per unit volume of solution $\left(\mathrm{M} = \frac{\mathrm{mol}}{\mathrm{L}}\right)$. Dilution is a way of preparing solutions with lower molarity (M) from a more concentrated stock solution.

As the number of moles of KOH is constant in both the stock solution (s_1) and the diluted solution (s_2), with V as the volumes we have

$$V_{s_1} \times M_{s_1} = V_{s_2} \times M_{s_2} \Longleftrightarrow \frac{40}{1000}\cancel{\mathrm{L}} \times 5\,\frac{\mathrm{mol}}{\cancel{\mathrm{L}}} = V_{s_2} \times \frac{2}{10}\,\frac{\mathrm{mol}}{\mathrm{L}}$$

$$\text{Thus, } V_{s_2} = \frac{40}{1000} \times 5\,\mathrm{mol} \div \frac{2}{10}\,\frac{\mathrm{mol}}{\mathrm{L}} = \frac{1}{25} \times 5\,\cancel{\mathrm{mol}} \times \frac{10}{2}\,\frac{\mathrm{L}}{\cancel{\mathrm{mol}}}$$

$$= \frac{50}{50}\,\mathrm{L} = 1\,\mathrm{L}.$$

Example in Biology: Under the lens of a light microscope, the unicellular ciliate *Paramoecium* is about 1.2 divisions long out of the 10 imaginary divisions into which the field diameter observed through the lens is divided. Knowing that the field diameter is $1500\,\mu\mathrm{m}$, estimate the length of the *Paramoecium.*

Solution: The light microscope is a common instrument to measure the sizes of microscopic objects, such as cells and organelles. The size of the field of view under the microscope decreases proportionately when the magnification is increased.

We can say that the ratio of imaginary divisions equals the ratio of real lengths: $\frac{1.2}{10} = \frac{x}{1500\,\mu\mathrm{m}}$, where x is the size of the *Paramoecium.*

So the *Paramoecium* length $x = \frac{1.2}{10} \times 1500\,\mu\mathrm{m} = \frac{1800}{10}\,\mu\mathrm{m}$ $= 180\,\mu\mathrm{m}$, which is within the typical size range of this species (*i.e.* 50–$330\,\mu\mathrm{m}$).

1.2.3 Percentages

Ratios, fractions, and percentages are related concepts. Many problems within Biosciences require students to have a good understanding of the relationships between these concepts and to use them in calculations.

Percentages are an alternative way to write fractions using 100 as the denominator.

For instance, 20% (namely, 'twenty per cent') equals the fraction $\frac{20}{100}$, so we can write $20\% = \frac{20}{100} = \frac{1}{5}$.

In general, if the quantity x is a part of the quantity y and we want to know what percentage x is of y (*i.e.* how large x is compared to y), we can use the formula

$$\text{percentage} = \frac{x}{y} \times 100\%$$

Example: What percentage is 21 of 84?

Solution: $\text{percentage} = \frac{21}{84} \times 100\% = \frac{1}{4} \times 100\% = 25\%$.

Example: What is 30% of 214?

Solution: We know the values of y (214) and the percentage (30%): $30\% = \frac{x}{214} \times 100\%$. We want to know the value of x:

$$x = \frac{30\%}{100\%} \times 214 = \frac{3}{10} \times 214 = 3 \times 21.4 = 64.2$$

Example in Chemistry: A sample of 3.5 g of carbon dioxide (CO_2) was obtained by heating 12.4 g of calcium carbonate ($CaCO_3$):

$$CaCO_3(s) \longrightarrow CaO(s) + CO_2(s)$$

What is the percentage yield for this reaction?

Solution: The amount of a product that can be produced by a reaction under specific conditions is called the *theoretical yield* of the reaction. However, the amount of product obtained, which is called the *actual yield*, is often less than the theoretical yield for a number of reasons (temperature, pressure, solvent, *etc.*).

In practice, the extent to which the theoretical yield of a reaction is achieved is commonly expressed as its

$$\textit{percentage yield} = \frac{\text{actual yield}}{\text{theoretical yield}} \times 100\%$$

12.4 g of calcium carbonate ($CaCO_3$) corresponds to

$$\frac{12.4\,\text{g}}{(40+12+16\times 3)\,\frac{\text{g}}{\text{mol}}} = \frac{12.4\,\cancel{\text{g}}}{100\,\frac{\cancel{\text{g}}}{\text{mol}}} = 0.124\,\text{mol}$$

As per the stoichiometry of the balanced chemical equation, the ratio of $CaCO_3$ to CO_2 is $1:1$. Thus, the expected yield of CO_2 should also be $0.124\,\text{mol}$, which corresponds to:

$$0.124\,\cancel{\text{mol}} \times (12+16\times 2)\,\frac{\text{g}}{\cancel{\text{mol}}} = (0.124\times 44)\,\text{g} = 5.5\,\text{g of } CO_2$$

Therefore, the percent yield of the reaction is: $\frac{3.5\,\cancel{\text{g}}}{5.5\,\cancel{\text{g}}} \times 100\% = 64\%$.

Example in Biology: Two wild-eyed fruit flies are crossed in one vial. The offspring are counted: 148 wild-eyes flies (w^+) and 47 white-eyed flies (w^-). Wild flies (normal phenotype) are red-eyed fruit flies, whereas white-eyed flies (mutated phenotype) have 'albino characteristics' with a genetic mutation on the gene '*white*'.

Calculate the percentage of each offspring compared to the total offspring and form an hypothesis on the genetic trait 'eye colour' in fruit flies.

Solution: Genetics is the science of heredity and of the mechanisms by which traits are passed from parent to offspring. Ratios and percentages are commonly used in genetics. The phenotype is an organism's appearance resulting from its genotype and/or the environment:

$$\text{percentage}^{w^+} = \frac{148}{148+47} \times 100\% = \frac{148}{195} \times 100\% = 0.758 \times 100\%$$
$$= 75.8\%$$

Percentage of white-eyed flies' (w^-) offspring can be calculated in two ways: $\text{percentage}^{w^-} = \frac{47}{148+47} \times 100\% = \frac{47}{195} \times 100\%$ $= 0.242\times 100\% = 24.2\%$, or $\text{percentage}^{w^-} = 100\% - \text{percentage}^{w^+}$ $= 100\% - 75.8\% = 24.2\%$.

Since both parental flies are wild-eyed and the percentage of the mutated offspring is 25% (*i.e.* the white phenotype is not dominant), this is consistent with the hypothesis that (w^-) is a recessive genetic trait and both parents are heterozygous (w^+/w^-) for the gene '*white*', that is to say that both parents have two different alleles for the gene '*white*'.

1.3 Elementary Tools of Algebra

As you have seen in the Biosciences examples examined so far, most of the physical, chemical, and biological quantities which have been calculated are not represented by their own English names, but abbreviations or single letters are used:
e.g. V = volume, n = number of moles, and x = size of the *Paramoecium*.

The use of these conventions is extremely helpful in Mathematics to simplify expressions and help handle calculations.

1.3.1 Algebraic representation

Numeric **constants** are represented by letters early in the alphabet: a, b, c, ... up to m, whereas algebraic **variables** are represented by the letters later in the alphabet: n, p, ... x, y, z (o is never used as a symbol since it can be confused with 0). This is simply convention to avoid confusion.

The identity $5 = 2+3$ can be represented as $z = x+y$, where $z = 5$, $x = 2$, and $y = 3$.

In the expression $3z = 4x + 6y$, possible values would be $z = 5$, $x = \frac{3}{4}$, and $y = 2$ to satisfy the equation. The **coefficient** of z is said to be 3, while the coefficient of x is 4 and that of y is 6. When no number is written beside a variable, the coefficient of the variable is 1.

In the identity $az = bx+cy$, there are a number of possible solutions, since a, b, and c can take any values from the set of real numbers ($\mathbb{R}$).

The **coefficient** of z is said to be a, while the coefficient of x is b and that of y is said to be c.

• Like and unlike terms

We can simplify like terms by combining them together but we cannot combine unlike terms.

Example: Collect and combine terms in x, y, and xy separately to simplify the expression $13x+xy-3y-5x+20y-5xy = 8x+17y-4xy$.

Solution: We can label like terms in the same way, so we can easily simplify the expression $\mathbf{13x}+xy-\underline{3y}-\mathbf{5x}+\underline{20y}-5xy = \mathbf{8x}+\underline{17y}-4xy$.

- **Brackets**

Everything inside the brackets is multiplied by what is immediately outside the bracket.

Example: Simplify the expression $5(x+4y-7)$.

Solution: We multiply the terms in the brackets by 5, so $5(x+4y-7) = 5x+20y-35$.

1.3.2 Formulae

A formula is an expression that connects two or more variable or constant quantities.

Example: Calculate the average cost per person of feeding a household for a week.

Solution: Let C = average cost, T = total bill, and n = number of people, so we divide the total food bill by the number of people living in the household: $C = \frac{T}{n}$.

- **Substitution**

If we know the values of some variables in a formula, such as x or y, we can put them into our formula and calculate the value of the mathematical expression.

To calculate a numerical value from a formula with n unknown variables, $(n-1)$ of them should be specified to determine the nth quantity.

Example: If $x = 6$, $y = 3$, and $z = -5$, find the value of the expression $z(y+2x)$.

Solution: Let us substitute x, y, and z and multiply the terms of the expression: $z(y+2x) = -5(3+2\times 6) = -5(3+12) = -5\times 15 = -75$.

Example in Chemistry: When one mole of gas is heated at constant pressure, the volume increases depending on the temperature (measured in Kelvin, where $273\,\mathrm{K} = 0\ (^\circ\mathrm{C})$ [Centigrade]). The gas volume equals $20.94\,\mathrm{L}$ at $255\,\mathrm{K}$ and increases by $83.0\,\mathrm{mL}$ per temperature degree. Write down the formula that represents the relationship between gas volume and temperature, and estimate the volume of one mole of gas at room temperature ($293\,\mathrm{K}$).

Solution: First define variables and constants with their corresponding measurement units: V = volume(L), t = temperature (K), a = initial constant (20.94 L), and b = gradient ($83.0\,\frac{\mathrm{mL}}{\mathrm{K}} = 0.083\,\frac{\mathrm{L}}{\mathrm{K}}$, as measurement units should be the same throughout the formula): $V = a + b \times t = 20.94\,\mathrm{L} + 0.083\,\frac{\mathrm{L}}{\mathrm{K}} \times t\,\mathrm{K}$.

We can substitute the value of t as $293 - 255 = 38\,\mathrm{K}$ in the formula and thus get $V = 20.94 + 0.083 \times 38 = 20.94 + 3.15 = 24.09\,\mathrm{L}$, which is the volume of a mole of gas at $293\,\mathrm{K}$.

1.3.3 Introducing indices and roots

The value of variables and constants in formulae might not often be a simple positive or negative integer, or a natural or rational number as you have seen in the examples so far, but might require superscript numbers to express very small or very large quantities.

Indices are used to condense multiple similar factors, simplify expressions, and help handle multiple products: *e.g.* $4 \times 4 \times 4 = 4^3$, or in symbols $a \times a \times a = a^3$.

In general,

$$a^n = \overbrace{a \times a \times a \times \cdots \times a}^{n \text{ times}}$$

This is called the **exponential** form: n is the **power** or **index**, and a is the **base**.

Warning! $a + a + a \neq a^3$, but $a + a + a = 3a$.

Example in Biology: *Escherichia coli* cells are usually cultured in an appropriate glucose-salts medium at $37\,^\circ\mathrm{C}$ to grow, and each bacterial cell on average divides about every 18 minutes under those conditions. Estimate the number of bacterial cells generated after 3 hours from a single bacterium.

Solution: Starting from a single cell, the bacteria subsequently generated will follow the sequence of numbers: $1, 2, 4, 8, 16, 32, \ldots$, which are, respectively, $2^0, 2^1, 2^2, 2^3, 2^4, 2^5, \ldots$. This can be formulated as 2^n with n being the number of cell cycle divisions.

In 3 hours, there will be 10 cell cycle divisions, as $\dfrac{3\,\mathrm{h} \times 60\frac{\mathrm{min}}{\mathrm{h}}}{18\,\mathrm{min}} = \dfrac{180}{18} = 10$. Thus, the number of bacterial cells generated from a single bacterium after 3 hours will be: $2^{10} = 1024$.

• Manipulating indices

Note that both '·' and '×' are used to represent the multiplication symbol.

The laws governing the operations with indices are as follows:

(i) When multiplying the powers of the same number together, we add the indices:

$$b^2 \cdot b^4 = (b \cdot b) \cdot (b \cdot b \cdot b \cdot b) = b^{2+4} = b^6$$

(ii) When dividing powers of the same number, we subtract the indices:

$$\frac{b^5}{b^3} = \frac{b \cdot b \cdot b \cdot b \cdot b}{b \cdot b \cdot b} = b^{5-3} = b^2$$

(iii) When taking powers of powers of a number, we multiply the indices together:

$$(b^3)^7 = b^{3 \cdot 7} = b^{21}$$

(iv) If a product is raised to a certain power, each factor of the product is raised to that power:

$$(a \cdot c)^6 = a^6 \cdot c^6$$

(v) If a quotient is raised to a certain power, both the numerator and denominator are raised to that power:

$$\left(\frac{m}{n}\right)^6 = \frac{m^6}{n^6}$$

(vi) If a number is raised to a negative number ($-n$, with $n \in \mathbb{N}$), we express the reciprocal fraction of the power:

$$b^{-3} = \frac{1}{b^3}$$

Important note: $a^0 = 1$, $a^1 = a$.

Warning! $a^2 + a^2 = 2a^2$. The expressions $a^4 + a^2 \neq a^6$, but $a^4 + a^2 = a^2(a^2 + 1)$ and $a^5 - a^3 \neq a^2$, but $a^5 - a^3 = a^3(a^2 - 1)$, as explained in Section 2.1.2.

Example: Multiply out and simplify the expression $2^x \cdot 4^{x+1} \div 16^{x-2}$.

Solution: Before simplifying the expression, first rewrite the expression to change all the bases to the same number:

$$2^x \cdot 4^{x+1} \div 16^{x-2} = 2^x \cdot (2^2)^{x+1} \div (2^4)^{x-2} = 2^x \cdot 2^{2x+2} \div 2^{4x-8}$$

$$= 2^{(x+2x+2)} \div 2^{4x-8} = 2^{3x+2} \div 2^{4x-8} = 2^{(3x+2-4x+8)} = 2^{10-x} = \frac{2^{10}}{2^x}$$

Warning! $(a+b)^n \neq a^n + b^n$ and similarly $(a-b)^n \neq a^n - b^n$.

Example: Calculate (i) $(1+3)^2$ and (ii) $1^2 + 3^2$.

Solution: (i) $(1+3)^2 = 4^2 = 16$ but (ii) $1^2 + 3^2 = 1 + 9 = 10$. These are different!

The brackets in (i) require that we add first and then square the result. However, in (ii), the order of operations requires that we square the individual numbers first, and then add the results together.

Example in Chemistry: Photon energy is the energy carried by a single photon. The amount of energy is calculated as the ratio between Planck's constant (h) multiplied by the speed of light (c) and the wavelength of the light (λ). Find the energy of a photon with light wavelength $= 400\,\mathrm{nm}$, knowing Planck's constant $= 6.63 \cdot 10^{-34}\,\mathrm{J\,s}$ and the speed of light $= 3.0 \cdot 10^{8}\,\mathrm{m\,s^{-1}}$.

Solution: The relationship connecting the constants and the variables can be expressed by the formula $E = \dfrac{h \cdot c}{\lambda}$.

Then, we can calculate E, using the International System of Units (*i.e.* metre (m) for length, (kg) for mass, (s) for time, (mol) for moles, and joule (J) for energy):

$$E = \frac{6.63 \cdot 10^{-34}\,\mathrm{J\,s} \cdot 3.0 \cdot 10^{8}\,\mathrm{m\,s^{-1}}}{400\,\mathrm{nm}} = \frac{19.89 \cdot 10^{-34} \cdot 10^{8} \cdot \mathrm{J}\,\cancel{\mathrm{s}}\,\cancel{\mathrm{m}}}{4 \cdot 10^{2} \cdot 10^{-9} \cdot \cancel{\mathrm{m}}\,\cancel{\mathrm{s}}}$$
$$= \frac{19.89}{4} \cdot \frac{10^{-34+8}}{10^{2-9}} = 4.97 \cdot \frac{10^{-26}}{10^{-7}}\,\mathrm{J} = 4.97 \cdot 10^{-26-(-7)}\,\mathrm{J}$$
$$= 4.97 \cdot 10^{-19}\,\mathrm{J}$$

- **Combining squares and higher-order roots**

Let a be a positive real number (≥ 0) and n be a positive integer number (>0),

$b = \sqrt[n]{a}$ is equivalent to $b^n = a$

The number b is called the nth **root** of a, n is the **root index**, and a is the **argument** of the root.

Important note: If the root index n is even, the argument a can only be ≥ 0, whereas if n is odd, a can be any real number.

The same rules as those for indices apply, we just have to deal with fractions instead of integers:

- **Square root:** $\sqrt{a} = a^{\frac{1}{2}}$
- **Cube root:** $\sqrt[3]{a} = a^{\frac{1}{3}}$
- **Fourth root:** $\sqrt[4]{a} = a^{\frac{1}{4}}$
- **nth root:** $\sqrt[n]{a} = a^{\frac{1}{n}}$

General rules of roots are as follows, with $a, b > 0$:

(i) $\sqrt{a} \cdot \sqrt{a} = a^{\frac{1}{2}} \cdot a^{\frac{1}{2}} = a^{\frac{1}{2}+\frac{1}{2}} = a$ (ii) $\sqrt{a} \cdot \sqrt{b} = \sqrt{ab}$

(iii) $\dfrac{a}{\sqrt{a}} = \dfrac{\sqrt{a} \cdot \cancel{\sqrt{a}}}{\cancel{\sqrt{a}}} = \sqrt{a}$ (iv) $\dfrac{\sqrt{a}}{\sqrt{b}} = \sqrt{\dfrac{a}{b}}$

(v) $\sqrt[n]{a^m} = \left(\sqrt[n]{a}\right)^m = a^{\frac{m}{n}}$ with $m, n \in \mathbb{N}$ (vi) $\sqrt[n]{\sqrt[m]{a}} = \sqrt[n \cdot m]{a}$ with $m, n \in \mathbb{N}$

To rationalise a denominator containing a single square root, multiply numerator and denominator by that square root:

$$\frac{b}{\sqrt{a}} = \frac{b \cdot \sqrt{a}}{\sqrt{a} \cdot \sqrt{a}} = \frac{b\sqrt{a}}{a}$$

Warning!:

- $\sqrt{a} + \sqrt{b} \neq \sqrt{a+b}$
- $\dfrac{1}{\sqrt{a} + \sqrt{b}} \neq \dfrac{1}{\sqrt{a+b}}$

Example: Simplify $\sqrt[3]{x^2 \cdot \sqrt[5]{\dfrac{1}{x}}}$.

Solution: Since the root indices are both odd (3 and 5), the arguments can be positive or negative, but $x \neq 0$ to allow the division $\dfrac{1}{x}$:

$$\sqrt[3]{x^2 \cdot \sqrt[5]{\frac{1}{x}}} = \sqrt[3]{x^2 \cdot \sqrt[5]{x^{-1}}} = \sqrt[3]{x^2 \cdot x^{-\frac{1}{5}}} = \sqrt[3]{x^{\left(2-\frac{1}{5}\right)}} = \sqrt[3]{x^{\frac{9}{5}}}$$
$$= x^{\left(\frac{9}{5} \cdot \frac{1}{3}\right)} = x^{\frac{3}{5}} = \sqrt[5]{x^3}$$

Warning! If the index is $\dfrac{1}{2}$, the square root is involved.

Example: Calculate (i) $\sqrt{9+16}$ and (ii) $\sqrt{9} + \sqrt{16}$.

Solution: (i) $\sqrt{9+16} = \sqrt{25} = 5$, but (ii) $\sqrt{9} + \sqrt{16} = 3 + 4 = 7$. These are NOT the same!

At some point in their life, almost everybody makes the mistake of assuming that $\sqrt{a^2 + b^2} = \sqrt{a^2} + \sqrt{b^2} = a + b$.

The **only** way this can be true is for a or b (or both) to be 0.

If neither a nor b is 0, then $\sqrt{a^2 + b^2}$ **must be different from** $a+b$.

Example in Chemistry: The ammonium acetate (CH_3OONH_4) salt hydrolyses into the cation CH_3COO^- (a *weak base*) and the anion NH_4^+ (a *weak acid*), both of which react with water to produce OH^- and H^+, respectively, thereby affecting the acidity of a solution: the higher $[H^+]\,(\mathrm{M})$, the more acidic the solution. A solution containing hydrolysed salt ions can be *acidic* or *alkaline*, depending on the strength of the base and the acid in solution.

The equilibria involved in this hydrolysis reaction are:

(i) $CH_3COO^- + H_2O \rightleftharpoons CH_3COOH + OH^-$

(ii) $NH_4^+ \rightleftharpoons NH_3 + H^+$

As both the acid and base of ammonium acetate are weak, $[H^+] = \sqrt{\frac{K_w \cdot K_a}{K_b}}$, where K_w is the dissociation constant of water, K_a is that of the conjugate acid (CH_3COOH) of the weak base, and K_b is that of the conjugate base (NH_3) of the weak acid. The Chemistry literature gives the value of those constants at $25°\mathrm{C}$: $K_w = 10^{-14}\,\mathrm{M}^2$, $K_a = 1.76 \cdot 10^{-5}\,\mathrm{M}$, and $K_b = 1.85 \cdot 10^{-5}\,\mathrm{M}$.

Estimate the concentration of H^+ in a water solution containing $0.01\,M$ ammonium acetate.

Solution: We can calculate $[H^+]$ by using the formula given above:

$$[H^+] = \sqrt{\frac{10^{-14}\,\mathrm{M}^2 \cdot 1.76 \cdot 10^{-5}\,\mathrm{M}}{1.85 \cdot 10^{-5}\,\mathrm{M}}} = \sqrt{10^{-14} \cdot 0.95}\,\mathrm{M}$$

$$= \sqrt{10^{-14}} \cdot \sqrt{\frac{95}{100}}\,\mathrm{M} = 10^{\frac{-14}{2}} \cdot \frac{\sqrt{95}}{\sqrt{10^2}}\,\mathrm{M} = 10^{-7} \cdot \frac{9.75}{10}\,\mathrm{M}$$

$$= 0.975 \cdot 10^{-7}\,\mathrm{M}$$

Given that a neutral solution has $[H^+] = 10^{-7}\,\mathrm{M}$, the solution containing $0.01\,\mathrm{M}$ CH_3COONH_4 has $[H^+] = 0.975 \cdot 10^{-7}\,\mathrm{M}$, which is slightly lower than $10^{-7}\,\mathrm{M}$; thus, the solution is slightly alkaline.

Important note: The initial concentration of the salt does not affect the estimation of $[H^+]$ in this case.

1.3.4 The modulus (or absolute value)

The **modulus** or the **absolute value** of a number is the size (magnitude) of a number (x), ignoring its sign and is written as $|x|$.

Let a be a real number, then $|a|$ is the **distance** between a and 0. Thus,

$$\text{if } a \geq 0, \text{ then } |a| = a$$
$$\text{if } a \leq 0, \text{ then } |a| = -a$$

The laws governing the operations with moduli (or 'mods') are

(i) $|ab| = |a| \cdot |b|$ (ii) $\left|\frac{a}{b}\right| = \frac{|a|}{|b|}$ (*if* $b \neq 0$)

(iii) $|a^n| = |a|^n$ (iv) $\sqrt{a^2} = |a|$

(v) $|a+b| \leq |a| + |b|$ (vi) $|a-b| \geq |a| - |b|$

Example: Evaluate $x|3-2y| + |x^3|$, when $x = -2$ and $y = 7$.

Solution: This exercise is actually no different from any other problem involving substitution of x and y values.

We should beware when we find the signs of the absolute values in the expression: if the number between the absolute value signs is negative, it changes to a positive one:

$$x|3-2y| + |x^3| = -2|3-2\cdot 7| + |(-2)^3| = -2|-11| + |-2|^3$$
$$= -2\cdot 11 + 2^3 = -14$$

1.4 Mathematical Operators

In science, some operators are used as short-hand for complex mathematical expressions; they are called **variable-binding operators**, as each of them binds (fixes) a variable, usually k, for some set of numbers, such as $\mathbb{N}$.

1.4.1 The operator Σ

The sum of multiple numbers can be shortened by using the **summation operator** $\sum$, which is the capital Greek letter sigma (Σ) enlarged.

The general form is

$$\sum_{k=a}^{n} X_k = X_a + X_{a+1} + X_{a+2} + \cdots + X_{n-1} + X_n$$

where,

- X is the *general term* of the summation (*i.e.* the numbers summed) and its expression is bound to the variable k;
- k is the *summation index*, identifies each term of the sum and indicates the first term of the sum (X_a);
- n denotes the last term of the sum, defining its range;
- k is a variable which is an integer, and it assumes the values $a, a+1, \ldots, n$.

General properties of the $\sum$ operator:

(i) $\sum_{k=1}^{n} c = \overbrace{c + c + \cdots + c + c}^{n \text{ times}} = \mathbf{n} \cdot \mathbf{c}$, where c is a positive constant independent of n;

(ii) $\sum_{k=0}^{n} c \cdot X_k = c \cdot X_0 + c \cdot X_1 + \cdots + c \cdot X_{n-1} + c \cdot X_n = \mathbf{c} \cdot \sum_{k=0}^{n} X_k$, where c is a constant independent of n;

(iii) $\sum_{k=0}^{n} (X_i \pm Y_i) = \sum_{k=0}^{n} X_i \pm \sum_{k=0}^{n} Y_i$.

Example: Calculate $\sum_{k=4}^{7} \left(\frac{k^2}{6} - \frac{7}{k} \right)$.

Solution: $\sum_{k=4}^{7} \left(\frac{k^2}{6} - \frac{7}{k} \right) = \frac{1}{6} \sum_{k=4}^{7} k^2 - 7 \cdot \sum_{k=4}^{7} \frac{1}{k}$

$= \frac{1}{6} (4^2 + 5^2 + 6^2 + 7^2) - 7 \cdot \left(\frac{1}{4} + \frac{1}{5} + \frac{1}{6} + \frac{1}{7} \right) = \frac{1}{6} \cdot 126 - 7 \cdot \frac{319}{420}$

$= 21 - \frac{319}{60} = \frac{1260 - 319}{60} = \frac{941}{60}$

Some of the summations can be calculated straightforwardly through a closed-form formula. A closed-form formula contains constants, variables and basic functions connected by arithmetic operations.

Important examples of summations to know are as follows:

- $\sum_{k=1}^{n} k = 1 + 2 + 3 + \cdots + (n-2) + (n-1) + n = \frac{\mathbf{n(n+1)}}{\mathbf{2}}.$
- $\sum_{k=1}^{n} k^2 = 1 + 4 + 9 + \cdots + (n-1)^2 + n^2 = \frac{\mathbf{n(n+1)(2n+1)}}{\mathbf{6}}.$
- $\sum_{k=0}^{n} c^k = c^0 + c^1 + c^2 + \cdots + c^{n-1} + c^n = \frac{\mathbf{1 - c^{(n+1)}}}{\mathbf{1 - c}}$ with $c \neq 1$, where c is a positive constant.

The **arithmetic mean** is commonly used in Mathematics and Statistics. It can be expressed using the summation notation, since it is the sum of a collection of values divided by the number of those values:

$$\text{arithmetic mean} = \frac{1}{n} \sum_{k=1}^{n} x_k$$

Examples: Find the closed-form formula to calculate the sum of the first n even natural numbers and then the sum of the first n odd natural numbers.

Solution: $\sum_{k=1}^{n} 2k = 2 \cdot \sum_{k=1}^{n} k = 2 \cdot \frac{n(n+1)}{2} = n(n+1)$

$$\sum_{k=1}^{n} (2k-1) = \sum_{k=1}^{n} 2k - \sum_{k=1}^{n} 1 = 2 \cdot \sum_{k=1}^{n} k - n \cdot 1 = 2 \cdot \frac{n(n+1)}{2} - n$$
$$= n^2 + n - n = n^2$$

which means that all the perfect squares of natural numbers can be generated by summing consecutive odd natural numbers starting from 1.

Example in Biology: The diversity of the ground flora in a sample of two different types of woodland was tested after sampling random quadrats (framed areas), that is, by noting both the number of plant species within each quadrat and the number of individuals (i) of each species. The following data were obtained:

Tree species	$i_{\text{Sample 1}}$	$i_{\text{Sample 2}}$
1. Maple	76	13
2. Poplar	52	23
3. Pine	72	158
4. Ash	0	26
Total	200	220

Use Simpson's Diversity Index to quantify and compare the biodiversity of the two habitats: $D = 1 - \sum_{k=1}^{n} \left(\frac{i_k}{N}\right)^2$, where N is the total number of species present and k is the number of individual of each species.

Simpson's Diversity Index is used in ecology as a measure of diversity, by taking into account the number of species present, as well as the abundance of each species. The value of this index (D) ranges from 0 and 1: the greater the value, the greater the sample diversity. The index then represents the probability that two individuals randomly selected from a sample will belong to different species.

Solution: In the first sample, $D_1 = 1 - \sum_{k=1}^{3} \left(\frac{i_k}{N_1}\right)^2$

$= 1 - \left(\frac{1}{N_1}\right)^2 \cdot \sum_{k=1}^{3} (i_k)^2$ using one of the properties of the Σ operator given above (ii).

Thus, $D_1 = 1 - \left(\frac{1}{200}\right)^2 \cdot \left[(76)^2 + (52)^2 + (72)^2\right] = 0.658.$

In the second sample, $D_2 = 1 - \sum_{k=1}^{4} \left(\frac{i_k}{N_2}\right)^2 = 1 - \left(\frac{1}{N_2}\right)^2 \cdot \sum_{k=1}^{4} (i_k)^2$

$= 1 - \left(\frac{1}{220}\right)^2 \cdot \left[(13)^2 + (23)^2 + (158)^2 + (26)^2\right] = 0.456.$

Finally, we can conclude that the biodiversity of the first habitat is greater than that of the second habitat.

1.4.2 The operator Π

The product of multiple numbers can be abbreviated by using the **product operator** $\prod$, which is the capital Greek letter sigma (Π) enlarged.

The general form is

$$\prod_{k=a}^{n} X_k = X_a \cdot X_{a+1} \cdot X_{a+2} \cdot \ldots \cdot X_{n-1} \cdot X_n$$

where:

- X is the *general term* of the product (*i.e.* the numbers to be multiplied) and its expression is fixed to the variable k;
- k is the *product index*, identifies each term in the product, and indicates the first term of the product (X_a);
- n denotes the last term of the product;
- k is a variable which is an integer, and it assumes the values a, $a+1$,.....n.

Example: Calculate $\prod_{k=2}^{6} \frac{n+1}{n-2}$.

Solution: $\prod_{k=3}^{6} \frac{n+1}{n-2} = \frac{4}{1} \cdot \frac{5}{2} \cdot \frac{6}{3} \cdot \frac{7}{4} = \cancel{4} \cdot \frac{5}{\cancel{2}} \cdot \cancel{2} \cdot \frac{7}{\cancel{4}} = 35.$

Important examples of product notation to know are as follows:

- $\prod_{k=1}^{n} k = 1 \cdot 2 \cdot 3 \cdot \ldots \cdot n-1 \cdot n = \mathbf{n}!$, which is called the **factorial** of a positive integer n or '$\mathbf{n}$ **factorial**'.
- $\prod_{k=1}^{n} c = \overbrace{c \cdot c \cdot c \cdot \ldots \cdot c}^{n \text{ times}} = \mathbf{c}^{\mathrm{n}}$, where c is a positive constant independent of n.

The **geometric mean** is also used in Mathematics and Statistics, and it can be expressed using the product notation, since it is the nth root of the product of n numbers:

$$\text{geometric mean} = \sqrt[n]{\prod_{k=1}^{n} x_k}$$

Example: Find the closed-form formula to calculate the following product $\prod_{k=1}^{n}\left(1+\frac{1}{k}\right)$.

Solution:

$$\prod_{k=1}^{n}\left(1+\frac{1}{k}\right) = \left(1+\frac{1}{1}\right)\cdot\left(1+\frac{1}{2}\right)\cdot\ldots\cdot\left(1+\frac{1}{n-1}\right)\cdot\left(1+\frac{1}{n}\right)$$

$$= \frac{1+1}{1}\cdot\frac{2+1}{2}\cdot\ldots\cdot\frac{n-1+1}{n-1}\cdot\frac{n+1}{n} = \frac{2}{1}\cdot\frac{3}{2}\cdot\ldots\cdot\frac{n}{n-1}\cdot\frac{n+1}{n}$$

$$= \frac{2\cdot 3\cdot\ldots\cdot n\cdot(n+1)}{1\cdot 2\cdot\ldots\cdot(n-1)\cdot n} = \frac{(n+1)!}{n!} = \frac{\cancel{n!}\cdot(n+1)}{\cancel{n!}} = (n+1)$$

Example in Chemistry: The chemical route to synthesise Diazepam (Valium), one of the oldest and most successful drugs for the treatment of a wide spectrum of central nervous system disorders, is accomplished through a variety of chemical steps to produce reactive intermediates. One of these routes starts with an aminobenzophenone and gives the final product Diazepam after six intermediate reactions. In one attempt of synthesis, a chemist reported the percentage yield of each intermediate reaction: $y_1 = 85\%$, $y_2 = 92\%$ $y_3 = 94\%$ $y_4 = 98\%$, $y_5 = 75\%$, $y_6 = 89\%$. Estimate the overall percentage yield for this synthesis.

Solution: The yield of multiple consecutive chemical reactions is given by the formula $Y = \prod_{k=1}^{n} y_k$.

Thus, in this case, $Y = \prod_{k=1}^{6} y_k = y_1 \cdot y_2 \cdot y_3 \cdot y_4 \cdot y_5 \cdot y_6$ $= 0.85 \cdot 0.92 \cdot 0.94 \cdot 0.98 \cdot 0.75 \cdot 0.89 = 0.48 = 48\%$.

Therefore, it can be concluded that this synthesis route seems cumbersome, quite expensive, and inefficient when considering the amount of final product resulting from the starting material.

1.4.3 The operator Δ

The difference between the initial and final values of a physical quantity can be abbreviated by using the **delta operator** Δ, which is the capital Greek letter delta. The delta operator is often used to express the change in a physical quantity with time.

The general form is

$$\Delta X = X_{\text{final}} - X_{\text{initial}}$$

where:

- X is the physical entity;
- X_{final} is the final value of X;
- X_{initial} is the initial value of X.

If the value ΔX is positive, then the physical quantity, X, has increased, whereas if the value ΔX is negative, then X has decreased.

Example: The volume (V) of a gas-filled balloon changes from $46.5\,\text{dm}^3$ to $45.0\,\text{dm}^3$ when the temperature decreases from $30\,°\text{C}$ to $20\,°\text{C}$. Calculate ΔV.

Solution: $\Delta V = V_{\text{final}} - V_{\text{initial}} = 45.0\,\text{dm}^3 - 46.5\,\text{dm}^3 = -1.5\,\text{dm}^3$, which means that the balloon volume has decreased, since $\Delta V < 0$.

Example in Biochemistry: In Biochemistry, the *Gibbs free energy change* (ΔG) is used as an indicator of both the usable energy and the direction of a biochemical reaction for particular values of pressure and temperature.

If $\Delta G < 0$, the reaction delivers energy to the system and spontaneously proceeds from left to right. Alternatively, if $\Delta G > 0$, the reaction requires energy to proceed from left to right and therefore occurs spontaneously from right to left. If $\Delta G = 0$, the reaction is exactly at equilibrium.

The reaction of ATP hydrolysis to ADP at typical cellular concentrations gives $\Delta G_1 = -53\,\text{kJ}\,\text{mol}^{-1}$:

$$\text{ATP} + H_2O \longrightarrow \text{ADP} + HPO_4^{2-}$$

Explain why a negative ΔG corresponds to energy delivery to the system and why non-spontaneous biochemical reactions, such as the biosynthesis of macro-molecules driven by enzymes in cells, often occur only if they are coupled to ATP hydrolysis.

Solution: A spontaneous reaction has $\Delta G < 0$, which means that $G_{\text{final}} - G_{\text{initial}} < 0 \Longleftrightarrow G_{\text{final}} < G_{\text{initial}}$; this is consistent with a loss of energy by the reaction to the system.

A non-spontaneous biochemical reaction in cells, such as $\text{A} \not\longrightarrow \text{B}$, should have $\Delta G_2 > 0$. As ATP hydrolysis provides a relatively large negative ΔG_1, coupling both reactions could result in an overall $\Delta G < 0$, if $\Delta G_2 < -53\,\text{kJ}\,\text{mol}^{-1}$ at typical cellular concentrations of its reactants.

Therefore, ATP hydrolysis will help drive the initial reaction in the non-spontaneous direction:

$$\text{A} + \text{ATP} + \text{H}_2\text{O(s)} \longrightarrow \text{B} + \text{ADP} + \text{HPO}_4^{2-}$$

1.5 Some Geometry

Knowledge of the elementary formulae of geometry is required to correctly estimate some physical quantities in the Biosciences (*e.g.* the volume of a cell). Table 1.1 shows some useful formulae to know in **planar geometry**.

Table 1.1: Formulae in planar geometry

Figure	Perimeter	Area	Parameter Definitions
Triangle	$a+b+c$	$\frac{b \cdot h}{2}$	$a, b, c = \text{sides};\ h = \text{height}$
Parallelogram	$2(a+b)$	$b \cdot h$	$a, b = \text{sides};\ h = \text{height}$
Circle	$2\pi \cdot r$	$\pi \cdot r^2$	$r = \text{radius}$

Table 1.2 shows some useful formulae to know in **solid geometry**.

Table 1.2: Formulae in solid geometry

Figure	Surface Area	Volume	Parameter Definitions
Cuboid	$2(a \cdot b + a \cdot c + b \cdot c)$	$a \cdot b \cdot c$	$a, b, c =$ edges
Cylinder	$2\pi \cdot r(r + l)$	$\pi \cdot r^2 \cdot l$	$r =$ radius; $l =$ length
Sphere	$4\pi \cdot r^2$	$\frac{4}{3}\pi \cdot r^3$	$r =$ radius

Example: A plastic drain pipe has a diameter of $5\,\text{cm}$ and a length of $3.5\,\text{m}$; the internal empty conduit of the pipe has a diameter of $4.5\,\text{cm}$. Estimate the volume of plastic used to make the pipe.

Solution: We can first calculate both the outer and inner radii: $r_{out} = \dfrac{5\,\text{cm}}{2} = 2.5\,\text{cm}$ and $r_{in} = \dfrac{4.5\,\text{cm}}{2} = 2.25\,\text{cm}$. We should also convert the pipe length to cm: $l = 3500\,\text{cm}$.

Thus, the volume of the plastic pipe is the difference between the outer cylinder volume and the inner empty cylinder volume:

$$V_{out} - V_{in} = r_{out}^2 \pi l - r_{in}^2 \pi l = 2.5^2 \cdot 3.14 \cdot 3500 - 2.25^2 \cdot 3.14 \cdot 3500$$
$$= 68687.5\,\text{cm}^3 - 55636.875\,\text{cm}^3 = 13050.625\,\text{cm}^3 \simeq 13\,\text{dm}^3 = 13\,\text{L}$$

Example in Biology: A human blood cell is around $0.008\,\text{mm}$ in diameter. What is its volume in μm^3?

Solution: Since the radius is half the diameter, the human blood cell is around $0.004\,\text{mm} = 4\,\mu\text{m}$ in radius. The volume (V) of a blood cell can be approximated with that of a sphere, thus $V = \frac{4}{3}\pi r^3 = \frac{4}{3} \cdot 3.14 \cdot 4^3\,\mu\text{m}^3 = 268\,\mu\text{m}^3$.

1.5.1 Scaling of quantities

In Geometry, sometimes it is useful to scale some quantities up or down.

If one-dimensional entities, such as lengths, are scaled up or down by a proportional factor, k, the final length is found by multiplying (scaling up) or dividing (scaling down) the original length by the factor k.

However, if two- or three-dimensional entities, such as areas or volumes, are scaled up or down, changing a single dimension of the area or volume by a factor k results in a scaling up or down of the original quantity by a factor of k^2 or k^3, respectively.

Key examples of this case are:

- a square of twice the side l has 4 times the area;
- a sphere of twice the radius r has 8 times the volume.

1.6 Manipulation of Physical Quantities

1.6.1 Dimensional analysis

We can analyse all formulae describing physical phenomena in terms of their 'dimensions'. When describing a quantity, units are needed to define what it means physically. Nevertheless, some derived quantities in Biochemistry do not have units, such as pH.

There are seven fundamental quantities in science from which all other quantities are derived, each with a fundamental measurement unit in the International System of Units (SI):

(i) **Length** $[\mathrm{L}]$ in metres (m).
(ii) **Mass** $[\mathrm{M}]$ in kilograms (kg).
(iii) **Time** $[\mathrm{T}]$ in seconds (s).
(iv) **Temperature** $[\Theta]$ in degrees Kelvin (K).
(v) **Amount of substance** $[\mathrm{N}]$ in moles (mol).
(vi) **Electric current** $[\mathrm{I}]$ in amperes (A).
(vii) **Luminous intensity** $[\mathrm{J}]$ in candelas (cd).

Whatever the mathematical law, usually formulated in an equation describing the particular phenomenon, these separate components MUST be the same on each side of the equation and also have the same order (*e.g.* linear or square dependence).

For instance, Newton's law of motion is $F = m \cdot a$, with F being the force, m the mass, and a the acceleration: m has dimension $[\mathrm{M}]$ and a has dimensions of $[\mathrm{L}][\mathrm{T}]^{-2}$, so F MUST have dimensions $[\mathrm{M}][\mathrm{L}][\mathrm{T}]^{-2}$ which gives the SI units of Force as $\mathrm{kg} \cdot \mathrm{m} \cdot \mathrm{s}^{-2} = \mathrm{N}$ (Newton).

If a unit has a negative exponent, then it is said to be '*per that unit*': *e.g.* Newton, N, can be said as 'kilogram metre *per second squared*'. Rates of change, such as speed which is defined as the change in linear distance per time $(\mathrm{m} \cdot \mathrm{s}^{-1})$, usually employ measurement units with negative powers. For example, in the Biosciences, common rates of change are:

- zeroth-order chemical reactions measured in [product $(\mathrm{M})] \cdot \mathrm{s}^{-1}$;
- bacterial growth measured in [bacteria] $\cdot \mathrm{min}^{-1}$;
- enzymatic decomposition of substrate measured in $-$[substrate $(M)] \cdot \mathrm{s}^{-1}$;
- patient's breathing measured in [breaths] $\cdot \mathrm{min}^{-1}$.

Also, any terms in a physical equation that are added or subtracted need to have the same units in order to produce an answer that makes sense. Thus, we cannot sum $5\,\mathrm{kg} + 12\,\mathrm{s}$, since this is physically meaningless.

Example: What are the dimensions of pressure, which is the force exerted on a surface?

Solution: As Pressure $= \dfrac{\text{Force}}{\text{Area}}$ and Force has dimensions $[\mathrm{M}][\mathrm{L}][\mathrm{T}]^{-2}$ and Area $[\mathrm{L}]^2$, the ratio of the dimensions gives $\dfrac{[\mathrm{M}][\mathrm{L}][\mathrm{T}]^{-2}}{[\mathrm{L}]^2}$ $= [\mathrm{M}][\mathrm{L}]^{-1}[\mathrm{T}]^{-2}$, as the dimensions of pressure, which then has the measurement units $\mathrm{kg} \cdot \mathrm{m}^{-1} \cdot \mathrm{s}^{-2}$, corresponding to the SI unit of pascal (Pa).

Example in Chemistry: Using the ideal gas equation $PV = nRT$, where p is pressure, V volume, n number of moles, and T temperature, find the dimensions of the constant $R = \dfrac{PV}{nT}$ and its measurement units.

Solution: We should first substitute the dimensions of each parameter into the formula: $R = \dfrac{[\mathrm{M}][\mathrm{L}]^{-1}[\mathrm{T}]^{-2} \cdot [\mathrm{L}]^3}{[\mathrm{N}] \cdot [\Theta]} = \dfrac{[\mathrm{M}][\mathrm{L}]^2[\mathrm{T}]^{-2}}{[\mathrm{N}][\Theta]}$. Then, we can substitute the fundamental SI units for each physical quantity: $R = \mathrm{kg} \cdot \mathrm{m}^2 \cdot \mathrm{s}^{-2} \cdot \mathrm{mol}^{-1} \cdot \mathrm{K}^{-1}$.

1.6.2 Scientific notation and unit prefixes

Most of the numbers used in Chemistry range very widely from extremely small values, such as the mass of electron, to remarkably large ones, such as Avogadro's number. In order to simplify their expression, a set of prefixes are conventionally associated with the various units (Table 1.3).

Also, to make their expression concise and enable their easier manipulation in calculations, powers of 10 are used in the **scientific notation**

Table 1.3: SI unit prefixes

Prefix	pico	nano	micro	milli	centi	deci	kilo	mega	giga	tera
Power of 10	10^{-12}	10^{-9}	10^{-6}	10^{-3}	10^{-2}	10^{-1}	10^{3}	10^{6}	10^{9}	10^{12}
Symbol	p	n	μ	m	c	d	k	M	G	T

(or **standard form**), where the number is written as a non-null digit followed by decimal figures, all multiplied by the appropriate power of 10 (*e.g.* Avogadro's number is written as $6.02 \cdot 10^{23}\,\mathrm{mol}^{-1}$).

The **order of magnitude** of a physical quantity is **the power of** 10 **closest** to that quantity. For instance, 8, 12, and 54 have the same order of magnitude, that is, 10; whereas, the order of magnitude of 55, 95, and 320 is $100 = 10^2$.

Example: Estimate, using the scientific notation, how much coffee is drunk over a 12-year period by an adult if he/she usually drinks three cups each of $0.25\,\mathrm{L}$ per day. Also give the order of magnitude of the estimated quantity.

Solution: The adult usually drinks about $3 \cdot 0.25\,\mathrm{L} = 0.75\,\mathrm{L}$ of coffee per day. As there are 365 days in normal years and in 12 years there are 3 leap years, the volume of coffee drunk in 12 years is $0.75 \cdot 365 \cdot 9 + 0.75 \cdot 366 \cdot 3 = 3287.25\,\mathrm{L}$, which can be rounded up to $3.3 \cdot 10^3\,\mathrm{L}$.

The order of magnitude is therefore 10^3.

Example in Chemistry: Express the speed of light in vacuum using scientific notation: $c = 299792458\,\mathrm{m\,s^{-1}}$.

Solution: The speed of light is $c = 299792458\,\mathrm{m\,s^{-1}} = 2.99792458 \cdot 10^8\,\mathrm{m\,s^{-1}}$, which can be rounded up to $3.0 \cdot 10^8\,\mathrm{m\,s^{-1}}$ for ease in calculations without losing significant precision.

1.6.3 Units and conversion

Bioscientists usually use different measurement units simultaneously in their experiments and need to recognise the correct units for common quantities.

Moreover, measurements sometimes need to be converted between different metric units to give the results according to SI conventions. In some formulae, dm^3 is used for volumes rather than L (litre), degrees (°) are required instead of radians for angles, or time can be expressed in either minutes (min) or hours (h) in addition to seconds (s). Careful manipulation of the units to be converted is essential.

Importantly, when substituting numerical values into formulae in order to find the value of another physical quantity, it is essential to substitute the values consistently in their SI form.

Thus, using measurement units within the same decimal scale may often require the conversion of the units into different multiples, and this involves multiplying or dividing by the appropriate factor, usually a power of 10, as shown in Table 1.4.

Table 1.4: Commonly used conversions between units

Length	$1\,\mathrm{m} = 10\,\mathrm{dm} = 10^2\,\mathrm{cm} = 10^3\,\mathrm{mm} = 10^6\,\mu\mathrm{m} = 10^9\,\mathrm{nm}$ $10^{-1}\,\mathrm{dm} = 1\,\mathrm{cm} = 10\,\mathrm{mm}$
Area	$1\,\mathrm{m}^2 = 10^2\,\mathrm{cm}\cdot 10^2\,\mathrm{cm} = 10^4\,\mathrm{cm}^2 = 10^3\,\mathrm{mm}\cdot 10^3\,\mathrm{mm} = 10^6\,\mathrm{mm}^2$ $1\,\mathrm{m}^2 = 10^6\,\mu\mathrm{m}\cdot 10^6\,\mu\mathrm{m} = 10^{12}\,\mu\mathrm{m}^2 = 10^9\,\mathrm{nm}\cdot 10^9\,\mathrm{nm} = 10^{18}\,\mathrm{nm}^2$
Volume	$1\,\mathrm{mL} = 1\,\mathrm{cm}^3 = 1\,\mathrm{cc}$ $1\,\mathrm{L} = 1000\,\mathrm{mL} = 1000\,\mathrm{cc}$ $1\,\mathrm{cm}^3 = 10^{-2}\,\mathrm{m}\cdot 10^{-2}\,\mathrm{m}\cdot 10^{-2}\,\mathrm{m} = 10^{-6}\,\mathrm{m}^3 = 10^{-3}\,\mathrm{dm}^3$
Density	$1\,\mathrm{g}\cdot\mathrm{mL}^{-1} = 1\cdot 10^3\,\mathrm{g}\cdot(1\cdot 10^3\,\mathrm{mL}^{-1}) = 1\,\mathrm{kg}\cdot\mathrm{dm}^{-3}$
Molarity	$1\,\mathrm{M} = 10^3\,\mathrm{mM} = 10^6\,\mu\mathrm{M} = 10^9\,\mathrm{nM}$ $1\,\mathrm{nM} = 10^{-3}\,\mu\mathrm{M} = 10^{-6}\,\mathrm{mM} = 10^{-9}\,\mathrm{M}$

Table 1.4 is particularly helpful for Biochemistry experiments, where volumes are usually measured in L or mL or μL and sometimes need to be readily converted to cm^{-3} or dm^{-3} or m^{-3} in formulae: thus, $1\,\mathrm{L} = 10^3\,\mathrm{cm}^3 = 1\,\mathrm{dm}^3 = 10^{-3}\,\mathrm{m}^3$, $1\,\mathrm{mL} = 10^{-3}\,\mathrm{dm}^3$, and $1\,\mu\mathrm{L} = 10^{-3}\,\mathrm{cm}^3$.

It should be noted that sometimes non-SI units are used in science for convenience:

- The **Angström** (abbreviated as Å) is a length unit with $1\,\text{Å} = 10^{-10}\,\mathrm{m}$ and is regularly used to describe inter-atomic distances.

- The **Dalton** (abbreviated as Da or u) is a unit of mass, commonly used for specifying the mass of proteins. One Dalton is a unified atomic mass unit and is equal in value to the mass of $\frac{1}{12}$ of a nuclide of ^{12}C:

$$1\,\mathrm{Da} = 1\mathrm{u} = 1.6605402 \cdot 10^{-27}\,\mathrm{kg}$$

Example: In an ancient Roman house (*domus*), 90,000 square tiles each of surface area $1.7\,\mathrm{cm}^2$ are used to restore a mosaic covering the floor of a bedroom (*cubiculum*). Estimate the surface area of the room floor in m^2.

Solution: We can multiply out $90{,}000 \cdot 1.7\,\mathrm{cm}^2 = 153{,}000\,\mathrm{cm}^2$ and convert the result to $15.3\,\mathrm{m}^2$ using Table 1.4.

Example in Biology: A $5\,\mathrm{mL}$ starter culture of *Escherichia coli* was left to incubate at $37\,°\mathrm{C}$ in a final $1\,\mathrm{L}$ volume of an appropriate glucose-salt medium. After $12\,\mathrm{h}$, a sample of $1\,\mathrm{mL}$ was collected and then serial dilutions were used to prepare a 100-times diluted sample of $1\,\mathrm{mL}$, of which $10\,\mu\mathrm{L}$ were used to streak an agar place to isolate individual colonies. Sixteen distinct colonies were finally found on the streak plate after $12\,\mathrm{h}$ incubation at $37\,°\mathrm{C}$. Estimate how many bacteria were in the original culture after a $12\,\mathrm{h}$ incubation at $37\,°\mathrm{C}$.

Solution: In Microbiology, dilutions are commonly used to work out how many bacteria are in the original culture.

Sixteen colonies were generated from the bacteria which were in the $10\,\mu\mathrm{L}$ sample. Therefore, we need to use dilution factors and unit conversions to work out the number of bacteria in the bulk culture after a $12\,\mathrm{h}$ incubation.

We can set up the following table to summarise the calculations:

Sample	Streaked	Volume factor	Diluted	Dilution factor	Collected	Volume factor	Culture
Volume	$10\,\mu\mathrm{L}$	$\xrightarrow{\times 100}$	$1000\,\mu\mathrm{L} = 1\,\mathrm{mL}$	$\xrightarrow{\times 100}$	$1\,\mathrm{mL}$	$\xrightarrow{\times 1000}$	$1000\,\mathrm{mL} = 1\,\mathrm{L}$
Bacteria	16	$\xrightarrow{\times 100}$	$1600 = 16 \cdot 10^2$	$\xrightarrow{\times 100}$	$16 \cdot 10^4$	$\xrightarrow{\times 1000}$	$16 \cdot 10^7$

Thus, 160,000,000 ($1.6 \cdot 10^8$) bacteria were in the $100\,\mathrm{mL}$ culture after a $12\,\mathrm{h}$ incubation at $37\,°\mathrm{C}$. This is just an estimate, as it would be impossible to count all the bacteria individually!

1.6.4 Errors

Errors occur in all physical measurements in science, and when a measurement is the result of a combination of measurements, the errors in the initial measurements are carried through to the final result.

Two different types of error can occur in measurements:

- **systematic error**, which occurs to the same extent in every measurement performed on the same instrument, *e.g.* when using a balance to weigh chemicals;
- **random error**, which occurs in the measurements as a result of variations in using the measurement technique, *e.g.* experimenter, instrument manipulation and limit of reading accuracy, and local temperature.

In science, either the **absolute error** or the **relative error** is quoted when the measured quantity is required.

The **absolute error** does not depend on the actual measurement. For example,

- the number of people in a room has no error possible;
- the length of a ruler, for which the absolute error is half of the smallest unit of measurement; thus, if the smallest measure is $0.1\,\mathrm{cm}$, then the error is $0.05\,\mathrm{cm}$ above and below that, *e.g.* $20 \pm 0.05\,\mathrm{cm}$.

The **relative error** is more useful since it is the absolute error in relation to the actual measurement and can be expressed as a fraction of the actual measured quantity or as percentage errors. Thus,

$$\text{relative error} = \frac{\text{absolute error}}{\text{measurement}}$$

For instance, if the instrument's absolute error is $0.05\,\mathrm{cm}$, then the relative error in $20\,\mathrm{cm}$ is $\frac{0.05}{20} = 0.25\%$, whereas for $0.2\,\mathrm{cm}$ it is $\frac{0.05}{0.2} = 25.0\%$!

Conventionally, it is always useful to calculate the maximum errors so that in practice the real errors are probably less than these.

Example: A set of weighing scales has $0.1\,\mathrm{g}$ as its smallest unit of measurement. What is the relative error when weighing out $5\,\mathrm{g}$ of substance?

Solution: The absolute error is half the value of the smallest measurement, *i.e.* $0.05\,\mathrm{g}$.

$$\text{Thus, relative error} = \frac{\text{absolute error}}{\text{measurement}} = \frac{0.05}{5} = 0.01 = 1\%.$$

Example in Medicine: Commercial glucose meters are medical devices for determining the approximate concentration of glucose in the blood and commonly have a maximum relative error of 15% according to the International Organisation for Standardisation (ISO) 15197. Estimate the maximum range of relative uncertainty for a blood glucose value of $105\,\mathrm{mg\,dl^{-1}}$ measured using a commercial glucose meter.

Solution: We can estimate the absolute error in the actual measurement:

$$\begin{aligned}\text{absolute error} &= \text{measurement} \cdot \text{relative error} = 0.15 \cdot 105\,\mathrm{mg\,dl^{-1}} \\ &= 16\,\mathrm{mg\,dl^{-1}}\end{aligned}$$

Therefore, the interval of uncertainty is $105 \pm 16\,\mathrm{mg\,dl^{-1}}$.

1.6.5 Significant figures

The final result of any calculation can have **no more significant figures** than ANY of the original measurements. It is therefore important not to quote excessive accuracy in the experimental results.

If a number needs to be rounded to have the correct number of significant figures, first the required number or decimal points after rounding should be defined. Then, the number should be either left the same if the next digit is < 5 (**rounding down**) or increased by 1 if the next digit is ≥ 5 (**rounding up**).

Example: Find the product of $73.24 \cdot 4.52$ using the correct number of significant figures.

Solution: The number 73.24 has 4 significant figures, so the error is one in ten thousand, while the number 4.52 has 3 significant figures, so the error is one in a thousand.

The product of these two numbers should therefore be recorded as $73.24 \cdot 4.52 = 331$, with 3 significant figures, NOT 331.0448 with 7 significant figures and an error of one in ten million, NOR 331.04 with 5 significant figures and an error of one in a hundred thousand.

The final result of 331 should be left as it is, since the next digit after 1 is 0, which is < 5.

Example in Chemistry: Calculate the number of moles of sodium chloride ($M_w = 58.4\,\mathrm{g\,mol^{-1}}$) present in $10\,\mathrm{mL}$ of a $0.5\,\mathrm{mg\,ml^{-1}}$ solution to the appropriate number of significant figures (sig. figs).

Solution: The value quoted with the smallest number of significant figures is the concentration of the solution (two sig. figs), therefore the final answer should also be quoted only to two sig. figs.

There are $0.5\,\mathrm{mg}$ NaCl per $1\,\mathrm{mL}$ of solution, therefore $0.5\,\mathrm{mg\,mol^{-1}} \cdot 10\,\mathrm{mL} = 5\,\mathrm{mg}$ in total.

One mole of NaCl weighs $58.4\,\mathrm{g}$.

Thus, $5\,\mathrm{mg} = 0.005\,\mathrm{g}$ correspond to $\dfrac{0.005\,\mathrm{g}}{58.4\,\mathrm{g\,mol^{-1}}}$ $= 8.86 \cdot 10^{-5}\,\mathrm{mol} = 88.6\,\mu\mathrm{mol} = 89\,\mu\mathrm{mol}$ to two sig. figs.

1.7 Exercises

(1) Calculate and express the answers as fractions in their simplest form for the following sums and differences. Do not use a calculator!

(a) $\frac{1}{2} + \frac{1}{3}$ (b) $\frac{1}{4} + \frac{2}{5}$

(c) $\frac{2}{3} + \frac{4}{7}$ (d) $\frac{19}{14} + \frac{1}{6}$

(e) $\frac{13}{16} + \frac{1}{32}$ (f) $\frac{1}{2} - \frac{1}{6}$

(g) $\frac{5}{6} - \frac{1}{12}$ (h) $\frac{3}{10} - \frac{2}{15}$

(i) $\frac{9}{11} - \frac{4}{5}$ (j) $\frac{27}{30} - \frac{1}{5}$

(k) $\frac{2}{13} - \frac{2}{169}$ (l) $\frac{1}{4} - \frac{1}{144}$

(2) Calculate and express the answers as fractions in their simplest form for the following products. Do not use a calculator!

(a) $\frac{2}{5} \times \frac{3}{8}$ (b) $\frac{11}{15} \times \frac{3}{4}$

(c) $\frac{1}{9} \times \frac{7}{6}$ (d) $\frac{16}{17} \times \frac{1}{4}$

(e) $\frac{-5}{3} \times \frac{6}{7}$ (f) $\frac{1}{2} \div \frac{1}{3}$

(g) $\frac{4}{9} \div \frac{2}{9}$ (h) $\frac{4}{9} \div \frac{1}{2}$

(i) $\frac{-4}{9} \div 2$ (j) $6 \times \frac{1}{3} \div 3 \times \frac{1}{4}$

(k) $\frac{8}{39} \div \frac{1}{3}$ (l) $\frac{5}{6} \div \frac{15}{9}$

(m) $\frac{1}{6} \div \frac{1}{5}$ (n) $\frac{-1}{3} \div \frac{-1}{15}$

(o) $\frac{99}{100} \div 33$ (p) $\frac{7}{-46} \div \frac{-5}{23}$

(3) Calculate the specified percentages of these numbers

(a) 16% of 50 (b) 5% of 120 (c) 25% of 144

(d) 0.7% of 84 (e) 1.3% of 1600 (f) 2.5% of 364

(4) Calculate the following percentages, rounded to two significant figures.

(a) 48 is ...% of 600 (b) 18 is ...% of 450 (c) 135 is ...% of 540

(d) 86 is ...% of 1155 (e) 0.9 is ...% of 18 (f) 4.5 is ...% of 360

(5) Calculate x if:

(a) 15 is 3% of x (b) 25 is 20% of x (c) 210 is 14% of x

(d) 125 is 2.5% of x (e) 24 is 0.12% of x (f) 75 is 0.3% of x

(6) Solve the following problems using percentages.

(a) The population of Oxford in 2011 was 150,200 and had increased to $159,950$ in 2015. What was the percentage increase?

(b) The diesel fuel price has increased from 126.7p to 128.9p per litre in the last couple of months in the UK. What has been the percentage increase?

(c) A pharmaceutical company used to sell a drug in packs of $100\,\text{g}$ at a price of $£10$ each; however, the marketing manager has now decided to reduce each pack to $80\,\text{g}$, keeping the price at $£10$. By what percentage has the drug price increased?

(d) A $2\,\mathrm{L}$ volume contains 75% ethanol and the remainder is water. How much ethanol should be added to have a final mixture containing 80% ethanol?

(e) Calculate the percentage % (weight/weight) of a solution prepared by dissolving $2.5\,\mathrm{g}$ of sodium chloride (NaCl) in $400\,\mathrm{g}$ of water.

(f) Estimate how much calcium chloride ($CaCl_2$) salt should be weighed out to prepare $650\,\mathrm{g}$ of a solution which is 24% (weight/weight) salt. In how much water (in g) should the salt be dissolved?

(7) Simplify the following expressions by collecting like terms.

(a) $6x + 15x + 7y + 9y$ (b) $15x - 6x - 3y + 10y$

(c) $-15x - 6x - 15y - 6y$ (d) $-15y + 4x - 6y + 7y + 3x$

(e) $6x^2 + 15x^2 + 7x^2$ (f) $3xy + 4z - yx + y^2 - z + 3zy - 3y^2$

(8) Simplify the following expressions by multiplying them out.

(a) $-3(z - y)$ (b) $4(2y - 6z)$

(c) $-y\left(4y - \dfrac{8}{y}\right)$ (d) $xy\left(\dfrac{2z}{x} - \dfrac{3z}{y}\right)$

(9) Simplify these fractions by cancelling terms.

(a) $\dfrac{6z}{zv}$ (b) $\dfrac{5xy}{xy}$ (c) $\dfrac{-10ab}{-5ab}$ (d) $\dfrac{yz}{xy}$

(10) Write down formulae to represent the following relationships. Remember to define all the variables and constants.

(a) The radius of a circle is half the diameter.

(b) A flea can jump 100 times higher in cm than its weight in g.

(c) The distance cycled by my daughter every week is: cycling to school and back each weekday, to her friend's house twice a week, and to the swimming pool once a week.

(d) The annual consumption of blades of grass by a horse if it eats for a quarter of the day and consumes as many blades of grass in an hour as hairs on its back.

(e) My net pay at the end of the month is the gross amount with my pension and national insurance contributions subtracted (these payments are untaxed), and then with 24% tax deducted from what remains after taking into account my tax free annual allowance.

(11) Calculate results for the following expressions, first if $x = 6$, $y = 3$, and $z = -5$, and then if $x = \dfrac{1}{3}$, $y = \dfrac{1}{4}$, $z = \dfrac{1}{5}$.

(a) $3xy + 5zy - 6xz$ (b) $z(y + 2x)$

(c) $\dfrac{xyz}{x + y + z}$ (d) $y\left(\dfrac{1}{x} - \dfrac{1}{z}\right)$

(12) Find the value of the variable in brackets below, using the data provided.

(a) $[P] = L + 2W$ when $L = 1\,\mathrm{m}$ and $W = 10\,\mathrm{cm}$

(b) $[V] = \pi \times l \times l \times d$ when $l = 10\,\mathrm{cm}$ and $d = 15\,\mathrm{cm}$

(c) $[v] = 3\left(\frac{1}{x} - \frac{1}{y}\right)$ when $x = 2$ and $y = 3$

(13) Write the numbers in exponent form, identifying the lowest base and the corresponding exponent.

(a) 125 (b) 81 (c) 169 (d) 64 (e) 400 (f) 256

(14) Calculate the value of the following expressions.

(a) 2^3 (b) 3^5 (c) 4^1 (d) 5^4 (e) 7^3 (f) 11^2

(15) Simplify the following expressions by manipulating the indices.

(a) $5^4 \cdot 5^2$ (b) $a^{12} \cdot a^{11}$ (c) $\left(3^0\right)^2$

(d) $\left(4x^5\right)^3$ (e) $\frac{x^3 \cdot x^2}{x^4}$ (f) $\frac{a^{13}}{a^{11}}$

(g) $\frac{12^2 \cdot 3}{2^3 \cdot 3^3}$ (h) $\frac{30^2 \cdot 5^3}{2^2 \cdot 3 \cdot 5^5}$ (i) $(3y)^4$

(j) $\left(s^3 r^4\right)^3$ (k) $(ab)^3 \cdot \left(a^2 b^2\right)^2$ (l) $2a + b^2 + a + a^2$

(m) $\frac{a^3 b^4 c^2 \cdot abc}{a^2 b^5 c^3}$ (n) $\left(\frac{b^4}{a^2}\right)^3$ (o) $\left(\frac{2x^2 y}{ab^2 c}\right)^2 \cdot \left(\frac{4a^2 bc}{3xy}\right)^3$

(16) Simplify the expressions by manipulating the indices and roots.

(a) $\frac{x^{\frac{1}{3}} \cdot x^{-\frac{4}{3}}}{x^{-1}}$ (b) $\frac{\left(x^{\frac{1}{2}}\right)^4}{x^3 \cdot x^2}$ (c) $\frac{\left(\sqrt{x}\right)^6}{x^5}$

(d) $\frac{x^2 \cdot \sqrt{x}}{x^{3.5}}$ (e) $\sqrt{\frac{x}{y}} \div \sqrt[6]{\frac{y}{x}} \cdot \sqrt[3]{y^2}$ (f) $\left(\frac{y}{8x^3}\right)^{\frac{1}{3}} \cdot \left(\frac{1}{4}x^{-2}y^4\right)^{\frac{1}{2}}$

(17) Calculate the results for the following expressions if $x = -2$, $y = \frac{1}{3}$, and $z = -\frac{1}{4}$.

(a) $-|-3x + |3 - 4x + y|| - |-5yz|$

(b) $|7 - 3yz| + |2 - |-2yz + 5|| \cdot \left|-\frac{3}{2}x\right|$

(c) $|xy - 1| - |-3yz + 7|$

(d) $\left|\frac{5}{2}x^2 y - 1\right| - \left|-2\frac{y}{z} + x\right|$

(18) Expand the following sum/product notations and then express them using closed-form formulae, *i.e.* without using the sum/product notations (n denotes a positive integer and c a positive constant).

(a) $\sum_{k=1}^{n}(n-2k)$ (b) $\sum_{k=1}^{n}(6k^2-1)$ (c) $\sum_{k=1}^{n}\left(c^k+\frac{1}{c^k}\right)$

(d) $\prod_{k=1}^{n}(k^2-9)$ (e) $\prod_{k=1}^{n}c^k$ (f) $\prod_{k=1}^{n}k^n$

(19) Solve the following problems in geometry.

(a) The average radius of a hydrogen atom is $25\,\mathrm{pm}$. Imagining that the orbital of its single electron around the nucleus is planar, calculate the circumference and the area of the orbital.

(b) The 2s electron orbital in an atom is spherical; the average distance of the 2s electron from the nucleus in a lithium atom is $145\,\mathrm{pm}$. Calculate the volume and the surface area of the 2s orbital.

(c) *Bacillus subtilis* is a genus of Gram-positive, rod-shaped bacteria and its form can be approximated as a cylinder $7\,\mu\mathrm{m}$ in length with a $0.5\,\mu\mathrm{m}$ radius. Estimate its volume and its outer membrane cell surface area.

(d) A cuboid protein crystal is $100\ \mu\mathrm{m}$ in length, $50\ \mu\mathrm{m}$ in width, and $140\ \mu\mathrm{m}$ in height. Find the volume of the crystal in m^3 and the external surface area in m^2.

(20) Calculate the dimensions of the physical quantities below (*abbreviations:* K = constant; E = energy; Kin. = kinetic; Pot. = potential).

(a) Speed $= \dfrac{\text{distance}}{\text{time}}$ (b) Acceleration $= \dfrac{\text{speed}}{\text{time}}$

(c) Density $= \dfrac{\text{mass}}{\text{volume}}$ (d) Force = mass $\cdot$ acceleration

(e) Kin. E $= \dfrac{1}{2}\text{mass}\cdot(\text{speed})^2$ (f) Pot. E = mass $\cdot$ acceleration $\cdot$ height

(g) Centrifugal $K = \dfrac{\text{force}\cdot\text{distance}}{\text{mass}\cdot(\text{speed})^2}$ (h) Gravity $K = \dfrac{\text{force}\cdot(\text{distance})^2}{(\text{mass})^2}$

(21) Solve the following problems using the appropriate measurement units and scientific notation.

(a) How many moles of ATP are there in $25\,\mu\mathrm{L}$ of a $3\,\mathrm{mM}$ solution?

(b) What is the weight of ATP in the $25\,\mu\mathrm{L}$ of solution, if the molecular weight (M_w) of ATP $= 573\,\mathrm{g\,mol^{-1}}$?

(c) The molecular weight of hen egg white lysozyme (HEWL) is $14.7\,\mathrm{kDa}$. How many moles of HEWL are there in $20\,\mathrm{mL}$ of a solution with a concentration of $30\,\mathrm{mg\,ml^{-1}}$?

(d) 10 mL of H_2O was then added to the HEWL solution above. What is the new concentration (in M)?

(e) 25 μmol of bovine serum albumin (BSA, $M_w = 69.3$ kDa) was added to 15 mL buffer. Assuming no change in volume after the BSA dissolved, express the concentration in mg ml^{-1}.

(f) To conduct an NMR study on protein structure, a protein needs to be concentrated to 10 mM. The protein of interest is dissolved in 2 mL buffer at a concentration of 50 $\mu\text{mol ml}^{-1}$. What volume of buffer would be needed to bring this amount of protein to the appropriate concentration?

(22) Solve the following problems on absolute and relative errors.

(a) A Gilson laboratory pipette has a maximum capacity of 200 μL and the smallest unit of measurement is 1 μL. What is the absolute error of the measurement? What is the relative error, in %, when pipetting 10μL of solution? What is the error for 150 μL of the solution?

(b) A different Gilson pipette has a maximum capacity of 20 μL and an absolute error of 0.05 μL. What is the smallest unit of measurement possible for this pipette? Which method produces more relative error: pipetting 100 μL in one go with the 200 μL pipette, or five times 20 μL with the 20 μL pipette?

(23) Use the appropriate number of significant figures for the calculation of the result and error for the following products:

(a) $819 \cdot 4.914$

(b) $81 \cdot \pi$

(c) $0.024 \cdot 361$

Answers

(1) (a) $\frac{5}{6}$ (b) $\frac{13}{20}$ (c) $\frac{5}{21}$ (d) $\frac{11}{21}$ (e) $\frac{27}{32}$ (f) $\frac{1}{3}$
(g) $\frac{3}{4}$ (h) $\frac{1}{6}$ (i) $\frac{1}{55}$ (j) $\frac{7}{10}$ (k) $\frac{24}{169}$ (l) $\frac{35}{144}$

(2) (a) $\frac{3}{20}$ (b) $\frac{11}{20}$ (c) $\frac{7}{54}$ (d) $\frac{4}{17}$ (e) $-\frac{3}{7}$ (f) $\frac{1}{2}$
(g) 2 (h) $\frac{8}{9}$ (i) $-\frac{2}{9}$ (j) $\frac{8}{3}$ (k) $\frac{8}{13}$ (l) $\frac{1}{2}$
(m) $\frac{5}{5}$ (n) 5 (o) $\frac{3}{100}$ (p) $\frac{7}{10}$

(3) (a) 8 (b) 6 (c) 36 (d) 0.588 (e) 20.8 (f) 9.1

(4) (a) 8% (b) 4% (c) 25% (d) 7.4% (e) 5% (f) 1.25%

(5) (a) 500 (b) 125 (c) 1,500 (d) 5,000 (e) 20,000 (f) 25,000

(6) (a) 6.5% (b) 1.7% (c) 2 £ each, 20% increase (d) 0.5 L (e) 0.6%
(f) 156 g $CaCl_2$; 494 g H_2O

(7) (a) $21x + 16y$ (b) $9x + 7y$ (c) $-21x - 21y$ (d) $-14y + 7x$
(e) $28x^2$ (f) $2xy + 3z - 2y^2 + 3zy$

(8) (a) $-3z + 3y$ (b) $8y - 24z$ (c) $-4y^2 + 8$ (d) $2zy - 3zx'$

(9) (a) $\frac{6}{v}$ (b) 5 (c) 2 (d) $\frac{z}{x}$

(10) (a) $r = d/2$ (b) $H = 100x$ (c) $D = 5x + 2y + z$ (d) $C = 2190h$
(e) $T = 0.76(G - P - I - A) + A$ (T = net pay, G = gross, P = pension, I = national insurance, A = tax free allowance)

(11) (a) $159; \frac{1}{10}$ (b) $-75; \frac{11}{60}$ (c) $22.5; \frac{1}{47}$ (d) $\frac{11}{10}; -\frac{1}{2}$

(12) (a) $120\,\mathrm{cm}$ (b) $4710\,\mathrm{cm}^3$ (c) $\frac{1}{2}$

(13) (a) 5^3 (b) 3^4 (c) 13^2 (d) 2^6 (e) 20^2 (f) 2^8

(14) (a) 8 (b) 243 (c) 4 (d) 625 (e) 343 (f) 121

(15) (a) 5^6 (b) a^{23} (c) 3^6 (d) 3^6 (e) x (f) a^2 (g) 2
(h) 3 (i) $81y^4$ (j) s^9r^{12} (k) $(a \cdot b)^7$ (l) $3a + b^2 + a^2$
(m) a^2 (n) $\frac{b^{12}}{a^6}$ (o) $\frac{256a^4cx}{27by}$

(16) (a) 1 (b) x^{-3} (c) x^{-2} (d) x^{-1} (e) $\sqrt[3]{x^2}$ (f) $4x^{-2}\sqrt[3]{y^7}$

(17) (a) $-\frac{71}{4}$ (b) $\frac{67}{4}$ (c) $-\frac{67}{12}$ (d) $\frac{5}{3}$

(18) (a) n (b) $2n^3 + 3n^2$ (c) $\frac{1-c^{n+1}}{1-c}\left(1 + \frac{1}{c^n}\right)$ (d) 0
(e) $c^{\frac{n(n+1)}{2}}$ (f) $n!^n$

(19) (a) $C = 157\,\mathrm{pm}$; $A = 1936\,\mathrm{pm}^2$ (b) $V = 0.013\,\mathrm{nm}^3$; $S = 0.26\,\mathrm{nm}^2$
(c) $V = 5.5\,\mu\mathrm{m}^3$; $S = 23.6\,\mu\mathrm{m}^2$ (d) $V = 7 \cdot 10^{-25}\,\mathrm{m}^3$; $S = 52\,\mathrm{m}^{-17}$

(20) (a) $[\mathrm{L}][\mathrm{T}]^{-1}$ (b) $[\mathrm{L}][\mathrm{T}]^{-2}$ (c) $[\mathrm{M}][\mathrm{L}]^{-3}$ (d) $[\mathrm{M}][\mathrm{L}][\mathrm{T}]^{-2}$
(e) $[\mathrm{M}][\mathrm{L}]^2[\mathrm{T}]^{-2}$ (f) $[\mathrm{M}][\mathrm{L}]^2[\mathrm{T}]^{-2}$ (g) No dimension (h) $[\mathrm{L}]^3[\mathrm{T}]^{-2}[\mathrm{M}]^{-1}$

(21) (a) $75\,\mathrm{mmol}$ (b) $43\,\mu\mathrm{g}$ (c) $408\,\mathrm{nmol}$ (d) $1.36 \cdot 10^{-25}\,\mathrm{M}$
(e) $120\,\mathrm{mg\,ml}^{-1}$ (f) $8\,\mathrm{mL}$

(22) (a) $0.5\,\mu\mathrm{L}$; 5%; 0.33% (b) $0.1\,\mu\mathrm{L}$; 0.5% *versus* 1.25%

(23) (a) 4020, 3 sig. figs, 0.1% error (b) 250, 2 sig. figs, 1% error
(c) 8.66, 3 sig. figs, 1% error

Chapter 2

Basic Algebra

Preamble

Algebra is the branch of Mathematics that uses alphabetical letters (usually called variables) to represent parameters with unknown values in a mathematical expression. The known values, such as numbers, are called constants.

Algebraic expressions usually contain variables and constants along with operations, such as addition, subtraction, multiplication, and division.

Algebra is a powerful tool for problem solving in Biosciences and many other disciplines, since using letters in place of words in mathematical expressions saves writing many pages for each problem and thus makes things succinct and avoids confusion.

2.1 Multiplication and Factorisation

2.1.1 Multiplying algebraic terms

- **Product of polynomials**

The product of $(a+b)(c+d)$ is calculated using the so called 'Binomial expansion' as follows:

$$(a+b)\cdot(c+d) = a\cdot(c+d) + b\cdot(c+d) = ac + ad + bc + bd$$

i.e. we perform four multiplications and we have four resulting terms.

Example: Multiply out $\left(\frac{x}{2}+7\right)(x-2)$.

Solution: $\left(\frac{x}{2}+7\right)(x-2) = \frac{x}{2}\cdot(x-2)+7\cdot(x-2)$
$= \frac{x}{2}\cdot x + \frac{x}{2}\cdot(-2)+7\cdot x+7\cdot(-2) = \frac{x^2}{2}+6x-14.$

• Quadratic expressions

Those expressions containing a term in x^2 are called **quadratic expressions**.

Generally, we have:

$$\text{(i)}\quad (a+b)^2 = a^2+2ab+b^2$$
$$\text{(ii)}\quad (a-b)^2 = a^2-2ab+b^2$$
$$\text{(iii)}\quad (a-b)(a+b) = a^2-b^2$$

Important note: $(a+b)^2 \neq a^2+b^2$

Example: Multiply out $(2x-3y)^2$ using the Binomial Expansion.

Solution: $(2x-3y)^2 = (2x)^2-2\cdot 2x\cdot 3y+(-3y)^2 = 4x^2-12xy+9y^2$.

• Binomial Series

A Binomial Series is the series of terms resulting from an expansion of expressions which have the form

$$(1+x)^n = 1+\frac{nx}{1}+\frac{n(n-1)}{2!}x^2+\frac{n(n-1)(n-2)}{3!}x^3+\cdots+nx^{n-1}+x^n$$

If n is a positive integer, this series has $(n+1)$ terms and will terminate; the expression is true for *any* value of x.

The Binomial Expansion can be generalised for the case where n is a positive integer:

$$(a+b)^n = a^n + na^{n-1}b + \frac{n(n-1)}{2!}a^{n-2}b^2 + \frac{n(n-1)(n-2)}{3!}a^{n-3}b^3 + \cdots + b^n$$

Example: Multiply out $(2-x)^3$.

Solution: $(2-x)^3 = 2^3 + 3\cdot 2^{3-1}(-x) + \dfrac{3(3-1)}{2!}2^{3-2}(-x)^2 + (-x)^3$
$= 8 - 12x + 6x - x^3$.

2.1.2 Factorisation

Factorisation is the breaking down of algebraic terms into components that have been multiplied together.

- **Common factors**

Often we want to reverse (or 'undo') the multiplication process and find the 'factors' which give the product.

We find the 'highest common factor' in the terms and take it out of all the terms:

$$ab + ca + ad = a\cdot(b+c+d)$$

However, the expression $ac + ad + bc + bd$ has no factor which is common in all four terms, but we notice a factor common (a) in the first two terms and another factor common (b) in the last two terms. So, we can partially collect the common factors and then further collect the common expression $(c+d)$:

$$ac + ad + bc + bd = a\cdot(c+d) + b\cdot(c+d) = (a+b)\cdot(c+d)$$

Example: Factorise $-5xyz - 10x^2yz^3 + 20xy^3z^2$.

Solution: $5xyz$ is the factor common in every term.

Thus: $-5xyz - 10x^2yz^3 + 20xy^3z^2 = 5xyz(4y^2z - 1 - 2xz^2)$.

Example: Factorise $cx + xz - 2az - 2ac$.

Solution: The expression has no factor which is common in all four terms, but we notice that x is a common factor in the first two terms and $-2a$ is a common factor in the last two terms.

So, $cx + xz - 2az - 2ac = x(c + z) - 2a(z + c) = (x - 2a)(z + c)$.

Warning! There is another way to factorise this expression: c is a common factor in the first and the fourth terms, and z is a common factor in the the second and the third terms. Whatever the way chosen, the result is the same and can easily be checked by multiplying out the expression again to ensure that it gives the same as the starting equation.

• Quadratic expressions

To factorise the quadratic expressions, two cases should be distinguished:

(i) Those where the coefficient of x^2 is 1:

$$x^2 + bx + c = (x + g)(x + h) = x^2 + (g + h)x + (g \cdot h)$$

so $c = g \cdot h$ and $b = g + h$: g and h are the factors of the constant term (which is c), whereas the algebraic sum of g and h is equal to the coefficient of x (which is b).

(ii) Those where the coefficient of x^2 is **NOT** 1.
We want to factorise the quadratic expression to give an expression of the form $(qx + r)(px + s)$:

$$ax^2 + bx + c = (q \cdot p)x^2 + (qs + rp)x + rs$$

So, we have $c = rs$, $b = qs + pr$, $a = qp$.

Thus, $ca = qprs$ and among the factors of ca are (qr and ps), (qp and rs), and (pr and qs). It is the last of these pairs which we want.

The method involves *finding the pair of factors of ac which will add up to b.*

Example: Factorise $x^2 + 10x + 24$.

Solution: Here $a = 1$ and $c = 24$.

The factors of 24 are (2 and 12) or (6 and 4) or (1 and 24). The coefficient of x is 10, which is the sum of 6 and 4.

Hence, the factorisation of $x^2 + 10x + 24$ is $(x + 4)(x + 6)$.

The answer should be checked by multiplying the brackets out again.

Example: Factorise $6x^2 + 19x + 10$.

Solution: Here $a = 6, b = 19, c = 10$. We notice that $ac = 60$ has factors (10 and 6), (15 and 4), (2 and 30), (12 and 5), (3 and 20), and (1 and 60).

Since $b = 19$ the pair required is (15 and 4) so that $4 = rp$ and $15 = qs$. The factors of 4 and 15 will now give r, p, q, and s individually.

So we can write:

$$\begin{aligned} 6x^2 + 19x + 10 &= 6x^2 + 4x + 15x + 10 = 2x(3x + 2) + 5(3x + 2) \\ &= (3x + 2)(2x + 5) \end{aligned}$$

• Difference of two squares

We can factorise the difference of two squares as follows:

$$x^2 - a^2 = (x + a)(x - a)$$

The *sum of two squares* has **NO real** factors: *e.g.* $x^2 + 9$. Complex numbers are required to factorise them (see Chapter 7, Section 7.3).

Example: Factorise $x^4 - 9$.

Solution: $x^4 - 9 = (x^2 + 3)(x^2 - 3) = (x^2 + 3)(x + \sqrt{3})(x - \sqrt{3})$.

• Perfect squares

Expressions containing the terms of a quadratic expression can be factorised into perfect squares:

$$\text{(i)} \quad a^2 + 2ab + b^2 = (a + b)^2$$

$$\text{(ii)} \quad a^2 - 2ab + b^2 = (a - b)^2$$

If the quadratic does not have exactly the above form, we can sometimes add another term to '*complete the square*' (as shown in the second example below).

Example: Factorise the expression $9b^2 - 12by + 4y^2$.

Solution: We notice that $9b^2 = (3b)^2$, $4y^2 = (2y)^2$, and $-12by = -2 \cdot 3b \cdot 2y$. So, we can factorise $9b^2 - 12by + 4y^2 = (3b - 2y)^2$ or $(2y - 3b)^2$, which are equivalent.

Example: Factorise $x^2 + 6x + 3$.

Solution: The expression can be made onto a perfect square by adding 6 and subtracting 6. So, we have $x^2 + 6x + 3 = x^2 + 6x + 3 + 6 - 6$ $= (x + 3)^2 - 6$.

• Combined factorisation

In algebraic expressions, we do not often have exactly the right terms for a single instance of factorisation, and thus the various methods of factorisation can be combined.

Example: Factorise $3ax^2 + 3ax - 6a$.

Solution: We notice that $3a$ is a common factor in all the terms. So, we first factorise: $3ax^2 + 3ax - 6a = 3a(x^2 + x - 2)$.

In the brackets, we have the terms of a quadratic expression where the coefficient of x^2 is 1. The factors of -2 are (± 2 and ± 1) and the coefficient of x is $+1$, which is the sum of $+2$ and -1.

Therefore, the complete factorisation is:

$$3ax^2 + 3ax - 6a = 3a(x^2 + x - 2) = 3a(x + 2)(x - 1).$$

2.2 Equations

2.2.1 Definition

An **equation** is a statement that two quantities are **equal (=)**: $5x - 1 = 6y + 3$. It is distinguished from an '*expression*' which does not contain an equals sign: $2x + 3 - 4y$.

2.2.2 Zero product

If the product of two numbers is zero, at least one of them must be zero: for instance, $3 \cdot 0 = 0$, $a \cdot 0 = 0$, $0 \cdot 0 = 0$.

The expression $ab = 0$ implies that either $a = 0$ or $b = 0$ or both a and b are 0. Similarly, if $abcd = 0$, at least one of a, b, c, and d must be zero.

The study of the zero products is helpful in finding out which real values make the denominator of rational expressions equal to zero, as a denominator of zero will not give a real number for the expression. Those values should be dealt with carefully in equations and inequalities.

Example: Find the real values which make the expression $x^2 - 2x - 3$ equal to zero.

Solution: We can factorise the expression using the method explained in Section 2.1.2: $x^2 - 2x - 3 = (x - 3)(x + 1)$. This expression equals 0 if at least one of the two factors is zero: *i.e.* $(x - 3) = 0$ or $(x + 1) = 0$.

Intuitively, the first factor $(x - 3) = 0$ when $x = 3$, while the second factor $(x + 1) = 0$ when $x = -1$.

2.2.3 Manipulation of equations

General rules for manipulating equations are as follows:

(i) An equation is unchanged by adding or subtracting the same number to/from both sides:

$$3x + 1 = 6 \text{ is equivalent to } 3x + 6 = 11 \text{ (adding 5 to both sides)}$$

(ii) An equation is unchanged by multiplying or dividing both sides by the same number $\neq 0$:

$$5x = 15 \text{ is equivalent to } 20x = 60 \text{ (multiplying both sides by 4)}$$

(iii) If $A = B$ and $B = C$, then $A = C$:

$$E = \frac{h\lambda}{c} \text{ and } \frac{h\lambda}{c} = \frac{h}{\nu}, \text{ then } E = \frac{h}{\nu}, \text{ with } c \neq 0 \text{ and } \nu \neq 0$$

(iv) If $A = B$ and $A = C$, then $B = C$:

$$V = \frac{RT}{P} \text{ and } V = \frac{M}{\rho}, \text{ then } \frac{RT}{P} = \frac{M}{\rho}, \text{ with } P \neq 0 \text{ and } \rho \neq 0$$

These rules are helpful in solving equations and handling scientific formulae.

2.2.4 Solving linear equations

In the above examples, the value of the unknown x could be deduced from the equations. This is known as **solving** or **finding the solution** of the equation.

A **linear equation** contains **one single** unknown (usually denoted x) and always has **one single** solution.

The general form of a linear equation is

$$a \cdot x = b \quad \text{with} \quad a \neq 0$$

and the solution is $\boxed{\mathrm{x} = \frac{\mathrm{b}}{\mathrm{a}}}$.

If $a = 0$ and $b \neq 0$, the equation has NO solutions, while if $a = 0$ and $b = 0$, the equation has unlimited solutions in the set of real numbers ($\mathbb{R}$), since x can take any value.

Example: Solve the equation $\frac{x+3}{2} = \frac{3-x}{3} + 3$.

Solution: We first find a common denominator and rewrite each expression with that denominator to handle fractions easily and to obtain the general form of a linear equation.

The common denominator is 6: $\frac{3x+9}{6} = \frac{6-2x}{6} + \frac{18}{6}$.

We can then multiply both sides of the equation by 6 to eliminate the denominators: $3x + 9 = 6 - 2x + 18$.

We can now solve for x: $3x + 2x = 6 + 18 - 9 \iff 5x = 15 \iff x = 3$.

Example in Medicine: A patient who is $1.80\,\mathrm{m}$ tall and weighs $95\,\mathrm{kg}$ suffers from cardiac dysfunction and is strongly recommended by physicians to lose weight, ideally to reach a body mass index (BMI) of 22.

BMI is an index indicating obesity, overweight, normal weight, or underweight in men and women, and is calculated according to the equation:

$$\text{BMI} = \frac{m}{h^2}$$

with m the weight of the patient in kg and h his/her height in m.

How much weight should the patient lose?

Solution: The equation to set up in order to find the ideal weight (the unknown x) is $22 = \frac{x}{1.80^2}$.

If we multiply both sides of the equation by 1.80^2, we have $x = 22 \cdot 1.80^2 = 71.3\,\mathrm{kg}$.

Therefore, the weight the patient should lose is $95\,\mathrm{kg} - 71.3\,\mathrm{kg} = 23.7\,\mathrm{kg}$, which is a very large amount!

2.2.5 Solving rational equations

Equations containing rational expressions with the unknown (usually x) in the denominator are called **rational equations**. We can solve these equations using the above techniques to perform operations with fractions and to solve algebraic equations.

Whenever there are unknowns in the denominator, we need to find any real values that should be excluded from the domain of solutions because they would make the denominator zero (see Section 1.2).

Example: Solve the rational equation: $\frac{2}{x-1} - \frac{3}{x^2-1} = \frac{1}{x+1}$.

Solution: First we can determine the values of x that result in a denominator of 0. In order to find out those values, we can factorise the denominator $(x^2 - 1) = (x-1)(x+1)$.

Thus, the values to be excluded from $\mathbb{R}$ are 1 and -1.

We can find the lowest common denominator and multiply both sides of the equation by it:

$$\left(\frac{2}{x-1} - \frac{3}{x^2-1}\right) \cdot (x-1)(x+1) = \frac{1}{x+1} \cdot (x-1)(x+1)$$

$$\iff 2(x+1) - 3 = x - 1$$

We now solve for x: $2x + 2 - 3 = x - 1 \iff 2x - x = -1 - 2 + 3 \iff x = 0$.

The solution $x = 0$ is acceptable as it is not a value initially excluded.

Example in Biochemistry: A protein in solution subjected to centrifugal force is known to move in the direction of the force at a velocity dependent on its mass. The *Svedberg equation* relates the molecular weight (M) of a protein to its sedimentation, S, and diffusion, D, coefficients: $M = \dfrac{SRT}{D(1-\overline{\nu}\rho)}$, where R is the ideal gas constant ($8.31\,\mathrm{J\,K^{-1}\,mol^{-1}}$), T is the temperature in K, ρ is the density $\mathrm{g\,mL^{-1}}$, and $\overline{\nu}$ is the partial specific volume.

Estimate the density of the solvent used to measure the molecular weight of the protein brain-derived neurotrophic factor (BDNF) using an ultracentrifuge at $20\,^{\circ}\mathrm{C}$. BDNF has a molecular weight of $27.3\,\mathrm{kDa}\ (=\mathrm{kg\,mol^{-1}})$, and previous experiments allowed the determination of the sedimentation coefficient $S = 2.5\cdot 10^{-13}\,\mathrm{s}$ ($= 2.5\,\mathrm{svedberg}$), the diffusion coefficient $D = 8.4\cdot 10^{-7}\,\mathrm{cm^2\,s^{-1}}$, and the partial specific volume ($0.7271\,\mathrm{mL\,g^{-1}}$).

Solution: In this problem, the unknown is $\rho = x$ and is in the denominator; thus the linear equation is rational and we should exclude the real values for x which make the denominator zero:

$$(1-\overline{\nu}x) \neq 0 \Longleftrightarrow x = \frac{1}{\overline{\nu}} \neq \frac{1}{0.7271\,\mathrm{mL\,g^{-1}}} \neq 1.375\,\mathrm{g\,mL^{-1}}$$

Using the general rules to manipulate the equation (see Section 2.2.3), we can solve it for x:

$$M = \frac{SRT}{D(1-\overline{\nu}x)} \Longleftrightarrow 1-\overline{\nu}x = \frac{SRT}{MD} \Longleftrightarrow \overline{\nu}x = 1-\frac{SRT}{MD}$$

$$\Longleftrightarrow x = \frac{1-\frac{SRT}{MD}}{\overline{\nu}}$$

Using the International System of measuring units and to simplify the task, we first calculate

$$\frac{SRT}{MD} = \frac{2.5\cdot 10^{-13}\,\mathrm{s}\cdot 8.31\,\mathrm{kg\,m^2\,K^{-1}\,mol^{-1}\,s^{-2}}\cdot 293\,K}{27.3\,\mathrm{kg\,mol^{-1}}\cdot 8.4\cdot 10^{-11}\,\mathrm{m^2\,s^{-1}}}$$

$$= 26.54\cdot 10^{-2}$$

Finally, we estimate $x = \dfrac{1 - 0.2654}{0.7271\,\mathrm{g\,mL^{-1}}} = 1.010\,\mathrm{g\,mL^{-1}}$, as the density of the solvent used for ultracentrifugation; this value is acceptable since it is not equal to $1.375\,\mathrm{g\,mL^{-1}}$.

2.2.6 Manipulation of formulae

The formula $y = 3x + 2$ is an **explicit** formula for y, as it explains how to calculate y without rearrangement. The variable y is said to be the **subject** of the equation.

However, the formula $\dfrac{1}{f} = \dfrac{1}{u} + \dfrac{1}{v}$ is said to be an **implicit** formula for f, since the relationship between f, u, and v is implied rather than explicit. Here $f = \dfrac{uv}{u+v}$ gives the explicit relationship of the last implicit formula.

Example: Make y the subject of the equation in the following implicit formula $\dfrac{x-y}{y+z} = 2w - x$.

Solution: First, we can multiply both sides of the equation by the denominator $(y+z)$: $x - y = (2w - x)(y + z)$.

We multiply out the expression in the brackets:

$$x - y = 2wy + 2wz - xy - xz.$$

Adding $xy - 2wy - x$ to both sides of the equation, we get

$$xy - y - 2wy = 2wz - xz - x.$$

We can now factorise y: $y(x - 1 - 2w) = 2wz - xz - x$, and then divide both sides by $(x - 1 - 2w)$ to finally get $y = \dfrac{2wz - xz - x}{x - 1 - 2w}$.

Example in Biochemistry: Make implicit the *Michaelis–Menten kinetics* formula $\left(v_i = \dfrac{V_{max} \cdot [s]}{K_m + [s]}\right)$, with $\dfrac{1}{v_i}$ being the subject of the equation: where V_{max} and K_m are constants

Solution: The Michaelis–Menten kinetics formula describes the rate of enzymatic reactions by relating the initial reaction rate v_i (rate of formation of product, $[P]$) to $[s]$, the concentration of substrate.

In order to make $\frac{1}{v_i}$ the subject of the equation, we can first invert both sides of the formula as equal fractions: $\frac{1}{v_i} = \frac{K_m + [s]}{V_{max} \cdot [s]}$.

Then, on the right side we divide each term of the numerator by the denominator and simplify the expressions appropriately:

$$\frac{1}{v_i} = \frac{K_m}{V_{max} \cdot [s]} + \frac{\cancel{[s]}}{V_{max} \cdot \cancel{[s]}}$$

We finally have a new form of the Michaelis–Menten kinetics formula: $\frac{1}{v_i} = \frac{K_m}{V_{max}} \cdot \frac{1}{[s]} + \frac{1}{V_{max}}$, which is easy to draw as a double reciprocal plot (better known as *the Lineweaver–Burk plot*) on a Cartesian graph with $\frac{1}{v_i}$ on the vertical (y) axis and $\frac{1}{[s]}$ on the horizontal (x) axis.

2.2.7 Elimination of a parameter in a formula

If we know how two variables (for instance, x and y) are related to a third variable (for instance, z), we can eliminate z and find out how x and y are related. This is called **eliminating a parameter**.

The method consists of the following steps:

(i) Make the term to be eliminated the subject of both equations.

(ii) Use Section 2.2.3 (point (iii)) to set the equations equal.

(iii) Manipulate the equation to give the required variable as the subject.

Example: Starting from the equations $P = \frac{RT}{V}$ and $\rho = \frac{M}{V}$, give the mathematical relationship between P and M.

Solution: We first notice that V is a common term in both formulae and can play the role of the term to be eliminated. Then we can make V the subject of both equations: $V = M/\rho$ and $V = RT/P$, so we can equate the two expressions: $\frac{M}{\rho} = \frac{RT}{P}$. Finally, multiplying both sides of the equations by $P\rho/M$, we have $P = \frac{RT\rho}{M}$.

Example in Medicine: Estimate the amount of sodium chloride (NaCl) required to prepare 2 litres of a saline solution at $0.154\,\mathrm{M}$ ($\mathrm{M} = \mathrm{mol\,L^{-1}}$).

Solution: The *saline solution*, also known as *physiological saline*, has a number of uses in medicine, for instance to clean wounds, to treat dehydration, or to dilute other medications to be given by injection. It is therefore critical that it is prepared accurately.

In Chemistry, we know that the number of moles (n in mol) is the ratio between the mass (m in g) of a species and its molecular weight (MW in $\mathrm{g\,mol^{-1}}$), *i.e.* $n = \dfrac{m}{MW}$, while the molarity (M) is expressed as the number of moles of solute (n) per volume litre (V in L) of solution, *i.e.* $M = \dfrac{n}{V}$.

The common term in both equations is n, which can be eliminated to solve the problem. Let us make n the subject of both equations: $n = \dfrac{m}{MW}$ and $n = M \cdot V$. Then set the equations equal:

$$\frac{m}{MW} = M \cdot V$$

We can now manipulate the resulting equation to have the unknown m as subject: $m = M \cdot V \cdot MW$, and substitute the variables with the known values: $M = 0.154\,\mathrm{mol\,L^{-1}}$, $V = 2\,\mathrm{L}$, while $MW_{\mathrm{NaCl}} = 58.5\,\mathrm{g\,mol^{-1}}$:

$$m = 0.154\,\mathrm{mol\,L^{-1}} \cdot 2\,\mathrm{L} \cdot 58.5\,\mathrm{g\,mol^{-1}} = 18.0\,\mathrm{g}$$

This mass of NaCl salt should then be put in a $2\,\mathrm{L}$ volumetric flask filled with sterile water up to the line.

2.3 Quadratic Equations

2.3.1 Solving quadratic equations

Such equations are usually of the form

$$ax^2 + bx + c = 0$$

Solving for the values of x which satisfy the equation is called **finding the roots** or **the zero values** of the quadratic equation.

For an equation containing x^2, there are **always two roots** (but they may sometimes be the same). The roots may be positive, negative, or complex numbers. Complex numbers are ones which have an imaginary part (see Chapter 7, Section 7.3).

Quadratic equations can also have the unknown parameter in the denominator, so we can first find out which real values to exclude from the domain of solutions, since they would make the denominator zero (see Section 2.2.5).

Three possible cases of quadratic equations can occur:

(i) If the constant term is missing ($c = 0$), we have $\mathbf{ax^2 + bx = 0} \Longleftrightarrow x(ax+b) = 0$, which implies that either $x = 0$ or $(ax + b) = 0$.

The two solutions are $\boxed{\mathbf{x_1 = 0}}$ and $\boxed{\mathbf{x_2 = -\frac{b}{a}}}$.

(ii) If the term in x is missing ($b = 0$) and $c \neq 0$, we have $\mathbf{ax^2 + c = 0}$, so $ax^2 = -c$.

Thus, the two solutions are

$$\boxed{\mathbf{x_{1,2} = \pm\sqrt{-\frac{c}{a}}}}$$

Only if $\frac{c}{a}$ is a negative number (*i.e.* $-\frac{c}{a}$ is a positive number), will the square roots of $-\frac{c}{a}$ be real numbers and the equation is said to have **real roots**.

If $\frac{c}{a}$ is a positive number, the square roots of $-\frac{c}{a}$ are not real numbers and the equation is said to have **imaginary roots**.

(iii) If all terms are present (a, b and c), there are two alternative methods of finding the roots of the quadratic equation: by factorisation (see Section 2.1.2) using the zero product rule (see Section 2.2.2) and by the quadratic equation formula (see Section 2.3.2).

Example: Solve the equations: $x^2 + 4x = 0$, $x^2 - 16 = 0$ and $x^2 + 5x + 6 = 0$, by recognising which case above (i, ii, or iii) applies.

Solution: The equation $x^2 + 4x = 0$ corresponds to case (i); so, $x^2 + 4x = 0 = x(x + 4) \Longleftrightarrow x = 0$ or $(x + 4) = 0$. The solutions are $x_1 = 0$ and $x_2 = -4$.

The equation $x^2 - 16 = 0$ corresponds to case (ii); so, $x^2 - 16 = 0 \iff x^2 = 16$ and thus the solutions are $x_{1,2} = \pm\sqrt{16} = \pm 4$.

The equation $x^2 + 5x + 6 = 0$ can be solved by factorisation (case (iii)); so, $x^2+5x+6 = 0 \iff (x+2)(x+3) = 0$, and the two roots are the real values that make the factors zero, giving $x_1 = -2$ or $x_2 = -3$.

2.3.2 The quadratic formula

An alternative method to factorising these is to manipulate the generalised quadratic equation to make x the subject (see Section 2.2.3 for equation manipulation). It is not necessary to know the derivation of the formula, but you should be able to follow it based on the previous sections of this chapter.

Let us start from $ax^2 + bx + c = 0$:

- dividing through by a: $x^2 + \frac{b}{a}x + \frac{c}{a} = 0$;
- subtracting $\frac{c}{a}$ from both sides: $x^2 + \frac{b}{a}x = -\frac{c}{a}$;
- completing the square by adding $\left(\frac{b}{2a}\right)^2$ to both sides (see Section 2.1.2): $x^2 + \frac{b}{a}x + \left(\frac{b}{2a}\right)^2 = \left(\frac{b}{2a}\right)^2 - \frac{c}{a}$;
- factorising the perfect square in the first side and finding a common denominator for the second side: $\left(x + \frac{b}{2a}\right)^2 = \frac{b^2 - 4ac}{4a^2}$;
- taking the square root of both sides: $x + \frac{b}{2a} = \pm\frac{\sqrt{b^2 - 4ac}}{2a}$;
- adding $-\frac{b}{2a}$ to both sides, so that we finally have:

$\boxed{\mathbf{x_{1,2} = \frac{-b \pm \sqrt{b^2 - 4ac}}{2a}}}$, where the expression $b^2 - 4ac$ is also defined as the **discriminant** (Δ) of the quadratic equation.

Three cases are possible:

(i) If $\Delta = b^2 - 4ac > 0$, the quadratic has **two, distinct, real roots** ($\mathbf{x_1}$ and $\mathbf{x_2}$).

(ii) If $\Delta = b^2 - 4ac = 0$, the square root term disappears and the quadratic has **a single root** ($\mathbf{x_1 = x_2}$).

(iii) If $\Delta = b^2 - 4ac < 0$, the square root term is imaginary and the quadratic has **no real roots**. The roots are complex numbers.

Example: Solve the following quadratic equations for x:

$$x^2 - 2x - 3 = 0,\ 4x^2 - 4x + 1 = 0,\ 2x^2 + 3x + 4 = 0$$

Solution: We can calculate the value of $\Delta = b^2 - 4ac$ for each equation: $\Delta_1 = 4 + 12 = 16$, $\Delta_2 = 16 - 16 = 0$, and $\Delta_3 = 9 - 32 = -23$.

Thus, the first equation has two distinct roots $\left(x_1 = \dfrac{2-4}{2} = -1\right.$ and $\left. x_2 = \dfrac{2+4}{2} = 3\right)$ and the second equation has two equal roots $\left(x_1 = x_2 = \dfrac{4 \pm 0}{8} = \dfrac{1}{2}\right)$, whereas the third equation has no real roots.

Example in Chemistry: A gas reaction vessel is filled with phosphorus pentachloride ($PCl_5(g)$, g = gas) at an initial pressure of $1.2\,\mathrm{atm}$, and PCl_5 undergoes dissociation into $PCl_3(g)$ and $Cl_2(g)$ at equilibrium. Calculate the concentrations of all chemical species at equilibrium, knowing that the equilibrium constant (K_p) is $0.5\,\mathrm{atm}$ at $260\,°\mathrm{C}$.

Solution: We first write out the balanced equation:

$$PCl_5(g) \rightleftharpoons PCl_3(g) + Cl_2(g)$$

In order to calculate gas pressures at equilibrium, we can track the known and unknown (x) pressures (in atm) of the chemical species throughout the reaction process using the following table:

	$PCl_5(g)$	$PCl_3(g)$	$Cl_2(g)$
Initial	1.2	0	0
Change	$-x$	x	x
Equilibrium	$1.2 - x$	x	x

The *equilibrium constant* (K_p) for gases is the ratio of the pressures of the products to the pressures of the reactants at the equilibrium, with each pressure being raised to the power of that species' stoichiometric coefficient in the balanced chemical equation.

Here, $K_p = \frac{(\text{PCl}_3)(\text{Cl}_2)}{(\text{PCl}_5)}$; thus, at equilibrium $0.5 = \frac{x \cdot x}{1.2 - x}$, which is a rational quadratic equation.

First, we can exclude $x = 1.2$ from the domain of real solutions, since this would make the denominator zero, and this is consistent with the hypothesis that the entire amount of reactant is converted to products, which is unlikely and also inconsistent with the equilibrium concept.

Let us then solve the equation for x by cross-multiplication and application of the quadratic formula:

$$0.5(1.2 - x) = \frac{x^2}{\cancel{1.2 - x}} \cdot \cancel{(1.2 - x)} \Longleftrightarrow 0.6 - 0.5x = x^2$$
$$\Longleftrightarrow x^2 + 0.5x - 0.6 = 0$$

$\Delta = (0.5)^2 - 4(-0.6) = 2.65$, which is > 0, and thus we have two possible solutions: $x_1 = \frac{-0.5 - 1.63}{2} = -1.565$ and $x_2 = \frac{-0.5 + 1.63}{2} = 0.565$. The former solution is non-physical, as it corresponds to a negative pressure.

Therefore, given $x = 0.565\,\text{atm}$, at the equilibrium we have $(\text{PCl}_5) = 1.2 - 0.565 = 0.635\,\text{atm}$, and $(\text{Cl}_2) = (\text{PCl}_3) = 0.565\,\text{atm}$.

• Factorisation of other quadratic expressions

So far, to factorise quadratic expressions $(ax^2 + bx + c)$, we have been looking for positive or negative natural numbers, which, upon multiplication, can give the quadratic coefficients a, b, and c (see Section 2.1.2).

However, this method is not always possible, since fractions or roots might be required, and with the above methods it can be almost impossible to factorise the quadratic expressions.

It can be demonstrated that

$$ax^2 + bx + c = a(x - x_1)(x - x_2)$$

where x_1 and x_2 are the roots of the associated quadratic equation $(ax^2 + bx + c = 0)$.

Therefore, this method is helpful, as it can be applied to any quadratic expressions with $\Delta > 0$, *i.e.* which have two distinct solutions (x_1 and x_2).

Example: Factorise the following quadratic expressions: $7x^2 - 3x - 1$.

Solution: Using the quadratic formula (see Section 2.3.2), we find that $x_1 = \dfrac{3+\sqrt{37}}{14}$ and $x_2 = \dfrac{3-\sqrt{37}}{14}$.
Thus, the factorisation is:

$$7x^2 - 3x - 1 = 7\left(x - \frac{3+\sqrt{37}}{14}\right) \cdot \left(x - \frac{3-\sqrt{37}}{14}\right)$$

These fractions and roots in brackets could not be guessed using the methods explained in Section 2.1.2 to factorise quadratic expressions.

2.3.3 Solving polynomial equations

Solving polynomial equations requires careful treatment. It is sometimes hard to solve them if they have a degree greater than 2, and for higher degrees that it can be impossible to solve them without special manipulations.

To solve a **polynomial equation**, the method consists of two steps:

(i) Factorise the polynomial (see Section 2.1.2).

(ii) Find the possible zero values of the polynomial by equating each factor in turn to 0, according to the zero-product property (see Section 2.2.2).

The **fundamental theorem of Algebra** states that any polynomial of degree n has n roots. However, some of the n roots might not be real solutions but may be complex numbers (see Chapter 7, Section 7.3).

Example: Find the values of x satisfying the polynomial equation:

$$x^6 - x^4 + 4x^2 - 4 = 0$$

Solution: First we notice that x^4 is a common factor in the first two terms, and $+4$ is a common factor in the last two terms.

So, $x^6 - x^4 + 4x^2 - 4 = x^4(x^2-1) + 4(x^2-1) = (x^2-1)(x^4+4)$.

(x^2-1) is a difference of two squares and can be further factorised as $(x+1)(x-1)$.

We can now equate each factor in turn to 0: $(x - 1) = 0$, $(x + 1) = 0$, and $(x^4 + 4) = 0$. Thus, the only real solutions of the initial equation are $x = 1$ and $x = -1$, as $(x^4 + 4)$ is always > 0. So, of the six possible roots of this polynomial equation, the other four solutions must be complex numbers.

Example in Medicine: The von Bertalanffy–Pütter equation has been widely used to describe growth models in Biology, including tumour growth examples. It states that

$$y = p \cdot m^a - q \cdot m^b$$

where y is the growth speed, m is the tumour size (in mm^3), and p, a, q, and b are positive constants, with $a < b$.

The term $p \cdot m^a$ describes the growth of the tumour due to abnormal cell replication, while the term $q \cdot m^b$ describes the degradation of the tumour in relation to therapies and apoptotic processes.

In general terms, what size will the tumour be when it stops growing?

If $p = \frac{1}{2}$, $a = 2$, $q = \frac{1}{1228250}$, and $b = 5$, solve the equation $y = 0$ for physically meaningful values of m.

Solution: First, we notice that since $a < b$, m^a is a common factor in the two terms, so we can factorise the equation as follows:

$$y = p \cdot m^a - q \cdot m^b = m^a \cdot (p - q \cdot m^{b-a})$$

In order to estimate how big the tumour would be when it stopped growing, we can now equate each factor in turn to 0: $m^a = 0$ and $p - q \cdot m^{b-a} = 0$.

The equation $m^a = 0$ has only one possible solution $m = 0$ but this is physically meaningless.

The equation $p - q \cdot m^{b-a} = 0$ can be rearranged as $m^{b-a} = \frac{p}{q}$ from which we obtain $m = \sqrt[(b-a)]{\frac{p}{q}}$ by taking the $(b-a)^{th}$ root of both sides of the equation. Obviously, only positive values of m are physically meaningful.

For the particular values of p, a, q, and b given above, we can solve the polynomial equation $\frac{m^2}{2} - \frac{m^5}{1228250} = 0$ to estimate the tumour volume at which the cancer would stop growing.

After factorising m^2 out of the two terms of the equation, we obtain $m^2 \cdot \left(\frac{1}{2} - \frac{m^3}{1228250}\right) = 0$.

We can now equate each factor in turn to 0, giving the solutions $m = 0$ which is not physically reasonable, and $m = \sqrt[3]{\frac{1228250}{2}}$ $= 85\,\mathrm{mm}^3$, which roughly corresponds to a cube with sides each of around $4.4\,\mathrm{mm}$.

2.4 Simultaneous Equations

Sometimes we have two equations which relate linear combinations of x and y and we can solve them for unique values of x and y (which is a pair of numbers x and y). These are called **simultaneous** equations, such as in the example below:

(i) $10x + y = 105$
(ii) $x - y = 5$

2.4.1 Method of elimination

The above pair of equations have the same coefficient of y but of opposite sign. If we add the equations together, y will be eliminated, and we get $11x = 110$, so $x = 10$. Substituting $x = 10$ into equation (ii) gives $10 - y = 5$ and $y = 5$.

If necessary, one equation can be multiplied through by a constant non-zero factor to give the same coefficients of x or y in both equations. The equations can then be added or subtracted from each other as appropriate.

It is usually necessary to multiply each of the pair of equations by a *different* factor for the elimination method.

Example: Solve the following pair of equations for a and b:

(i) $5a - 4b = 6$
(ii) $a + b = 3$

Solution: We can multiply every term in equation (ii) by 4: $4a + 4b = 12$, and add it to equation (i): $9a = 18$, so $a = 2$.

Thus, the solutions are $a = 2$ and $b = 1$.

We can check this by substituting these values back into equations (i) and (ii): (i) $5 \cdot 2 - 4 \cdot 1 = 6$ as expected and (ii) $2 + 1 = 3$ as expected. It is **always** wise to do this check!

Example in Chemistry: The vapour pressure above a mixture containing water and a 0.7 mole fraction of ethanol is $4.75\,\mathrm{atm}$ and falls to $4.40\,\mathrm{atm}$ when the mole fraction of ethanol is decreased to 0.6. Estimate the vapour pressures of pure ethanol and water as solvents.

Solution: First, we can set up the equations with the information given:

(i) $4.75\,\mathrm{atm} = 0.7\,p^*_{Et} + 0.3\,p^*_{H_2O}$
(ii) $4.40\,\mathrm{atm} = 0.6\,p^*_{Et} + 0.4\,p^*_{H_2O}$

with p^*_{Et} = vapour pressure of ethanol and $p^*_{H_2O}$ = vapour pressure of water.

If we multiply the first equation by 6 and the second by 7, we obtain a pair of equations in which the coefficients of p^*_{Et} are equal:

(iii) $28.5\,\mathrm{atm} = 4.2\,p^*_{Et} + 1.8\,p^*_{H_2O}$
(iv) $30.8\,\mathrm{atm} = 4.2\,p^*_{Et} + 2.8\,p^*_{H_2O}$

If we now subtract equation (iii) from equation (iv), we have $(30.8 - 28.5)\,\mathrm{atm} = (2.8 - 1.8)\,p^*_{H_2O}$, which simplifies to $p^*_{H_2O} = 2.3\,\mathrm{atm}$.

Substituting $p^*_{H_2O} = 2.3\,\mathrm{atm}$ into equation (i) and using the general rules to manipulate equations, we have $4.75\,\mathrm{atm} = 0.7\,p^*_{Et} + 0.3 \cdot 2.3$, which gives:

$$p^*_{Et} = \frac{4.75 - 0.3 \cdot 2.3}{0.7} = 5.8\,\mathrm{atm}$$

2.4.2 Method of substitution

An alternative method of elimination is to manipulate the equations to make x or y the subject of both and then equate them.

Example: Solve the following pair of equations for x and y:

(i) $5x + 5 = 10$

(ii) $2x - 3y = 14$

Solution: We can initially rearrange (i):

$5x = 10 - 5y$, so $x = \dfrac{10 - 5y}{5} \iff x = 2 - y$ (iii);

then we rearrange (ii): $2x = 14 + 3y$, so $x = \dfrac{14 + 3y}{2}$ (iv).

Then, by equating (iii) and (iv) we obtain $2 - y = \dfrac{14 + 3y}{2}$

$\iff 4 - 2y = 14 + 3y \iff 4 - 14 = 2y + 3y \iff -10 = 5y$

$\iff y = -2.$

This value for y is now substituted back into (iii), giving $x = 4$.

2.4.3 Solution with one nonlinear equation

If the equations contain terms in x^2, y^2, or xy, the method of substitution *must* be used. The resulting expression can then be factorised (see Section 2.1) and will usually give more than one pair of solutions.

The problem is equivalent to finding the points of intersection of a line (represented by the linear equation) and a curve (represented by the nonlinear equation).

Example: Solve the following pair of equations for x and y:

(i) $x^2 + y^2 = 13$

(ii) $x - y = 5$

Solution: If we substitute $x = 5 + y$ from (ii) into (i), we obtain $(5+y)^2 + y^2 = 13 \iff 25 + y^2 + 10y + y^2 = 13$.

Now, we should factorise: $2y^2 + 10y + 12 = 0 \iff y^2 + 5y + 6 = 0 \iff (y+2)(y+3) = 0$, so $y = -2$ or -3.

Putting $y = -2$ into (ii), the equation gives $x = 3$ and putting in $y = -3$ gives $x = 2$.

Example in Biology: A population of butterflies can have black, grey, or white wings, and the black allele (B) has complete dominance over the white allele (b). The homozygous dominant genotype BB provides the butterflies with black wings, the heterozygous genotype Bb grey wings, and the homozygous recessive genotype bb white wings.

The allele B gives the butterflies a higher probability of surviving from zygote to adult and a greater ability to survive and transmit its genes to the next generation, which is called *fitness*. The relative fitness values (w) of the genotypes BB and Bb are higher than that of the genotype bb: $w_{BB} = 1$, $w_{Bb} = 0.8$ and $w_{bb} = 0.4$, with an average fitness of the population $\overline{w} = 0.9$. As this population is in a state of *Hardy–Weinberg equilibrium*, determine the frequency of the dominant allele B and the recessive allele b in the population.

Solution: According to the Hardy–Weinberg law, which provides an equilibrium baseline for scientists to measure gene evolution in a given population, there are two equations needed to solve a Hardy–Weinberg equilibrium question:

(i) $p + q = 1$

(ii) $\overline{w} = p^2 w_{BB} + 2pq w_{Bb} + q^2 w_{bb}$

where the unknowns p and q are respectively the frequency of the dominant allele B and the frequency of the recessive allele b.

If we substitute $p = 1 - q$ from (i) into (ii), we obtain

$$\begin{aligned} \overline{w} &= (1-q)^2 w_{BB} + 2(1-q)q w_{Bb} + q^2 w_{bb} \\ &\Longleftrightarrow \overline{w} = q^2 w_{BB} - 2q w_{BB} + w_{BB} + 2q w_{Bb} - 2q^2 w_{Bb} + q^2 w_{bb} \\ &\Longleftrightarrow q^2(w_{BB} - 2w_{Bb} + w_{bb}) + 2q(-w_{BB} + w_{Bb}) + w_{BB} - \overline{w} = 0 \end{aligned}$$

which is a quadratic equation for q.

We can now substitute the fitness values into the quadratic equation: $q^2(1 - 2 \cdot 0.8 + 0.4) + 2q(-1 + 0.8) + 1 - 0.9 = 0$ which can be expressed as $-2q^2 - 4q + 1 = 0$ by multiplying both sides by 10.

We can use the quadratic formula to find the solutions: $q_1 = \dfrac{-2-\sqrt{6}}{2} = -2.22$, which is non-physical as it is a negative root, and $q_2 = \dfrac{-2+\sqrt{6}}{2} = 0.22$, which is then the frequency of the recessive allele b.

Finally, putting $q = 0.22$ into equation (i) gives $p = 1 - 0.22 = 0.78$, which is the frequency of the dominant allele B.

Important note: In the Hardy–Weinberg equilibrium equation (ii), p^2 is the frequency of individuals with the homozygous dominant genotype, $2pq$ is the frequency of individuals with the heterozygous genotype, and q^2 is the frequency of individuals with the homozygous recessive genotype. Therefore, if the total number of a population of butterflies was given, the frequency of individuals per genotype and number of individuals per genotype could also be determined for that population.

2.5 Inequalities

Inequality links two mathematical expressions by one of the symbols: $>$, $<$, $\geq$, $\leq$:

- $a > b$ is read as 'a *is greater than* b'.
- $a \geq b$ is read as 'a *is greater than or equal to* b'.
- $a < b$ is read as 'a *is less than* b'.
- $a \leq b$ is read as 'a *is less than or equal to* b'.

2.5.1 Manipulation of inequalities

The general rules for manipulating inequalities are as follows:

(i) An inequality is unchanged by adding or subtracting the same number to/from both sides:

$$2x - 1 > 3 \text{ is equivalent to } 2x + 4 > 8$$
$$\text{(adding 5 to both sides)}$$

(ii) An inequality is unchanged by multiplying or dividing both sides by a **positive** number (> 0):

$$5x < 15 \text{ is equivalent to } 20x < 60$$
$$\text{(multiplying both sides by 4)}$$

(iii) An inequality is reversed by multiplying or dividing both sides by a **negative** number (< 0):

$$3x < 4 \text{ is equivalent to } -9x > -12$$
$$\text{(multiplying both sides by } -3\text{)}$$

(iv) If $A < B$ and $B < C$, then $A < C$. This is also equivalent to saying that if $A > B$ and $B > C$, then $A > C$:

$$2 < 4 \text{ and } 4 < 7, \text{ then } 2 < 7$$

These rules are helpful in solving inequalities and handling scientific inequalities.

Warning! If $A < B$ and $A < C$, we cannot draw any conclusions *a priori* about the relationship between B and C if the values of both B and C are unknown.

However, we can only state that $2A < B + C$ by summing the two relationships. This is also valid for the inequality symbol $>$.

Important note: We must never subtract inequalities! For instance, using the inequalities (i) $-3 < 4$ and (ii) $-3 < 1$ we can state that $1 < 4$ (since we know the numbers involved in the inequalities).

Also, if we subtract (i)–(ii), we obtain:
$-3 - (-3) < 4 - 1 \Longleftrightarrow 0 < 3$, which is true.

However, if we subtract (ii)–(i) we obtain:
$-3 - (-3) < 1 - 4 \Longleftrightarrow 0 < -3$, which is NOT true.

This illustrates why we must never subtract inequalities.

2.5.2 Interval notations

Different mathematical notations can be used to represent an interval or a continuous set of numbers. For instance, to indicate all the numbers less than or equal to a real value a, a panel of diverse expressions can be adopted:

- **Inequality**: $x \leq a$.
- **Number line**: If the extreme value of the interval is included, a **black filled circle** is drawn (●), whereas if the extreme value of the interval is excluded, an **open circle** is drawn (○).

- **Brackets**: $(-\infty, a]$, where **( )** are used for the ∞ symbol (**unlimited interval**) or to exclude an endpoint (**open limited interval**), while **[ ]** are used to include an endpoint (**closed limited interval**).

2.5.3 Solving linear inequalities

In the above examples, a range of values for the unknown x satisfying the initial inequality could be deduced. This is known as **solving** the inequality, or **finding the solutions** of the inequality.

A **linear inequality** contains **one single** unknown (usually x) and usually has **a range of solutions** in the set of real numbers ($\mathbb{R}$), which can be represented on a number line.

The range of solutions can be represented on a **number line** to help visualise the solutions of the inequality.

Example: Solve the inequality $4 - 3x < -2$.

Solution: We first subtract 4 from both sides: $-3x < -6$.

Then, we divide both sides by -3, which is a negative number and reverses the inequality symbol from $<$ to $>$: $x > 2$.

This means that any real value greater than 2 satisfies the initial inequality.

For instance, if we put $x = 5$ into the initial inequality, we have $4-3 \cdot 5 < -2 \iff -11 < -2$ which is true, whereas if we put $x = -2$ into the initial inequality, we have $4 - 3 \cdot (-2) < -2 \iff 2 < -2$ which is false.

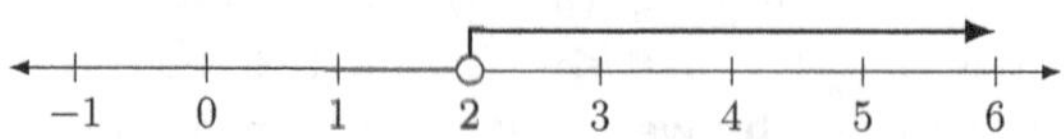

The **thick arrowed line** shows the range of values that x can take to satisfy the initial inequality. An **open circle** at 2 is drawn to show that, although the thick line goes from 2 onwards, x cannot equal 2.

Important note: If the extreme value of the range of solutions is also a solution of the inequality, as per the $\geq$ or $\leq$ symbols, a **black filled circle** is drawn ($\bullet$) on the graph.

Example in Chemistry: Knowing that the enthalpy, H, and entropy, S, change their values ($\Delta H = 177.8\,\mathrm{kJ\,mol^{-1}}$ and $\Delta S = 160.5\,\mathrm{J\,K^{-1}\,mol^{-1}}$) at $1\,\mathrm{atm}$ and $25\,°\mathrm{C}$ for the following reaction, at what temperature T is the reaction spontaneous?

$$CaCO_3(s) \longrightarrow CaO(s) + CO_2(s)$$

Solution: In thermodynamics, the Gibbs function ΔG helps us predict whether a reaction is possible or not at a particular temperature T. If ΔG is negative at that temperature, the reaction is spontaneous.

It is known that $\Delta G = \Delta H - T\Delta S$. Thus, we can solve the inequality $\Delta H - T\Delta S < 0$ to find out the range of temperatures for which the reaction is spontaneous:

$$177.8 \cdot 10^3\,\mathrm{J\,mol^{-1}} - x \cdot 160.5\,\mathrm{J\,K^{-1}\,mol^{-1}} < 0$$

Let us subtract $177.8 \cdot 10^3\,\mathrm{J\,mol^{-1}}$ from both sides of the inequality: $-x \cdot 160.5\,\mathrm{J\,K^{-1}\,mol^{-1}} < -177.8 \cdot 10^3\,\mathrm{J\,mol^{-1}}$.

Then, we divide both sides by $-160.5\,\mathrm{J\,K^{-1}\,mol^{-1}}$, thereby reversing the inequality symbol from $<$ to $>$:

$$x > \frac{-177.8 \cdot 10^3\,\mathrm{J\,mol^{-1}}}{-160.5\,\mathrm{J\,K^{-1}\,mol^{-1}}}$$

Thus, $x > 1.108 \cdot 10^3\,\mathrm{K} \iff x > 1108\,\mathrm{K}$, that is $T > 835\,°\mathrm{C}$, since $0\,°\mathrm{C} = 273\,\mathrm{K}$.

2.5.4 Solving compound inequalities

Sometimes we have two inequalities which relate to each other by a logical conjunction, which can be either **AND** or otherwise **OR**.

If two or more inequalities are linked by **AND**, the problem is equivalent to finding the ranges of solutions where the two inequalities are **both** satisfied. The inequalities should be first solved separately and then their corresponding solutions should be represented on a number line to find out the ranges of x values where the two inequalities are **both** verified.

If two or more inequalities are linked by **OR**, the problem is equivalent to finding the ranges of solutions where **at least one** of the inequalities is satisfied. The inequalities should be first solved separately and then their corresponding solutions should be represented on a number line to find out the ranges of x values where **at least one** of the two inequalities is verified.

Example: Solve the inequalities:

$$\text{(i) } 2 - x < 4 + x \text{ AND (ii) } 5 - 2x \geq x - 4$$

Solution: Let us first solve the inequalities separately:

$$\text{(i)} \quad x > -1 \quad \text{AND} \quad \text{(ii)} \quad x \leq 3$$

Both ranges of solutions should be plotted on the same number line:

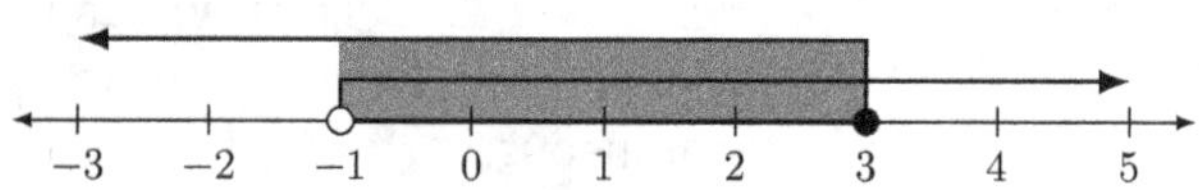

The range of values satisfying both inequalities is coloured in grey: $-1 < x < 3$. All values within this interval fulfil both initial inequalities: *i.e.* if we substitute x with 1 in both (i) and (ii), we have simultaneously $1 < 5$ and $3 \geq -3$, which are both true.

Example: Solve the inequalities (i) $3 + x \geq 2$ OR (ii) $5 - 2x < 3$

Solution: Let us first solve the inequalities separately:

$$\text{(i)} \quad x \geq -1 \quad \text{OR} \quad \text{(ii)} \quad x > 1$$

Both ranges of solutions should be plotted on the same number line:

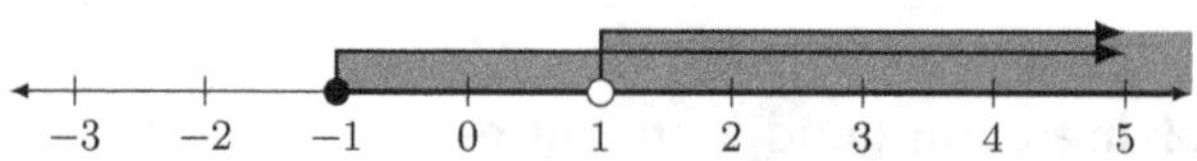

The range of values satisfying at least one of the two inequalities is coloured in grey: $x \geq -1$, *i.e.* all values of $x \geq -1$ fulfil at least one of the two initial inequalities. If we substitute x with 0 in both (i) and (ii), we have $3 \geq 2$ in (i), which is true, but $5 < 3$ in (ii), which is false. However, $x = 0$ is still acceptable since it at least satisfies the inequality (i).

Example: Solve the inequalities (i) $3x > 6$ OR (ii) $2x \leq 2$

Solution: Let us first solve the inequalities separately:

$$\text{(i)} \quad x > 2 \quad \text{OR} \quad \text{(ii)} \quad x \leq 1$$

Both ranges of solutions should be plotted on the same number line:

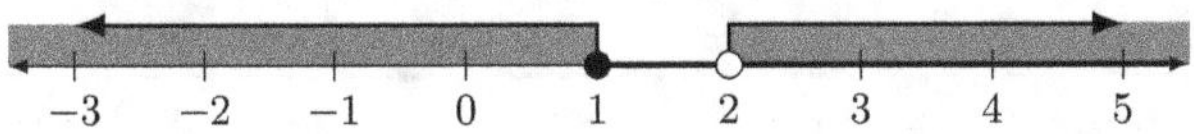

The range of values satisfying at least one of the two inequalities is coloured in grey: $x > 2$ OR $x \leq 1$, *i.e.* $x > 2$ OR $x \leq 1$ fulfils at least one of the two initial inequalities.

Example in Chemistry: A precious biological sample has been sent from a laboratory at the Columbia University in New York to the University of Oxford in the UK. Among the precautions suggested by the sender to keep the sample in its best condition is to maintain the sample frozen at a temperature between $-22\,°\mathrm{F}$ and $-4\,°\mathrm{F}$. What is the appropriate temperature range on the Celsius ($°\mathrm{C}$) scale?
Solution: The relationship that transforms the temperature values from Celsius scale (t in $°\mathrm{C}$) to Fahrenheit scale (T in $°\mathrm{F}$) is $\frac{9}{5}t + 32 = T$.

Thus, we are looking for temperature values in $°\mathrm{C}$ so that $-22 < \frac{9}{5}t + 32 < -4$, which is equivalent to solving two compound inequalities:

$$\text{(i)} \quad -22 < \frac{9}{5}t + 32 \quad \text{AND} \quad \text{(ii)} \quad \frac{9}{5}t + 32 < -4$$

First let us solve the inequalities separately:

$$\begin{array}{llll} \text{(i)} & -22-32<\frac{9}{5}t & \text{AND} & \text{(ii)} \quad \frac{9}{5}t<-4-32 \\ \text{(i)} & -54\cdot\frac{5}{9}<t & \text{AND} & \text{(ii)} \quad t<-36\cdot\frac{5}{9} \\ \text{(i)} & t>-30 & \text{AND} & \text{(ii)} \quad t<-20 \end{array}$$

Both ranges of solution should be plotted on the same number line to visualise the temperature range suitable for keeping the sample safe in the freezer:

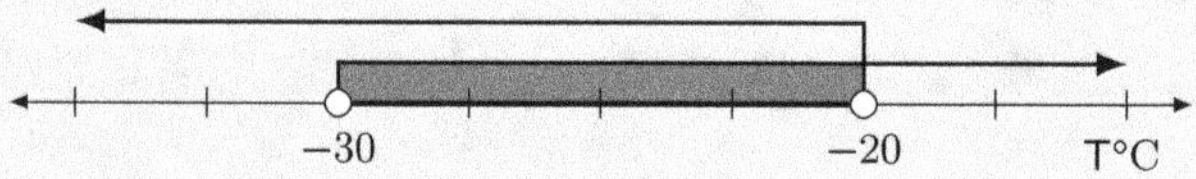

2.5.5 Solving quadratic inequalities

Such inequalities are usually of the form

$$ax^2+bx+c>0 \quad \text{or} \quad ax^2+bx+c<0$$

They can also have the inequality symbols $\leq$ or $\geq$, respectively, and need handling with care.

We can solve for the ranges of values of x satisfying the inequality, called finding the **ranges of solution** of the inequality.

We know, so far, that for a quadratic equation containing x^2 there are **always two roots**, which may sometimes be the same, may be positive, negative, or complex numbers, and can be found either by recognising one of the three canonical cases (see Section 2.3.1) or by using the quadratic formula (see Section 2.3.2).

The method used to solve quadratic inequalities consists of the following steps:

(i) Only if the term in x^2 (*i.e.* a) is negative, multiply both sides of the inequality by -1 to obtain a positive a. The inequality symbol should then be reversed.

(ii) Solve the **associated quadratic** ($ax^2+bx+c=0$) of the inequality and find the possible real solutions (*i.e.* x_1 and x_2, with $x_1 \leq x_2$).

Table 2.1: Solution ranges of quadratic inequalities

Δ value	$\Delta > 0$	$\Delta = 0$	$\Delta < 0$
$ax^2 + bx + c > 0$	$x < x_1$ OR $x > x_2$	$\mathbb{R} - \{x_1 = x_2\}$	$\mathbb{R}$
$ax^2 + bx + c < 0$	$x_1 < x < x_2$	$\varnothing$	$\varnothing$

Notes: $\mathbb{R}$ is the set of real numbers, $\varnothing$ indicates no real solution.

(iii) Define the ranges of solution using Table 2.1, in which the entries come from expressing $ax^2 + bx + c$ as $a(x - x_1)(x - x_2)$ with $a > 0$:

- $(x - x_1)(x - x_2) > 0$ when $x > x_1$ and $x > x_2$ OR when $x < x_1$ and $x < x_2$;
- $(x - x_1)(x - x_2) < 0$ when $x < x_1$ and $x > x_2$ OR when $x > x_1$ and $x < x_2$.

The ranges of solution of the quadratic inequalities can be represented on a number line as follows:

- In the case of $x < x_1$ OR $x > x_2$,

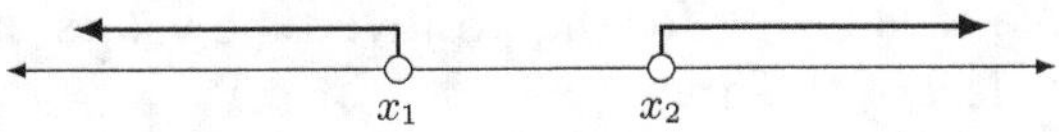

- In the case of $x_1 < x < x_2$,

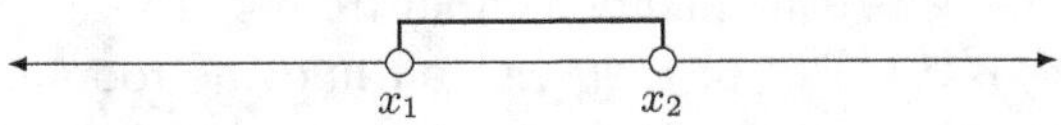

Important note: If $\Delta > 0$ and the inequality symbols are $\geq$ or $\leq$, then the roots x_1 and x_2 are included and the circle should be black-filled (•) on the number line.

Additionally, if $\Delta = 0$ and the inequality symbol is $\geq$, then the solution range is $\mathbb{R}$, whereas if the inequality symbol is $\leq$, the unique solution is $x_1 = x_2$.

Example: Solve for x: $2 + x(1 - x) \leq 2(4x + 7)$.

Solution: We first multiply out and simplify the inequality using the rules for manipulating inequalities:

$$2 + x - x^2 \leq 8x + 14 \Longleftrightarrow -x^2 - 7x - 12 \leq 0$$

We can multiply both sides of the inequality by -1 to get $a > 0$ and simultaneously reverse the inequality symbol:

$$x^2 + 7x + 12 \geq 0$$

Then we factorise the polynomial $x^2 + 7x + 12 = (x + 3)(x + 4)$ and find its zero values $x_1 = -4$ and $x_2 = -3$.

Following Table 2.1, we can define the ranges of solution as $x \leq -4$ OR $x \geq -3$ and plot them on the number line:

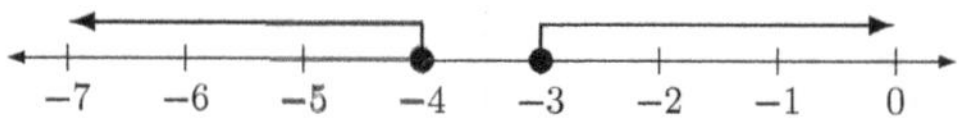

Example in Medicine: Following cutaneous injection of a medicine in a patient, the amount of medicine in the patient's blood follows the time law: amount $= -x^2 + 7.017x - 0.117$, with time (x) measured in hours (h), before complete degradation. How many hours after injection is the medicine in the patient's blood?

Solution: We are looking for the positive time values of the pharmacokinetics law of the drug:

$$\text{amount} > 0 \Longleftrightarrow -x^2 + 7.017x - 0.117 > 0$$

We first multiply both sides of the quadratic inequality by -1, thereby reversing the inequality symbol from $>$ to $<$: $x^2 - 7.017x + 0.117 < 0$; then we can find the roots of the polynomial using the quadratic formula:

$$x_{1,2} = \frac{7.017 \pm \sqrt{7.017^2 - 4 \cdot 0.117}}{2}$$

thus $x_1 = 0.017\,\text{h} = 1\,\text{min}$ and $x_2 = 7\,\text{h}$.

Following Table 2.1, we can define the ranges of solution as $0.017 < x < 7$ and plot them on the number line:

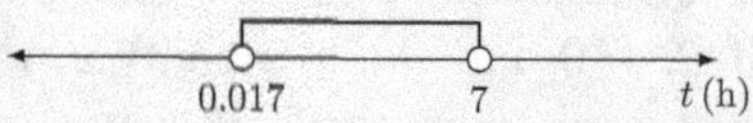

Therefore, the medicine is in the patient's blood 1 minute after injection and persists for 6 hours and 59 minutes in peripheral blood circulation.

2.5.6 Solving polynomial and rational inequalities

Solving rational inequalities and polynomial inequalities is very similar, and needs handling with care.

To solve a **polynomial inequality**, the method consists of the following steps:

(i) Factorise the polynomial (see Section 2.1.2).
(ii) Find the zero values of the polynomial (see Sections 2.2.4 and 2.3.1).
(iii) Divide the number line into intervals using the zero values found.
(iv) Find the sign of each factor on the real number line using inequalities (factor > 0), since whether or not the polynomial is greater than zero depends strictly upon the sign of the factors composing the polynomial.

The method for solving a **rational inequality** similarly consists of the following steps:

(i) Factorise both the numerator and the denominator (see Section 2.1.2).
(ii) Find the zero values of the numerator and the denominator (see Sections 2.2.4 and 2.3.1); however, because rational expressions have denominators, their zeroes are places where the fractions are not defined (known as **undefined points**, see Section 1.2), and thus we have to be a little more careful.
(iii) Divide the number line into intervals using the zero values found.
(iv) Find the sign of each factor on the real number line using inequalities (factor > 0), since whether or not the fraction is greater than zero depends strictly upon the sign of the factors composing the numerator and the denominator.

An example of the solution of a rational inequality is presented below, since polynomial inequalities are solved by similar methods and are also fractions, but with 1 as the denominator which is positive and thus does not affect the sign of the entire polynomial.

Example: Solve the rational inequality: $\dfrac{2x - x^2}{x + 2} \leq 0.$

Solution: We can factorise the numerator as $x(2 - x)$ to have $\dfrac{(2 - x)x}{x + 2} \leq 0.$

Next, we investigate the sign of each factor on the real number line using the chart below.

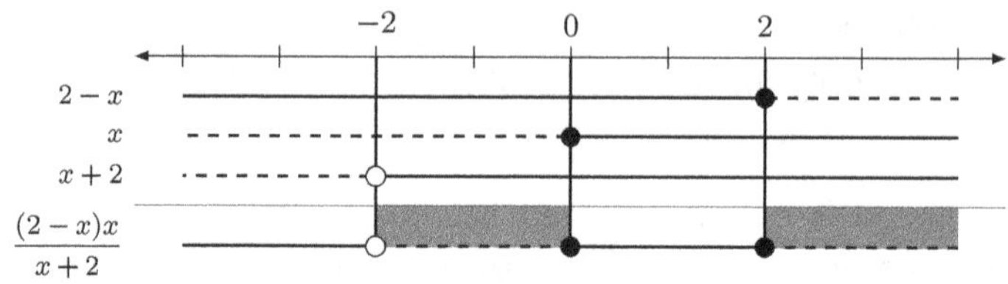

On the top of the number line, we write the places where both the numerator and the denominator can be equal to zero (*i.e.* when $x = -2$, 0 or 2) and list the factors on the left, with the fraction (or the polynomial) at the bottom left.

We then study the sign of each factor by investigating where each factor is > 0: **continuous line** corresponds to a *positive* factor while a **dashed line** to a *negative* factor.

In the bottom row, the sign of the final fraction is obtained by multiplying the sign of the factors within the interval and then is represented using the same conventions.

Since the fraction can also be equal to 0 (as the inequality symbol is $\leq$), the factors in the numerator can equal zero (*i.e.* its zero values can be solutions of the inequality), whereas the zeroes of the denominator factors should be excluded (see Section 1.2).

Full black circles ($\bullet$) indicate possible solutions, while for the empty circles ($\circ$), the values are excluded.

Since, we are looking for where the fraction $\dfrac{2x - x^2}{x + 2} \leq 0$ (*i.e.* where the line is dashed), our grid shows that this is true when $-2 < x \leq 0$ OR $x \geq 2$, as shown in grey.

We can then plot the ranges of solution on another line number for better visualising the solution intervals:

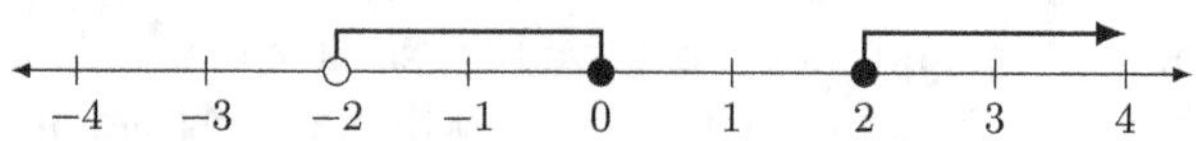

Example in Biology: A new kit for polymerase chain reaction (PCR) needs to be tested in the laboratory. The manufacturer's protocol states that the primer should have a minimum 40% content of guanine (G) and cytosine (C) nucleotides, should terminate in one or more C or G bases, and should be between 25 and 45 bases in length. The equation suggested by the manufacturer for the melting temperature (T_m) of the primer is $T_m = 81.5 - \frac{675}{N+150}$, where $N =$ total number of bases.

However, the machine for carrying out the polymerase chain reaction (PCR) in the laboratory only allows melting temperatures (T_m) above $78°\text{C}$.

Estimate the possible numbers of bases necessary for a DNA primer according to the manufacturer's protocol for the new kit.

Solution: We should first set up the inequality to model the problem:

$$T_m \geq 78 \Longleftrightarrow 81.5 - \frac{675}{N+150} \geq 78$$

which is a rational inequality as the unknown $N = x$ is in the denominator.

We can simplify the expression using the rules for manipulating inequalities: $-\frac{675}{x+150} \geq 78 - 81.5 \Longleftrightarrow \frac{675}{x+150} \leq 3.5.$

Let us then find a common denominator and simplify appropriately in order to investigate the sign of the final fraction:

$$\frac{675}{x+150} - \frac{(x+150)\cdot 3.5}{x+150} \leq 0 \Longleftrightarrow \frac{675 - 3.5x - 525}{x+150} \leq 0$$
$$\Longleftrightarrow \frac{150 - 3.5x}{x+150} \leq 0$$

We can plot a chart of the signs of the fraction:

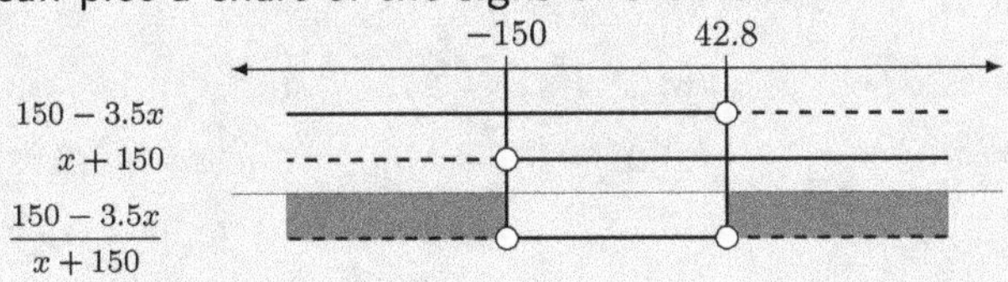

We are looking for where the fraction $\frac{150 - 3.5x}{x + 150} \leq 0$ (*i.e.* where the line is dashed), and our grid shows that this is true when $x \leq -150$ OR $x \geq 42.8$ in grey.

Therefore, the number of bases of the primer could be ≤ -150 OR ≥ 42.8; however, the manufacturer suggested using from 25 up to 45 bases per primer and the number of bases can only be positive. Thus, using the new PCR kit, the possible number of bases available could be 43 or 44 in order to have a primer performing with a $T_m \geq 78$.

2.6 Exercises

(1) Multiply out the following expressions.

(a) $(x - 4)(y + 3)$ (b) $(x + 2)^2$

(c) $(x - 4)^2$ (d) $(3x + 1)(3x - 1)$

(e) $6(4x + 2y)(c + d)$ (f) $-\frac{1}{2}(3 - 2x)(2x + 3)$

(2) Use the Binomial Expansion to find the terms for the following expressions.

(a) $(1 + x)^7$ (b) $(1 - 3x)^4$

(c) $(4 + x)^6$ (N.B.): $= 4^6\left(1 + \frac{x}{4}\right)^6$ (d) $(2 - x)^5$ (N.B.): $= 2^5\left(1 + \left(\frac{-x}{2}\right)\right)^5$

(3) If possible, factorise the following expressions.

(a) $rx + sx + tx + ux$ (b) $2\pi r + \pi r^2$

(c) $12x + 15y$ (d) $cx - dy - dx + cy$

(e) $x^2 + dx + cx + cd$ (f) $1 - y + z - y^2z^2 + y^3z^2 - y^2z^3$

(g) $2x^4y^3 + 6x^2y^5 - 8x^3y^4$ (h) $\frac{3}{2}ab + a - 1 - \frac{3}{2}b$

(4) If possible, factorise the following quadratic expressions, using an appropriate method.

(a) $x^2 + 12x + 27$ (b) $x^2 + 14x + 24$ (c) $x^2 - 4x - 12$

(d) $x^2 - 4x + 3$ (e) $2x + 15 - x^2$ (f) $6x - 8 - x^2$

(g) $3x^2 + 5x + 2$ (h) $10x^2 + 24x + 8$ (i) $6x^2 + 7x + 2$

(j) $6x^2 - 7x + 2$ (k) $6x^2 + x - 2$ (l) $6x^2 - x - 2$

(5) If possible, factorise the following quadratic expressions appropriately.

(a) $x^2 - 81$ (b) $x^2 + 36$

(c) $121 - x^2$ (d) $4x^2 - 9y^2$

(6) Factorise the following perfect squares. If necessary, add or subtract the extra term that completes the square in the following expressions and then factorise them.

(a) $x^2 + 12x + 36$ (b) $x^2 - 8x + 16$ (c) $9x^2 + 12x + 4$

(d) $x^2 + 4x + 2$ (e) $4x^2 + 12x$ (f) $x^2 + 10x + 15$

(7) Factorise the following algebraic expressions using and combining appropriate methods.

(a) $4x - 4y + ya^4 - xa^4$ (b) $2xy^2 - 3xya + xa^2$ (c) $0.16x^5y^3 - x^3y$

(d) $x^5 - x^4 + x^3 - x^2$ (e) $x^2 + 4y^2 - 4xy - z^2$ (f) $x^4 - 7x^2 + 10$

(8) By applying the rules for manipulating of equations, solve the following equations for x.

(a) $x + 6 = 15$ (b) $3x + 5 = 17$ (c) $16x - 4 = 28$

(d) $9x - 2 = -6$ (e) $\frac{15}{x} = 3$ (f) $\frac{x}{9} = 10$

(g) $\frac{x}{6} = \frac{5}{3}$ (h) $\frac{42}{x} = \frac{6}{7}$ (i) $15 - x = -3$

(j) $4(x + 7) = 6(x - 3)$ (k) $6(x + 1) = 18$ (l) $\frac{5x}{4} + \frac{1}{2} = 3$

(m) $\frac{3x - 1}{4} = \frac{3}{5}$ (n) $\frac{x}{4} + 6 = \frac{x}{3} - 4$ (o) $\frac{x - 6}{7} = \frac{x + 4}{10}$

(9) Solve the following rational equations for x. Be careful to exclude from the solutions the values which make the denominators 0.

(a) $\frac{5x - 3}{x - 1} = 3$ (b) $\frac{x - 3}{x - 7} = \frac{x - 7}{x - 3}$

(c) $\frac{x - 1}{3 + x} + \frac{2 - x}{x - 3} = \frac{5}{x^2 - 9}$ (d) $1 - \frac{2}{x^2 - x} = \frac{x}{x - 1} + \frac{2}{x}$

(10) Manipulate the following equations to change the subject as specified. Also make explicit the conditions for the existence of the subject variable in $\mathbb{R}$.

(a) $y = 3x + 2$ for x (b) $x = u + at$ for t

(c) $y = cx - a$ for x (d) $F = mg + m\omega^2 r$ for r

(e) $E = hc/\lambda$ for λ (f) $y = bx^3 + d$ for x

(g) $\dfrac{x^2}{\sqrt{a}} = 3q$ for a

(h) $y = \dfrac{3x-2}{3x+1}$ for x

(i) $a = \dfrac{2bc^2}{b+c}$ for b

(j) $\dfrac{1}{a} + \dfrac{1}{b} = \dfrac{c-2}{x}$ for x

(k) $y = \dfrac{1-x^2}{1+x^2}$ for x

(l) $\dfrac{a}{A} = \sqrt{\dfrac{b+d}{b-d}}$ for b

(11) Use the elimination method to obtain the required subject for the following cases.

(a) Eliminate b, making y the subject, from $x = 3y + 5b$ and $b = 6xy$

(b) Eliminate ν, making λ the subject, from $E = h\nu$ and $\lambda\nu = c$

(c) Eliminate t, making u the subject, from $v = u + at$ and $s = ut + \dfrac{1}{2}at^2$

(d) Eliminate r, making A the subject, from $V = 4\pi r^3/3$ and $A = 4\pi r^2$

(12) Find the roots of the following equations, by recognising the appropriate quadratic cases.

(a) $x^2 - 6x = 0$ (b) $4x^2 - 2x = 0$ (c) $x^2 - x = 0$

(d) $25x^2 - x = 0$ (e) $x^2 - 49 = 0$ (f) $16 - 4x^2 = 0$

(g) $100 + x^2 = 0$ (h) $13x^2 - 21 = 0$ (i) $x^2 + 9x + 20 = 0$

(j) $x^2 + 4x - 21 = 0$ (k) $4x^2 + 6x + 2 = 0$ (l) $x^2 - 5x - 50 = 0$

(13) By using the quadratic formula, find the roots of the following quadratics.

(a) $5x^2 - 13x + 6 = 0$ (b) $10x^2 + x - 2 = 0$ (c) $3 - x^2 - 2x = 0$

(d) $x^2 + x + 12 = 0$ (e) $4 - 2x - 6x^2 = 0$ (f) $2x^2 + 4x + 2 = 0$

(14) Solve the following rational equations for x. Be careful to exclude values from the solutions that make the denominators 0.

(a) $-\dfrac{1}{2x} = \dfrac{3}{x^2+3x} + \dfrac{1}{x^2-4x}$

(b) $1 - x = \dfrac{x^2+5-6x}{x-1}$

(c) $\dfrac{5}{x+5} - 1 = -\dfrac{4}{x+4}$

(d) $\dfrac{5x}{x-1} - \dfrac{6}{1+x} = \dfrac{2}{x^2-1}$

(15) Solve the following quadratics using the quadratic formula and then factorise the expressions appropriately.

(a) $8x^2 - 12x + 3 = 0$ (b) $3x^2 + 2x - 3 = 0$ (c) $2x^2 - 3x - 4 = 0$

(16) Solve the following polynomial equations for x in the set of real numbers.

(a) $x^3 - 4x^2 + 6x - 24 = 0$ (b) $x^6 - 9x^4 + 5x^2 - 45 = 0$

(c) $-x^4 + 13x^2 - 36 = 0$ (d) $-\frac{1}{12} - \frac{x}{4} + \frac{x^2}{3} + x^3 = 0$

(17) Solve the following pairs of equations for x and y using the elimination method.

(a) $20x + y = 81$ $2x - y = 7$ (b) $2x + 5y = 16$ $5x + 2y = 19$

(c) $2x - 7y = 8$ $4x - 6y = 0$ (d) $2y - x = 1$ $5y + \frac{x}{2} = 1$

(e) $2y - 6x = 6$ $20y + 12x = 0$ (f) $3y + 2x = 9$ $5x - y = -3$

(18) Solve the following pairs of equations for x and y using the substitution method.

(a) $4y + x = 15$ $9y - x = 37$ (b) $3x + 2y = 7$ $2x + 3y = 3$

(c) $2y - 2x = 6$ $3y + 3x = -3$ (d) $2x - 20y = 15$ $20x + 4y = 48$

(e) $3y - 6x = 36$ $13y + 3x = -18$ (f) $3y - 10x = 2$ $15y + 50x = 0$

(19) Solve the following problems using simultaneous equations.

(a) A mixed pile of 2 kg and 4 kg bricks weighs a total of 1000 kg and contains 300 bricks. How many of each weight brick are there?

(b) A car averages 28.3 miles/gallon in town and 32.4 miles/gallon on the motorway. On a recent 300 mile trip, fuel consumption averaged 31.2 miles/gallon. How many miles of the journey were on the motorway?

(20) Solve the following pairs of equations for x and y.

(a) $x^2 + y^2 = 2$ $2x - y = 1$

(b) $y^2 - x^2 = 7$ $4x - 3y = 0$

(c) $x^2 + 2x + y^2 = 17$ $5x - 3y = 1$

(d) $y^2 + xy = 6$ $5y + 2x = 15$

(e) $x^2 + y^2 + 3x - 2y = 4$ $13x - 6y = 1$

(f) $25x^2 + 9y^2 - 5x - 3y = 6$ $5x - 3y = 2$

(21) Solve the following linear inequalities for x and plot the solutions on a number line.

(a) $x - 4(x - 1) < -4$ (b) $-\frac{1}{3}x > -2 - (4 + x)$

(c) $\frac{3x - 1}{4} - \frac{2 - 4x}{2} \leq \frac{2}{3}x$ (d) $\frac{1}{2}\left(\frac{x}{3} - 2\right) - x \geq \frac{4 - x}{3}$

(e) $0.4 - 1.2x < 0.6(x + 1.1)$ (f) $7 - \sqrt{2}x > -2(\sqrt{2}x - 3)$

(22) Solve the following linear compound inequalities for x.

(a) $3(2x-1) \leq \frac{1}{2}(x-3)$ AND $2x-3>5$

(b) $-4(x-2)<0$ AND $\frac{x+4}{2}-\frac{x-2}{4} \geq 1$

(c) $3(0.5x-1.2)>-2.1x$ AND $2.5-0.3x \geq -0.4$

(d) $\frac{x-3}{3^{-1}} \leq \frac{4-x}{\left(\frac{1}{2}\right)^{-1}}$ AND $\frac{x}{2}<-\frac{x}{3}$

(e) $3+x<2(x-1)+1$ OR $\frac{4-x}{3}-\frac{x}{2} \geq 0$

(f) $7x-4 \geq 0$ OR $4>-7x+2$

(23) Solve the following quadratic inequalities for x.

(a) $x^2+x-6<0$

(b) $x^2-2x-8>0$

(c) $x(x+3) \geq -2x$

(d) $-x^2+3x \geq 2$

(e) $x^2+3x+5<1$

(f) $9+x(x-5)>0$

(g) $x(x-6)+1>7$

(h) $4x^2+20x+25 \leq 0$

(i) $-\frac{5}{2}x+\frac{3}{4} \leq 2x^2$

(j) $-x^2-2+\frac{5}{\sqrt{2}}x<0$

(k) $0.6x^2-1.2x+0.5 \geq 0$

(l) $-1.8>x(0.02+x)$

(24) Solve the following polynomial and rational inequalities for x.

(a) $(4x^2-1)(x-3)>0$

(b) $x^3-25x<0$

(c) $(x-5) \geq \frac{3}{x-3}$

(d) $\frac{x+3}{1-x} \leq 0$

(e) $\frac{x+2}{x-3}>4$

(f) $\frac{x-2}{5+x}-1>0$

(g) $\frac{x}{3}+\frac{8}{3} \geq \frac{3}{x}$

(h) $\frac{4x}{x+1}-\frac{x+1}{x}<0$

(i) $\frac{3x^2+x-4}{x^2-3x+4} \leq 0$

Answers

(1) (a) $xy+3x-4y-12$ (b) x^2+4x+4 (c) $x^2-8x+16$ (d) $9x^2-1$
(e) $24cx+24dx+12cy+12dy$ (f) $2x^2-\frac{9}{2}$

(2) (a) $x^7+7x^6+21x^5+35x^4+35x^3+21x^2+7x+1$
(b) $81x^4-108x^3+54x^2-12x+1$
(c) $x^6+24x^5+240x^4+1280x^3+3840x^2+6144x+4096$
(d) $-x^5+10x^4-40x^3+80x^2-80x+32$

(3) (a) $x(r+s+t+u)$ (b) $\pi r(r+2)$ (c) $3(4x+5y)$
(d) $(x+y)(c-d)$ (e) $(d+x)(c+x)$ (f) $(1-y+z)(1-yz)(1+yz)$
(g) $2x^2y^3(x-y)(x-3y)$ (h) $(a-1)\left(\frac{3b}{2}+1\right)$

(4) (a) $(x+3)(x+9)$ (b) $(x+2)(x+12)$ (c) $(x+2)(x-6)$
(d) $(x-1)(x-3)$ (e) $-(x+3)(x-5)$ (f) $-(x-2)(x-4)$
(g) $(x+1)(3x+2)$ (h) $2(x+2)(5x+2)$ (i) $(2x+1)(3x+2)$
(j) $(2x-1)(3x-2)$ (k) $(2x-1)(3x+2)$ (l) $(2x+1)(3x-2)$

(5) (a) $(x+9)(x-9)$ (b) Impossible
(c) $(11+x)(11-x)$ (d) $(2x+3y)(2x-3y)$

(6) (a) $(x+6)^2$ (b) $(x-4)^2$ (c) $(3x+2)^2$
(d) $(x+2)^2-2=(x+2+\sqrt{2})(x+2-\sqrt{2})$
(e) $(2x+3)^2-9=4x(x+3)$
(f) $(x+5)^2-10=(x+5+\sqrt{10})(x+5-\sqrt{10})$

(7) (a) $-(x-y)(a^2+2)(a+\sqrt{2})(a-\sqrt{2})$ (b) $x(a-y)(a-2y)$
(c) $x^3y(0.4xy+1)(0.4xy-1)$ (d) $x^2(x-1)(x^2+1)$
(e) $(x-2y+z)(x-2y-z)$ (f) $(x+\sqrt{2})(x-\sqrt{2})(x+\sqrt{5})(x-\sqrt{5})$

(8) (a) $x=11$ (b) $x=4$ (c) $x=2$ (d) $x=-\frac{4}{9}$ (e) $x=5$
(f) $x=90$ (g) $x=10$ (h) $x=49$ (i) $x=18$ (j) $x=23$
(k) $x=2$ (l) $x=2$ (m) $x=\frac{17}{15}$ (n) $x=120$ (o) No solution

(9) (a) $x=0$ (b) $x=5$ (c) $x=\frac{4}{5}$ (d) No solution

(10) (a) $x=\frac{y-2}{3}$ (b) $t=\frac{x-u}{a};\ a\neq 0$ (c) $x=\frac{y+a}{c};\ c\neq 0$
(d) $r=\frac{F-mg}{m\omega^2};m\neq 0$ (e) $\lambda=\frac{ch}{E};\ E\neq 0$ (f) $x=\sqrt[3]{\frac{y-d}{b}};b\neq 0$
(g) $a=\frac{x^4}{9q^2};q\neq 0$ (h) $x=\frac{-2-y}{3(-1+y)};\ y\neq 1$ (i) $b=-\frac{ac}{a-2c^2};\ a\neq 2c^2$
(j) $x=\frac{abc-2ab}{a+b};\ a\neq -b$ (k) $x=\pm\sqrt{\frac{1-y}{y+1}};\ y\neq -1$ (l) $b=\frac{-a^2d-A^2d}{A^2-a^2};\ A\neq a$

(11) (a) $y=\frac{x}{3(10x+1)};\ x\neq -\frac{1}{10}$ (b) $\lambda=\frac{ch}{E};\ E\neq 0$
(c) $u=\pm\sqrt{v^2-2as}$ (d) $A=\sqrt[3]{36\pi V^2}$

(12) (a) $x=6,\ x=0$ (b) $x=\frac{1}{2},\ x=0$ (c) $x=1,\ x=0$
(d) $x=\frac{1}{25},\ x=0$ (e) $x=7,\ x=-7$ (f) $x=2,\ x=-2$
(g) Imaginary roots (h) $x=\sqrt{\frac{21}{13}},\ x=-\sqrt{\frac{21}{13}}$ (i) $x=-4,\ x=-5$
(j) $x=3,\ x=-7$ (k) $x=-\frac{1}{2},\ x=-1$ (l) $x=10,\ x=-5$

(13) (a) $x=2,\ x=\frac{3}{5}$ (b) $x=\frac{2}{5},\ x=-\frac{1}{2}$ (c) $x=-3,\ x=1$
(d) No real roots (e) $x=-1,\ x=\frac{2}{3}$ (f) $x=-1$

(14) (a) $x=-10,\ x=3$ (b) $x=3$
(c) $x=2\sqrt{5},\ x=-2\sqrt{5}$ (d) No real solution

(15) (a) $x=\frac{3+\sqrt{3}}{4},\ x=\frac{3-\sqrt{3}}{4}$ (b) $x=\frac{\sqrt{10}-1}{3},\ x=-\frac{1+\sqrt{10}}{3}$
(c) $x=\frac{3+\sqrt{41}}{4},\ x=\frac{3-\sqrt{41}}{4}$

(16) (a) $x=4$ (b) $x=3,\ x=-3$
(c) $x=2,\ x=-2,\ x=3,\ x=-3$ (d) $x=-\frac{1}{3},\ x=\frac{1}{2},\ x=-\frac{1}{2}$

(17) (a) $x = 4,\ y = 1$ (b) $x = 3,\ y = 2$ (c) $x = -3,\ y = -2$
(d) $x = -\frac{1}{2},\ y = \frac{1}{4}$ (e) $x = -\frac{5}{6},\ y = \frac{1}{2}$ (f) $x = 0,\ y = 3$

(18) (a) $x = -1,\ y = 4$ (b) $x = 3,\ y = -1$ (c) $x = -2,\ y = 1$
(d) $x = \frac{5}{2},\ y = -\frac{1}{2}$ (e) $x = -6,\ y = 0$ (f) $x = -\frac{1}{10},\ y = \frac{1}{3}$

(19) (a) $x + y = 300$ and $2x + 4y = 1000$; 100 $2\,\text{kg}$ bricks and 200 $4\,\text{kg}$ bricks
(b) $x + y = 300$ and $\dfrac{x}{28.3} + \dfrac{y}{32.4} = \dfrac{300}{31.2}$; 220.4 miles on the motorway

(20) (a) $(x = 1,\ y = 1)$ $\left(x = -\frac{1}{5},\ y = -\frac{7}{5}\right)$
(b) $(x = 3,\ y = 4)$ $(x = -3,\ y = -4)$
(c) $(x = 2,\ y = 3)$ $\left(x = -\frac{38}{17},\ y = -\frac{69}{17}\right)$
(d) $\left(x = -\frac{5}{2},\ y = 4\right)$ $(x = 5,\ y = 1)$
(e) $(x = 1,\ y = 2)$ $\left(x = -\frac{131}{205},\ y = -\frac{318}{205}\right)$
(f) $\left(x = \frac{3}{5},\ y = \frac{1}{3}\right)$ $\left(x = 0,\ y = -\frac{2}{3}\right)$

(21) (a) $x > \frac{8}{3}$ (b) $x > -9$ (c) $x \leq \frac{3}{5}$
(d) $x \leq -\frac{14}{3}$ (e) $x > -0.0875$ (f) $x > -\frac{\sqrt{2}}{2}$

(22) (a) No real solution (b) $x > 2$ (c) $1 < x \leq \frac{29}{3}$
(d) $x < 0$ (e) $x \leq \frac{8}{5}$ OR $x > 4$ (f) $x > -\frac{2}{7}$ OR $x \geq \frac{4}{7}$

(23) (a) $-3 < x < 2$ (b) $x < -2$ OR $x > 4$
(c) $x \leq -5$ OR $x \geq 0$ (d) $1 \leq x \leq 2$
(e) No real solution (f) $\mathbb{R}$
(g) $x < 3 - \sqrt{15}$ OR $x > 3 + \sqrt{15}$ (h) $x = -\frac{5}{2}$
(i) $x \leq -\frac{3}{2}$ OR $x \geq \frac{1}{4}$ (j) $x < \frac{\sqrt{2}}{2}$ OR $x > 2\sqrt{2}$
(k) $x \leq \frac{6-\sqrt{6}}{6}$ OR $x \geq \frac{6+\sqrt{6}}{6}$ (l) No real solution

(24) (a) $-\frac{1}{2} < x < \frac{1}{2}$ OR $x > 3$ (b) $x < -5$ OR $0 < x < 5$
(c) $2 \leq x < 3$ OR $x \geq 6$ (d) $x \leq -3$ OR $x > 1$
(e) $3 < x < \frac{14}{3}$ (f) $x < -5$
(g) $-9 \leq x < 0$ OR $x \geq 1$ (h) $-1 < x < -\frac{1}{3}$ OR $0 < x < 1$
(i) $-\frac{4}{3} \leq x \leq 1$

Chapter 3

Graphs

Preamble

In Biosciences, there are many situations for which data can be presented graphically. Thus, it is essential to be familiar with the types of graphs and the conventions used, and how to interpret the information provided by the graphical representation.

Physical, chemical, and biological quantities are often linked by equations. Converting graphs into algebraic equations and analysing the shape of the plot are major goals which allow the scientist to make a critical judgement about the possible relationships between the variables plotted on the graph and to obtain analytical expressions to describe physical phenomena. This then allows extrapolation beyond the ranges covered by the experiments, providing predictive power.

3.1 Graph Etiquette

3.1.1 Titles, labels, and units

Plotting data on a graph has the major advantage that trends and patterns can be represented and recognised.

In scientific experiments, we usually have the following:

- an **independent variable**, X or x, sometimes called the *abscissa* or *control variable*, marked on the *horizontal axis*;
- a **dependent variable**, Y or y, sometimes called the *ordinate* or *measured variable*, marked on the *vertical axis*.

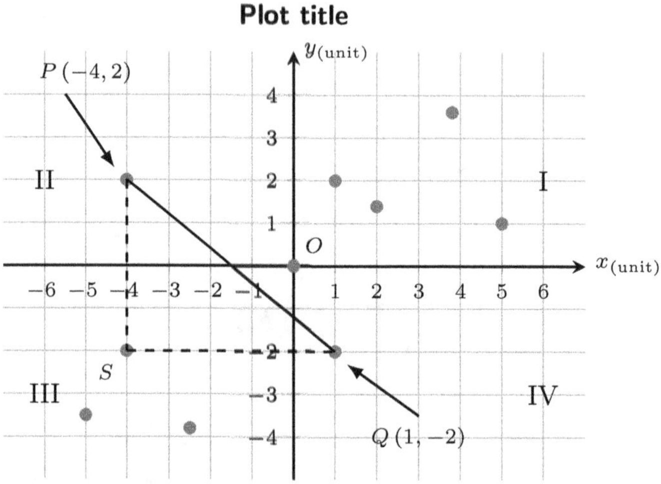

Figure 3.1: Example of graph

Essential features are required to plot a graph in an understandable way (Figure 3.1):

- an informative title;
- two perpendicular number lines (axes) usually crossing at a point O with coordinates $(0, 0)$;
- axis labels with appropriate measurement units included, as often graphs have some physical significance;
- a sensible linear scale on axes so that all points to be plotted fall within the graph area;
- adequate space on the page.

The axes y and x divide the Cartesian plane (named after the mathematician and philosopher René Descartes) into four infinite regions, called **quadrants**, which are numbered using Roman numerals, as shown in Figure 3.1.

The dependent variable y is said to be graphed *against* the independent variable x.

Each point (*e.g.* P and Q in Figure 3.1) has to be defined using Cartesian coordinates (x, y).

The intersection of the axes x and y is called the **origin** of the Cartesian plane and has coordinates $(0, 0)$.

The distance between two points, such as $P(-2,1)$ and $Q(1,-2)$ in Figure 3.1, can be calculated by applying **Pythagoras' Theorem**, which states that for a right angled triangle with perpendicular sides a and b, the length of the hypotenuse, c, is $c=\sqrt{a^2+b^2}$. Thus,

$$\overline{PQ}=\sqrt{\overline{PS}^2+\overline{SQ}^2}=\sqrt{(x_Q-x_P)^2+(y_Q-y_P)^2} \tag{3.1}$$

which gives $\overline{PQ}=\sqrt{(2+2)^2+(-4-1)^2}=\sqrt{16+25}=\sqrt{41}$.

3.2 Straight Lines

3.2.1 General equation

The equation of a **straight line**, also known as *slope-intercept* equation, is

$$y=m\cdot x+c \tag{3.2}$$

where (Figure 3.2):

- x is the controlled or *independent* variable;
- y is the observed or *dependent* variable;
- c is a constant, called the *intercept* of the line on the y-axis;
- m is also a constant, called the *slope*, and describes the steepness of the line.

The **intercept** of a straight line with the y axis is the point $\boxed{(\mathbf{0},\mathbf{c})}$ since at $x=0$, $y=c$ (Figure 3.2).

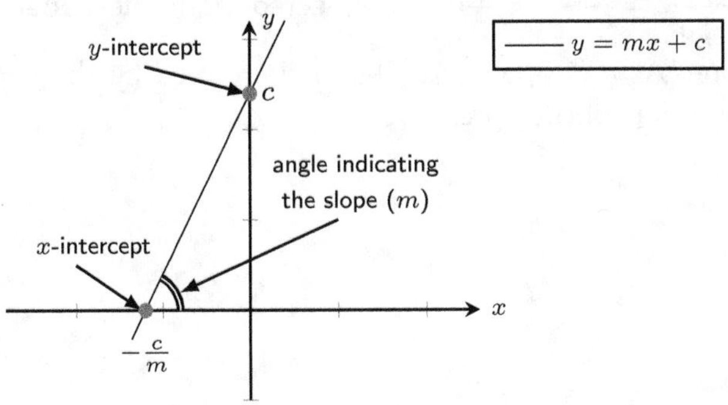

Figure 3.2: Plot of a straight line

The **slope** of a straight line (also called its '*gradient*') can be positive or negative and is given by $\boxed{\mathbf{m} = \frac{\mathbf{\Delta y}}{\mathbf{\Delta x}} = \frac{\mathbf{y_B - y_A}}{\mathbf{x_B - x_A}}}$, where $A\ (x_A, y_A)$ and $B\ (x_B, y_B)$ are two points on the straight line.

A *positive* gradient means that the quantity y increases as x increases, while a *negative* gradient means y is a decreasing quantity as x increases (Figure 3.2).

The **intercept** of a straight line with the x axis (also known as the *zero* of the line) is the point $\boxed{\left(-\frac{\mathbf{c}}{\mathbf{m}}, 0\right)}$ since at $y = 0$ we can solve $mx + c = 0$ (Figure 3.2).

A point $P(x_P, y_P)$ is situated on a certain straight line if the slope-intercept equation is satisfied when the coordinates are substituted into it.

Example: $y = 3x + 7$ and $y = -3x + 21$. For each equation find the y-intercept, the slope m, and the x-intercept. Draw both straight lines and mark the angles they make with the x-axis.

Solution: If $y = 3x + 7$, $c\,(y\text{-intercept}) = 7$. If we take two points on the line $y = 3x + 7$: $A\,(0, 7)$ and $B\,(5, 22)$, the gradient $m = \frac{22 - 7}{5 - 0} = \frac{15}{5} = 3$, as expected from the equation of this straight line. At $y = 0$, $x = -\frac{7}{3}$. The angle formed by the line with the positive x-axis is acute (*i.e.* $<90°$).

If $y = -3x + 21$, $c\,(y\text{-intercept}) = 21$. If we take two points on the line $y = -3x + 21$: $C\,(1, 18)$ and $D\,(5, 6)$, the gradient $m = \frac{18 - 6}{1 - 5} = \frac{12}{-4} = -3$, as expected from the equation of this straight line. At $y = 0$, $x = 7$. The angle formed by the line with the positive x-axis is obtuse (*i.e.* $<90°$).

This can all be visualised in the graph below:

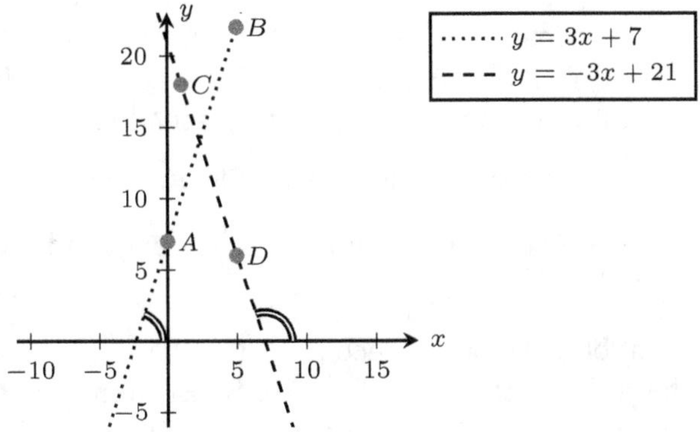

3.2.2 Finding the equation of a straight line

Two pieces of data are always required to write the equation of an unknown straight line. The most common cases are as follows:

(i) The coordinates of two points $A(x_A, y_A)$ and $B(x_B, y_B)$ are known, and if $x_A \neq x_B$,

$$y - y_A = \frac{y_B - y_A}{x_B - x_A} \cdot (x - x_A) \tag{3.3}$$

(ii) The coordinates of a single point $P(x_P, y_P)$ and the value of the slope m are known:

$$y - y_P = m \cdot (x - x_P) \tag{3.4}$$

In both cases, the expression can be rearranged to have the form of the slope-intercept equation (Equation 3.2), using the rules for manipulating equations (see Section 2.2.3).

The **distance** from a point $Q(x_Q, y_Q)$ to a straight line is the length of the segment beginning at Q and ending perpendicularly on the straight line. Let $Q(x_Q, y_Q)$ be a point and $y = mx + c$ be the equation of a certain line. It can be demonstrated that the distance d from Q to the line is

$$d = \frac{|c + m \cdot x_Q - y_Q|}{\sqrt{1 + m^2}} \tag{3.5}$$

Example: Find the equation of the straight line which passes through points $A(-12, 7)$ and $B(-8, -5)$. Also find the equation of the straight line which passes through $P(3, 2)$ and has $m = 4$, and then verify whether $Q(10, 7)$ is situated on this second line. If it is not, calculate the distance from point Q to the second line.

Solution: Using equation 3.3, for the first line we have $y - 7 = \frac{-5 - 7}{-8 + 12}(x + 12)$, which can be rearranged to give $y = -3x - 29$.

Using equation 3.4, for the second line we have $y - 2 = 4(x - 3)$, which can be rearranged to $y = 4x - 10$. Substituting the coordinates of point Q into $y = 4x - 10$, we have $7 = 4 \cdot 10 - 10$, giving $7 = 30$ which is incorrect.

Therefore, point Q is not situated on the line represented by the equation $y = 4x - 10$. Thus, using equation 3.5, we can calculate the distance from Q to the straight line:

$$d = \frac{|-10 + 4 \cdot 10 - 7|}{\sqrt{1 + 4^2}} = \frac{23}{\sqrt{17}} = \frac{23\sqrt{17}}{17} \approx 5.5$$

This can all be visualised in the graph below:

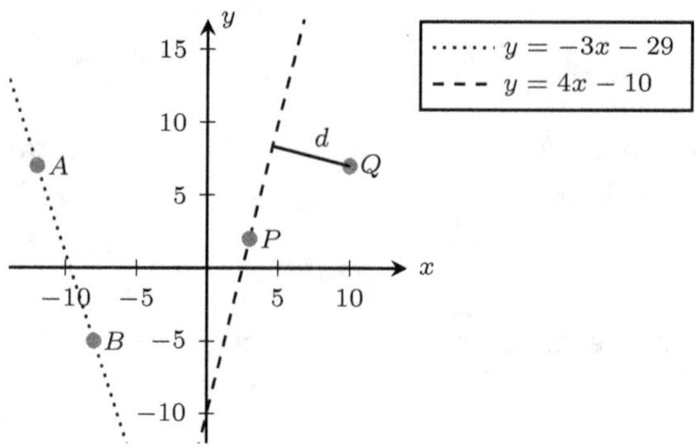

3.2.3 Special cases

Particular attention should be paid to special cases of straight lines (Figure 3.3):

(i) When $c = 0$, then the equation is $y = mx$ and the line passes through the point $(0, 0)$.

(ii) When $m = 0$, then the equation is $y = c$ and the line is **horizontal**.

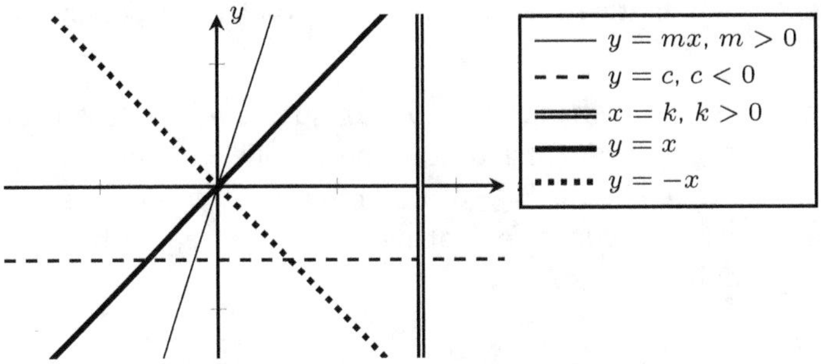

Figure 3.3: Special cases of straight lines

(iii) When the line equation is $x = k$ with $k =$ constant, the line is **vertical**.

(iv) The line bisecting (dividing into two equal parts) the right angle formed by the positive x- and y-axes (usually called quadrant I) has equation $y = x$, whereas the line bisecting the right angle formed by the negative x-axis and the positive y-axis (usually called quadrant II) has equation $y = -x$.

(v) Two lines with equations $y = m_1x + a$ and $y = m_2x + b$, respectively, are **parallel** if and only if their slopes are the same (*i.e.* $\boxed{\mathbf{m_1 = m_2}}$) or if they have no slope.

(vi) Two lines with equations $y = m_1x + a$ and $y = m_2x + b$, respectively, are **perpendicular** if and only if $\boxed{\mathbf{m_1 \cdot m_2 = -1}}$, or if one is vertical and the other is horizontal.

When the equation of a straight line is of the form $y = mx$ with $m > 0$, the dependent variable y is said to be **directly proportional** to the independent variable x.

Since a straight line consists of all the solutions for a linear equation in x and y, finding the point where two straight lines intersect is equivalent to solving both equations simultaneously (see Section 2.4) for the unique values of x and y.

Example: Draw the straight lines of equations $y = 3x$, $y = 7$, and $x = -5$, and comment on their position in relation to the Cartesian axes. Then, draw the lines of equations $y = 2x + 15$ and $y = 2x + 10$ and comment on their relative positions in the Cartesian coordinate system. Draw the lines of equations $y = -2x + 9$ and $y = \frac{1}{2}x + 1$, comment on their relative positions, and find their possible point of intersection.

Solution: If $y = 3x$ when $x = 0$, $y = 0$, so the y-intercept is at $(0, 0)$. If $y = 7$, its graph is a horizontal line parallel to the x-axis, with a y-intercept at $(0, 7)$. If $x = -5$, its graph is a vertical line parallel to the y-axis, with an x-intercept at $(-5, 0)$. This can all be seen in the following graph:

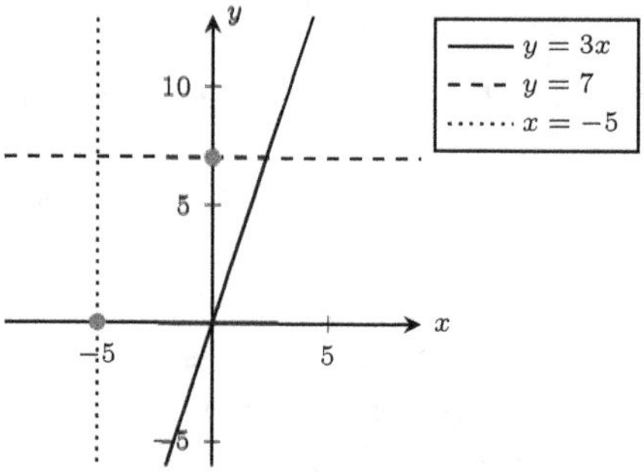

The lines $y = 2x + 15$ and $y = 2x + 10$ are parallel, as the values of m are equal ($m_1 = m_2 = 2$). The lines $y = -2x + 9$ and $y = \frac{1}{2}x + 1$ are perpendicular, as $m_1 \cdot m_2 = -2 \cdot \frac{1}{2} = -1$.

The intersection of the straight lines $y = -2x + 9$ and $y = \frac{1}{2}x + 1$ is found by solving both equations simultaneously using the substitution method: $P\left(\frac{16}{5}, \frac{13}{5}\right)$. These lines can be visualised in the following graph:

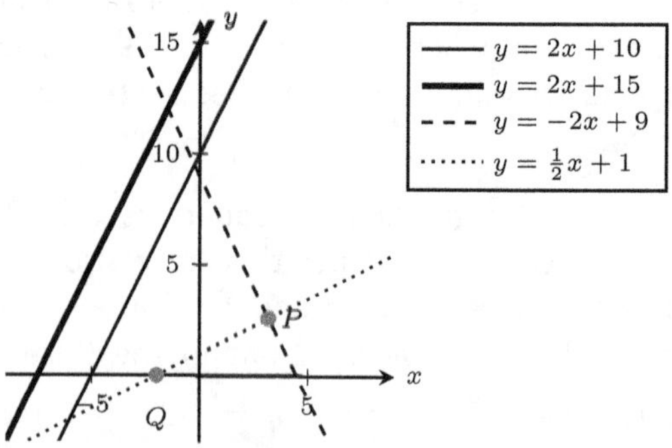

Important note: In all the Biosciences examples, the **grey areas** of graphs relate to the **non-physical solutions** of the problem, and these meaningless solutions should ALWAYS be excluded.

Example in Biochemistry: The rate at which a given enzyme catalyses a reaction depends on the substrate concentration following the equation:

$$\frac{1}{v_i} = \frac{K_m}{V_{max}} \cdot \frac{1}{[s]} + \frac{1}{V_{max}}$$

where v_i is the initial rate of the reaction, $[s]$ is the substrate concentration, and K_m and V_{max} are positive constants. We can derive a straight line graph from the above formula by plotting $\frac{1}{v_i}$ against $\frac{1}{s}$, as explained earlier in Section 2.2.6. This particular plot, known as the *Lineweaver–Burk plot*, is used to investigate enzyme kinetics.

Five enzymatic assays each using a different substrate concentration were performed to measure their corresponding initial velocities. Plotting $\frac{1}{v_i}$ against $\frac{1}{s}$ for each assay allowed the researcher to write the equation of the straight line:

$$y = 0.125\,\text{min} \cdot x + 2.5\,\text{min}\,\text{mM}^{-1}$$

Plot the straight line in an appropriate Cartesian coordinate system, and mark the slope, the y-intercept, and the x-intercept on the graph. Then, estimate the values of V_{max} (the maximum rate achievable by the system) and K_m (the substrate concentration at which the reaction rate is half of V_{max}) from the given straight line equation.

Solution: The straight line representing the equation $y = 0.125\,\text{min} \cdot x + 2.5\,\text{min}\,\text{mM}^{-1}$ can be appropriately plotted in the Cartesian coordinate system as follows:

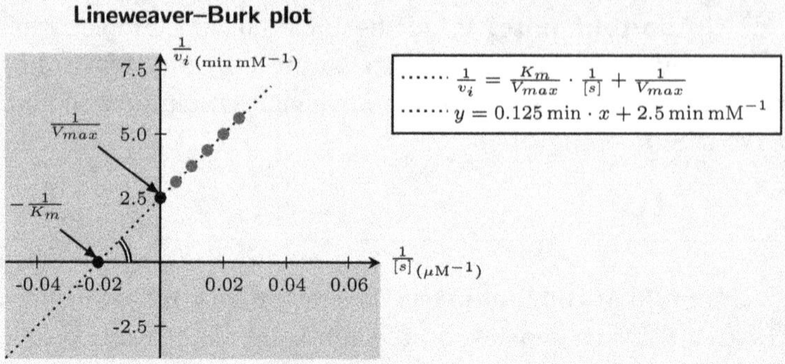

Since the y-intercept $= c = \dfrac{1}{V_{max}} = 2.5\,\text{min mM}^{-1}$, we can use it to estimate $V_{max} = \dfrac{1}{2.5}\,\text{mM min}^{-1} = 0.4\,\text{mM min}^{-1}$ $= 400\,\mu\text{M min}^{-1}$.

Using the Lineweaver–Burk equation, we can also find the x-intercept $= -\dfrac{1}{V_{max}} \cdot \dfrac{V_{max}}{K_m} = -\dfrac{1}{K_m}$.

Since x-intercept $= -\dfrac{c}{m} = -\dfrac{2.5\,\text{min mM}^{-1}}{0.125\,\text{min}} = 20\,\text{mM}^{-1}$ $= 0.02\,\mu\text{M}^{-1}$, we can use the last equation to estimate that $K_m = \dfrac{1}{0.02}\,\mu\text{M} = 50\,\mu\text{M}$.

The expression $mx + c$ involves only a single power of the variable x. It cannot model any of the 'curvature' that a data set might possess. Therefore, it is sometimes useful to introduce mathematical equations with negative or higher powers of x as well, such as the *rectangular hyperbola* $\left(y = \dfrac{k}{x} = kx^{-1}\right)$ or the *parabola* $(y = ax^2 + bx + c)$.

3.3 Rectangular Hyperbolae

3.3.1 General equation

The equation of a particular type of hyperbola, known as a **rectangular (or equilateral) hyperbola**, is

$$x \cdot y = k \tag{3.6}$$

which is equivalent to

$$y = \frac{k}{x} \tag{3.7}$$

with constant $k \neq 0$.

The Cartesian axes are the **asymptotes** (*i.e.* at very large positive or negative values of x or y, the curve approaches the Cartesian axes) and $\mathbf{O(0,0)}$ is the **centre of symmetry** (Figure 3.4).

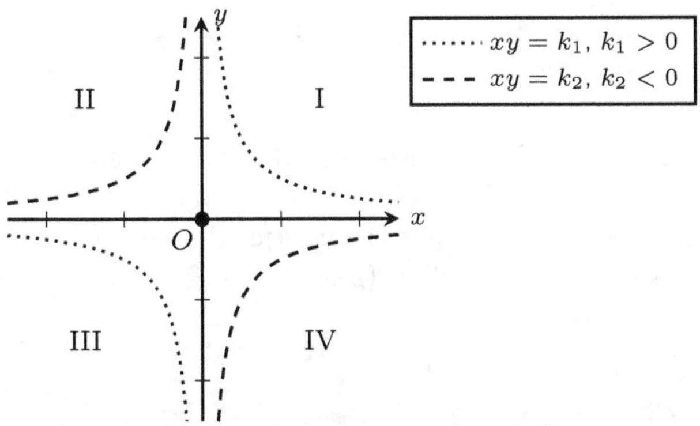

Figure 3.4: Plot of a rectangular hyperbola
Note: The Cartesian axes are the asymptotes, and quadrants are numbered

When $k > 0$ in the equation of a rectangular hyperbola, the dependent variable y is said to be **inversely proportional** to the independent variable x in quadrant I.

A point $P(x_P, y_P)$ belongs to a certain hyperbola if, upon substitution, its coordinates satisfy the hyperbolic equation.

Finding the common points of a line (represented by the linear equation $y = mx + c$) and a rectangular hyperbola $\left(\text{represented by the nonlinear equation } y = \frac{k}{x}\right)$ is equivalent to solving both equations simultaneously (see Section 2.4) for the unique values of x and y.

Three cases are possible:

- two distinct intersection points;
- one single shared point;

- no intersection, so the line does not meet the hyperbola at all.

Example: Draw the hyperbola H described by the equation $xy = -3$. Then, find the possible common points between the straight line $y = 2x + 1$ and the hyperbola H.

Solution: Let us find some points belonging to the hyperbola, that is, those points with coordinates which satisfy the hyperbolic equation when substituted. We can then draw the curve.

x	-5	-3	$-\frac{3}{2}$	-1	$-\frac{3}{5}$	$\frac{3}{5}$	1	$\frac{3}{2}$	3	5
y	$\frac{3}{5}$	1	2	3	5	-5	-3	-2	-1	$-\frac{3}{5}$

In order to determine the Cartesian coordinates of the possible common points between the parabola H and the straight line $y = 2x + 1$, we can solve their equations simultaneously for the unique values of x and y using the substitution method (see Section 2.4.2):

(i) $y = -\dfrac{3}{x}$

(ii) $y = 2x + 1$

which gives no possible solutions, as shown in the plot below.

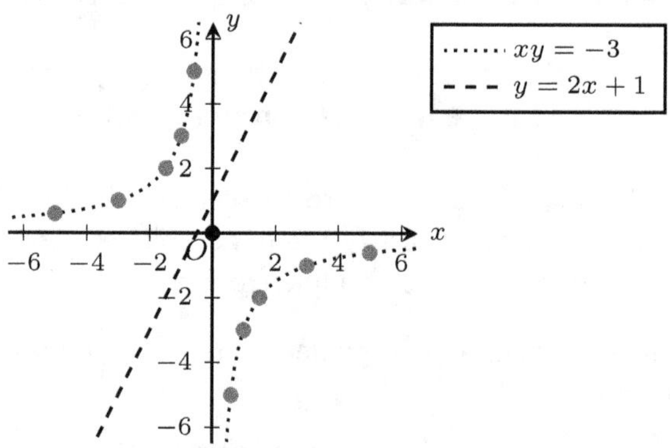

Example in Chemistry: The solubility equilibrium constant of $CaCrO_4$ (molecular weight: $156.07\,\mathrm{g\,mol^{-1}}$) is $K_{ps} = 7.1 \cdot 10^{-4}\,\mathrm{M}^2$ at $25\,°\mathrm{C}$.

Determine whether a $200\,\mathrm{mL}$ solution containing each of the following concentrations of Ca^{2+} and CrO_4^{2-} ions is saturated or whether the ions are completely dissolved or precipitated:

(a) $[Ca^{2+}] = 13\,\mathrm{mM}$, $[CrO_4^{2-}] = 55\,\mathrm{mM}$
(b) $[Ca^{2+}] = 65\,\mathrm{mM}$, $[CrO_4^{2-}] = 11\,\mathrm{mM}$
(c) $[Ca^{2+}] = 11\,\mathrm{mM}$, $[CrO_4^{2-}] = 32\,\mathrm{mM}$
(d) $[Ca^{2+}] = 50\,\mathrm{mM}$, $[CrO_4^{2-}] = 40\,\mathrm{mM}$

If the solution is supersaturated, estimate the concentration of both the dissolved ions (in mM) and the mass of the precipitated salt (in mg).

Solution: A solubility equilibrium exists when a salt in the solid state is in chemical equilibrium with a solution of its ions. The dissociation reaction of $CaCrO_4$ is as follows:

$$CaCrO_4 \rightleftharpoons Ca^{2+} + CrO_4^{2-}$$

The equilibrium constant $K_{ps} = [\mathrm{Ca}^{2+}]\cdot[\mathrm{CrO}_4^{2-}] = 7.1\cdot10^{-4}\,\mathrm{M}^2 = 710\,\mathrm{mM}^2$ is known as the solubility product.

Plotting $[CrO_4^{2-}](\mathrm{mM})$ on the y-axis against $[Ca^{2+}](\mathrm{mM})$ on the x-axis, we can derive a rectangular hyperbola of equation $x \cdot y = 710\,\mathrm{mM}^2$:

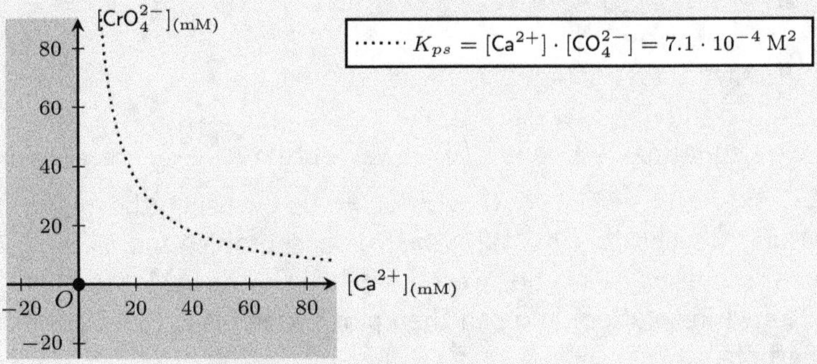

(a) As $[Ca^{2+}] \cdot [\mathrm{CrO}_4^{2-}] = 13\,\mathrm{mM} \cdot 55\,\mathrm{mM} = 710\,\mathrm{mM}^2 = K_{ps}$, the solution is saturated at equilibrium. Thus, we can plot the point $A\,(13, 55)$ on the hyperbola.

(b) In a similar way, as $[Ca^{2+}] \cdot [CrO_4^{2-}] = 65\,\text{mM} \cdot 11\,\text{mM} = 710\,\text{mM}^2 = K_{ps}$, the solution is saturated at equilibrium. Thus, we can plot the point $B\,(65, 11)$ on the hyperbola.

(c) Since $[Ca^{2+}] \cdot [CrO_4^{2-}] = 11\,\text{mM} \cdot 32\,\text{mM} = 352\,\text{mM}^2$ is $< K_{ps}$, the ions are completely dissolved. The point $C\,(11, 32)$ cannot be plotted on the hyperbola.

(d) Since $[Ca^{2+}] \cdot [CrO_4^{2-}] = 50\,\text{mM} \cdot 40\,\text{mM} = 2000\,\text{mM}^2$ is $> K_{ps}$, the solution is supersaturated and part of the Ca^{2+} and CrO_4^{2-} ions will be precipitated in the form of solid $CaCrO_4$ salt. The point $D\,(50, 40)$ cannot be plotted on the hyperbola.

To estimate the amount of salt precipitated in condition (d), it should be noted that the ratio of the Ca^{2+} and CrO_4^{2-} ions precipitating is $1 : 1$ to form the salt $CaCrO_4$. Thus, the equation of the straight line representing the salt precipitation in the Cartesian coordinate system will have a gradient $m = \dfrac{\Delta y}{\Delta x} = 1$ and pass through the point $D\,(50, 40)$: $y = x - 10$.

Thus, calculating the coordinates (x_E, y_E) of the intersection point between the straight line and the hyperbola gives the concentrations of the Ca^{2+} and CrO_4^{2-} ions in solution, respectively:

(i) $y = x - 10$

(ii) $y = \dfrac{710}{x}$

By equating (i) and (ii), we obtain $\dfrac{710}{x} = x - 10 \iff x^2 - 10x - 710 = 0$ with $x \neq 0$; by using the quadratic formula, we obtain $x_1 = 32$, which is acceptable, and $x_2 = -22$, which is non-physical. So, we have $[Ca^{2+}] = 32\,\text{mM}$ and $[CrO_4^{2-}] = 22\,\text{mM}$ in solution, and can then plot the point $E\,(32, 22)$ on the hyperbola.

Finally, we can estimate the amount of salt that precipitates by subtraction: $18\,\mathrm{mM}$ of salt, which corresponds to $562\,\mathrm{mg}$ in a $200\,\mathrm{mL}$ volume.

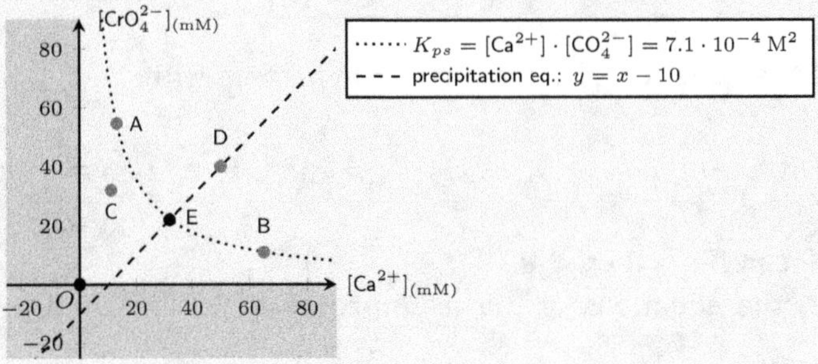

3.3.2 Translated rectangular hyperbolae

The equation of a rectangular hyperbola with asymptotes parallel to the Cartesian axes (*i.e.* at very large positive or negative values of x or y, the curve approaches particular straight lines parallel to the Cartesian axes, Figure 3.5) is:

$$y = \frac{ax + b}{cx + d} \tag{3.8}$$

where a, b, c and d can be any real number, with the conditions that b is NOT 0, and c AND d CANNOT both be equal to 0.

Equation 3.8 can be rearranged using the rules for manipulating equations (Section 2.2.3) as follows:

$$y = \frac{ax + b}{cx + d} \Longleftrightarrow y(cx + d) = ax + b \Longleftrightarrow cxy + dy - ax = b$$

We can manipulate the formula by applying several different operations: multiplying both sides by c, subtracting ad from both sides, and then factorising the result as shown below:

$$\begin{aligned} cxy + dy - ax = b &\Longleftrightarrow (cxy + dy - ax)c = bc \\ &\Longleftrightarrow c^2xy + cdy - cax - ad = bc - ad \\ &\Longleftrightarrow (cy - a)(cx + d) = bc - ad \end{aligned}$$

$$\iff \left(y - \frac{a}{c}\right)\left(x + \frac{d}{c}\right) = \frac{bc - ad}{c^2}$$

$$\iff \boxed{\mathbf{Y} \cdot \mathbf{X} = \mathbf{K}}$$

which is the equation of a rectangular hyperbola with

$$\mathbf{Y} = y - \frac{a}{c}, \quad \mathbf{X} = x + \frac{d}{c}, \quad \text{and} \quad \mathbf{K} = \frac{bc - ad}{c^2}$$

where $\mathbf{K}$ can be > 0 or < 0.

Thus, the equations of the **asymptotes** of a translated hyperbola (Figure 3.5) are $\boxed{\mathbf{y} = \frac{\mathbf{a}}{\mathbf{c}}}$ and $\boxed{\mathbf{x} = -\frac{\mathbf{d}}{\mathbf{c}}}$, whereas the **centre of symmetry** has coordinates $\boxed{\mathbf{C}'\left(-\frac{\mathbf{d}}{\mathbf{c}}, \frac{\mathbf{a}}{\mathbf{c}}\right)}$ (Figure 3.5).

The x-**intercept** of a translated hyperbola with the x-axis (also known as the *zero* of the fraction) is given by $\boxed{\left(\mathbf{x} = -\frac{\mathbf{b}}{\mathbf{a}}\right)}$ since at $y = 0$ we can solve $\dfrac{ax + b}{cx + d} = 0$ (Figure 3.5).

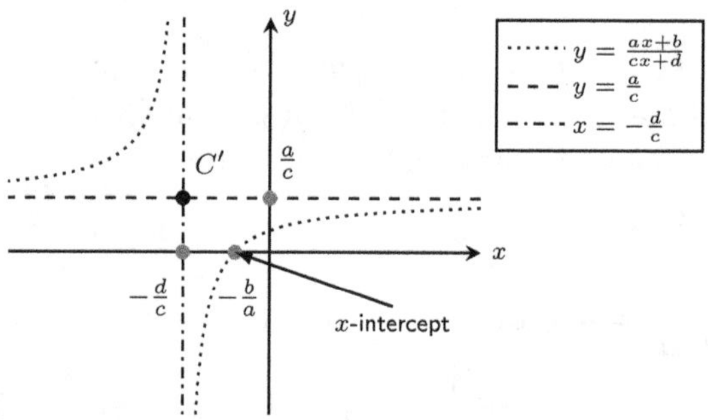

Figure 3.5: Plot of a translated hyperbola

A point $P(x_P, y_P)$ belongs to a translated hyperbola if its coordinates, when substituted, satisfy the equation of the translated hyperbola.

Finding the common points for a line (represented by the linear equation $y = mx + c$) and a translated hyperbola (represented by the nonlinear equation $y = \dfrac{ax + b}{cx + d}\Big)$ is equivalent to solving the two equations simultaneously (see Section 2.4) for the unique values of x and y.

As for an equilateral hyperbola, three cases are possible for a translated hyperbola: two, one, or no common points can be calculated and visualised.

Example: Plot the curve of the translated rectangular hyperbola $y = \dfrac{20x+3}{4x-8}$ and determine its asymptotes, centre of symmetry, and x-intercept.

Solution: Let us first find some points belonging to the translated hyperbola to draw the curve.

x	-11	-5	0	1	3	5	11
y	$\frac{217}{52}$	$\frac{97}{28}$	$-\frac{3}{8}$	$-\frac{23}{4}$	$\frac{63}{4}$	$\frac{103}{12}$	$\frac{223}{36}$
$(\simeq y)$	(4.2)	(3.5)	(−0.4)	(−5.8)	(15.8)	(8.6)	(6.2)

The equations of the asymptotes are $y = 5$ and $x = 2$, and thus the centre of symmetry is $C'\,(2, 5)$.

The x-intercept point of the translated hyperbola has coordinates $\left(-\dfrac{3}{20}, 0\right)$.

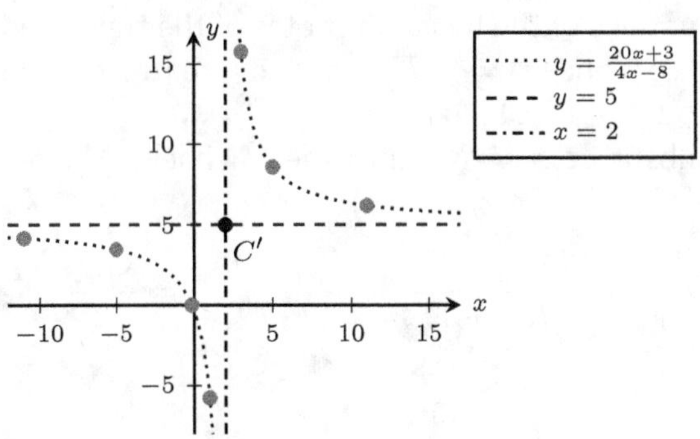

Example in Biochemistry: Show that the Michaelis–Menten equation $v_i = \dfrac{V_{max} \cdot [s]}{[s] + K_m}$, which describes the rate of enzymatic reactions by relating the reaction rate of formation of product (v_i) to the concentration of a substrate $[s]$, is a rectangular hyperbola and describe its characteristics.

Show that K_m corresponds to the substrate concentration at which the reaction velocity v_i is equal to one half of the maximal velocity V_{max} for the reaction.

Solution: Using the rules for manipulating equations (Section 2.2.3), the Michaelis–Menten formula can be rearranged as follows:

$$v_i = \frac{V_{max} \cdot [s]}{[s] + K_m} \Longleftrightarrow v_i \cdot ([s] + K_m) = V_{max} \cdot [s]$$
$$\Longleftrightarrow v_i \cdot ([s] + K_m) - V_{max} \cdot [s] = 0$$

We are aiming to rearrange the formula to have a constant on the right-hand side, and to facilitate this, we add and subtract $K_m \cdot V_{max}$ to/from the left-hand side and then rearrange:

$$v_i \cdot ([s] + K_m) - V_{max} \cdot [s] - K_m \cdot V_{max} + K_m \cdot V_{max} = 0$$
$$\Longleftrightarrow v_i \cdot ([s] + K_m) - V_{max} \cdot ([s] + K_m) = -K_m \cdot V_{max}$$
$$\Longleftrightarrow (v_i - V_{max}) \cdot ([s] + K_m) = -K_m \cdot V_{max} \Longleftrightarrow Y \cdot X = K$$

where $Y = v_i - V_{max}$, $X = [s] + K_m$, and $K = -K_m \cdot V_{max}$ which is < 0, as K_m and V_{max} are both positive physical entities.

The equations of the two asymptotes of the translated hyperbola are $y = V_{max}$ and $x = -K_m$, and its centre of symmetry has coordinates $C'(-K_m, V_{max})$.

If we substitute $s = K_m$ into the Michaelis–Menten equation, we obtain

$$v_i = \frac{V_{max} \cdot K_m}{K_m + K_m} = \frac{V_{max} \cdot \cancel{K_m}}{2 \cdot \cancel{K_m}} = \frac{V_{max}}{2}$$

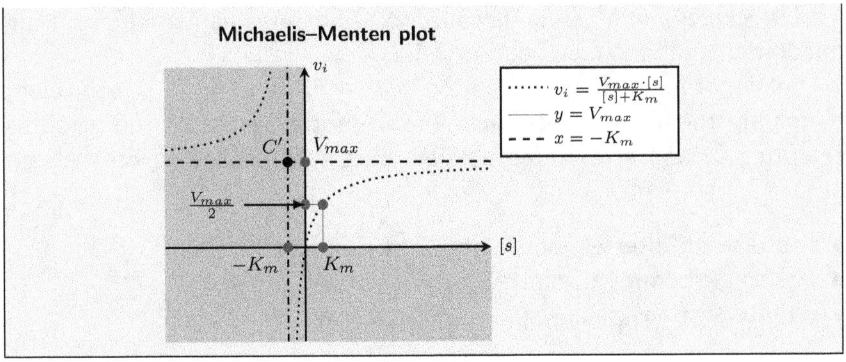

3.4 Parabolae

3.4.1 General equation

The equation of a **parabola** with the axis of symmetry parallel to the y-axis is a quadratic equation of the form

$$y = a \cdot x^2 + b \cdot x + c \tag{3.9}$$

where (Figure 3.6):

- a is a constant giving information on the *concavity* of the curve;
- b is a constant providing part of the information on the position of the *axis of symmetry* towards the y-axis;
- c is a constant known as the *intercept* of the parabola on the y-axis.

In particular, the coefficient a CANNOT be equal to 0, and if $a > 0$, the parabola is **concave upwards**, whereas if $a < 0$, the parabola is **concave downwards**.

The equation of the axis of **symmetry** (Figure 3.6) can be shown to be $\boxed{\mathbf{x = -\frac{b}{2a}}}$.

Thus, the x-coordinate of the vertex V is $x_V = -\frac{b}{2a}$, which implies that $y_V = -\frac{b^2 - 4ac}{4a}$ upon substitution of x_V into the parabolic equation. Finally, the vertex has coordinates $\boxed{\mathbf{V\left(-\frac{b}{2a}, -\frac{b^2 - 4ac}{4a}\right)}}$ (Figure 3.6).

The expression $b^2 - 4ac$ is called Δ, as defined earlier when solving quadratic equations (see Section 2.3.2).

Solving the quadratic equation $ax^2 + bx + c = 0$ is equivalent to finding the x-intercepts (also known as the *zeros*) of the parabola (Figure 3.6) and gives three possible outcomes depending on the sign of Δ:

- two distinct intersection points $X_1(x_1, 0)$ and $X_2(x_2, 0)$, if $\Delta > 0$;
- one single common point $X(x_1 = x_2, 0)$, if $\Delta = 0$;
- no intersections, if $\Delta < 0$.

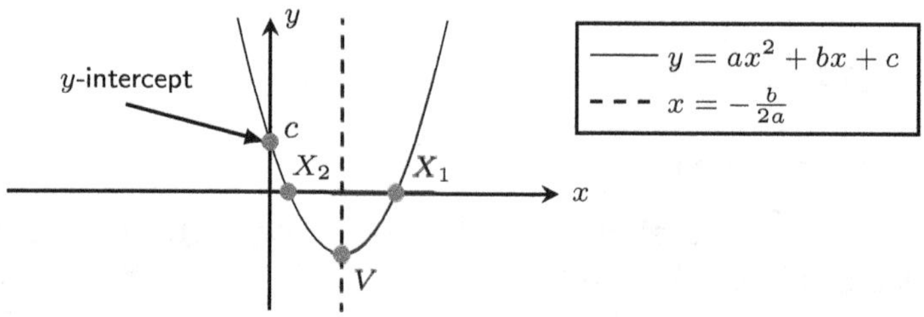

Figure 3.6: Plot of a parabola

Example: Find the equation of the axis of symmetry and the Cartesian coordinates of the vertex, y-intercept, and possible x-intercepts of the parabola P with equation $y = \frac{1}{2}x^2 - \frac{3}{2}x - 9$; then draw the curve with appropriate ranges for the x- and y-axes.

Solution: Using the formulae derived from the equation of a parabola, the axis of symmetry of P is $x = -\frac{-\frac{3}{2}}{2 \cdot \frac{1}{2}} = \frac{3}{2}$.

Thus, the coordinates of the vertex V are

$$\left(-\frac{-\frac{3}{2}}{2 \cdot \frac{1}{2}}, -\frac{\left(-\frac{3}{2}\right)^2 - 4 \cdot \frac{1}{2} \cdot (-9)}{4 \cdot \frac{1}{2}}\right) \iff \left(\frac{3}{2}, -\frac{81}{8}\right)$$

The y-intercept of the parabola has coordinates $(0, -9)$, while the possible x-intercepts can be obtained by solving the quadratic $\frac{1}{2}x^2 - \frac{3}{2}x - 9 = 0$, which has roots of $x_1 = -3$ and $x_2 = 6$.

Thus, $X_1(-3,0)$ and $X_2(6,0)$ are the coordinates of the two distinct points of intersection between the parabola P and the x-axis, respectively.

The following plot summarises the characteristics of this parabola P.

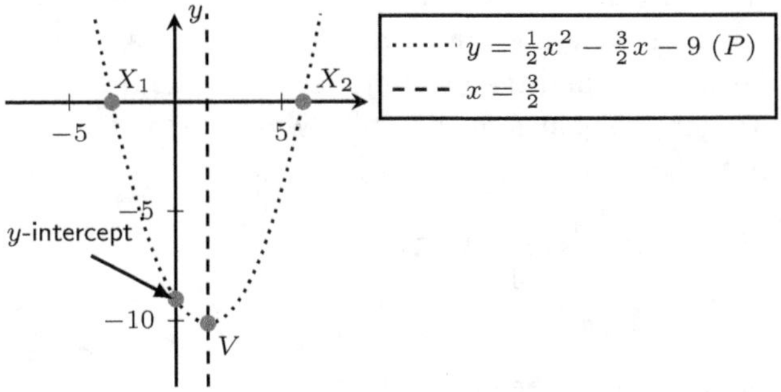

3.4.2 Finding the equation of a parabola

Three sets of point coordinates or two pieces of other data are always required to find the equation of an unknown parabola; the most common cases of available data are as follows:

(i) The coordinates of three points $A(x_A, y_A)$, $B(x_B, y_B)$, and $C(x_C, y_C)$ can be separately substituted into the general equation of a parabola to get three different equations which have a, b, and c as unknowns and can be solved simultaneously using the elimination or substitution method (see Section 2.4.2).

(ii) The coordinates of the vertex $V(x_V, y_V)$ and the value of a:

$$y - y_V = a \cdot (x - x_V)^2 \tag{3.10}$$

In both cases, the expression can then be arranged in the form of the general equation (Equation 3.9), using the rules for manipulating equations (see Section 2.2.3).

A point $P(x_P, y_P)$ belongs to a certain parabola if its coordinates satisfy the parabolic equation upon substitution.

Example: (a) Find the equation of the parabola P_1 passing through the points $A(-4,2)$, $B(0,4)$, and $C(1,7)$, find its intersections with the x and y axes, and then draw the curve.

(b) Find the equation of the parabola P_2 with vertex $V(5,0)$ and $a = -\frac{1}{2}$, find its intersections with the x and y axes, and then draw the curve.

Solution (a): In order to find the equation of the parabola P_1 passing through the points A, B, and C, we should substitute their coordinates separately into the general equation of a parabola $y = a \cdot x^2 + b \cdot x + c$ and then simultaneously solve the three equations obtained to get the values of the unknowns a, b, and c.

The three equations are:

(i) $2 = a \cdot (-4)^2 + b \cdot (-4) + c$
(ii) $4 = a \cdot 0^2 + b \cdot 0 + c$
(iii) $7 = a \cdot 1^2 + b \cdot 1 + c$

These equations can be rearranged as:

(i) $2 = 16a - 4b + c$
(ii) $4 = c$
(iii) $7 = a + b + c$

Using the elimination or substitution method (see Section 2.4.2), we obtain the values $a = \frac{1}{2}$, $b = \frac{5}{2}$, and $c = 4$, which define the equation of P_1: $y = \frac{1}{2}x^2 + \frac{5}{2}x + 4$.

Solving this quadratic equation: $\frac{1}{2}x^2 + \frac{5}{2}x + 4 = 0$ gives no real solutions because the determinant, $\Delta = b^2 - 4ac = 6.25 - 8 < 0$, is negative, therefore the parabola P_1 has no intersections with x-axis, as shown in the plot below.

(b): In order to find the equation of P_2, the coordinates of the vertex V and the a value can be substituted into equation 3.10, giving $y - 0 = -\frac{1}{2} \cdot (x-5)^2$ which can be rearranged as $y = -\frac{x^2}{2} + 5x - \frac{25}{2}$.

Solving the quadratic $-\frac{x^2}{2} + 5x - \frac{25}{2} = 0$ gives a single repeated real solution $x = 5$, thus there is a single common point between the parabola P_2 and x-axis, and this corresponds to its vertex V, as shown in the plot below.

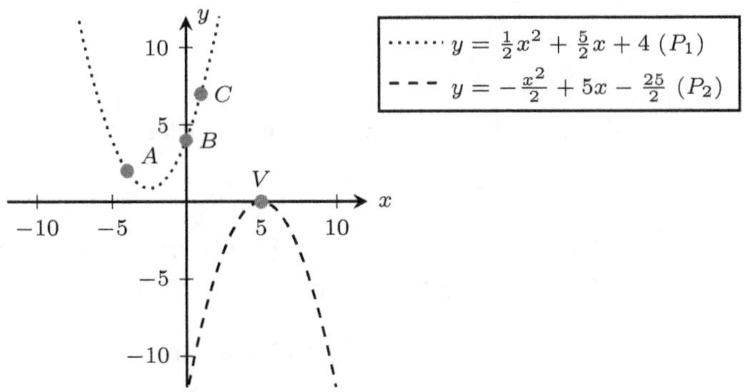

3.4.3 Special cases

Particular care should be taken with special cases of parabolas (Figure 3.7):

(i) When $c = 0$, the general equation is $y = ax^2 + bx$ and the parabola passes through the point $(0, 0)$.
(ii) When $b = 0$, the general equation is $y = ax^2 + c$ and the equation of the axis of symmetry is $x = 0$, which is the y-axis.
(iii) When $c = b = 0$, then the general equation is $y = ax^2$, so the parabola passes through the point $(0, 0)$ and its axis of symmetry is the y-axis.

Finding the possible common points for a line (represented by the linear equation $y = mx + c$) and a parabola (represented by the quadratic equation $y = ax^2 + bx + c$) is equivalent to solving the equations simultaneously (see Section 2.4) for the unique values of x and y.

Three cases are possible:

(i) two distinct intersection points;
(ii) one single shared point;
(iii) no intersection, so the line does not meet the parabola at all.

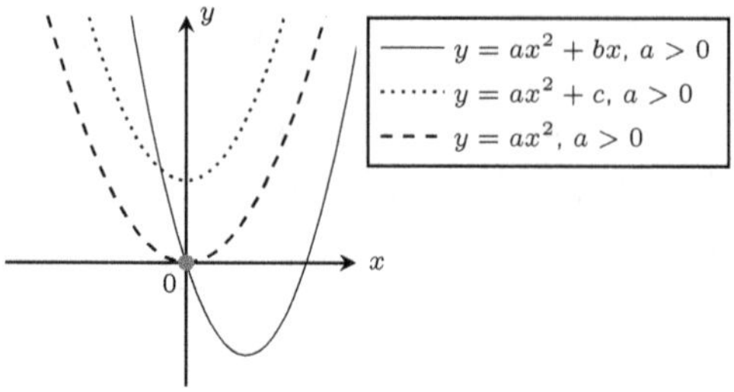

Figure 3.7: Special cases of parabolae

Example: Find the possible intersection points between the parabola $y = x^2 + x - 6$ (P) and the straight lines $y = x + 3$ (L_1), $y = 3x - 7$ (L_2), and $y = -\dfrac{1}{4}x - 8$ (L_3), and then plot all the curves.

Solution: In order to determine the Cartesian coordinates of the possible intersection points between the parabola P and the straight line L_1, we should solve their equations simultaneously for the unique values of x and y using the substitution method (see Section 2.4.2):

(i) $y = x^2 + x - 6$
(ii) $y = x + 3$

which gives two possible solutions: $A\,(-3, 0)$ and $B\,(3, 6)$, as represented in the plot below.

Solving the equations for P and L_2 simultaneously gives a single solution $C\,(1, -4)$.

Finally, solving the equations for P and L_3 simultaneously gives no real solutions, thus the line L_3 does not meet the parabola P.

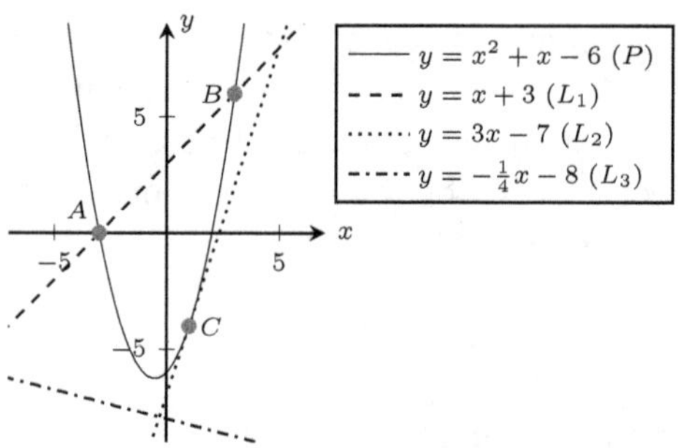

Example in Biochemistry: The effects of pH on the initial activity rate (v_i) of an enzyme are investigated. The enzymatic rate is measured as μM of substrate transformed to product per minute ($\mu\text{M}\,\text{min}^{-1}$) and is acquired for five different pH values.

pH	5	6	7.3	8	8.5
$v_{i\,(\mu\text{M}\,\text{min}^{-1})}$	0.40	0.60	0.56	0.40	0.23

After plotting the five values on appropriate x and y axes, if the points apparently follow a parabolic trend, estimate the possible equation of the parabola and plot the curve.

Then, verify that the curve also passes through a fourth point, and estimate at which pH there is either the highest activity rate or no possible activity at all.

Finally, estimate the pH values for which $v_i = 0.3\,\mu\text{M}\,\text{min}^{-1}$.

Solution: The five experimental values obtained can be plotted with the pH values on the x-axis and the initial activity rates on the y-axis, giving a parabolic trend.

Therefore, substituting the points $(5, 0.40)$, $(6, 0.60)$, $(8, 0.40)$ into the general equation of a parabola (Equation 3.9), we get three equations with a, b, and c as unknowns:

(i) $0.40 = 25a + 5b + c$

(ii) $0.60 = 36a + 6b + c$

(iii) $0.40 = 84a + 8b + c$

Using appropriate elimination and/or substitution methods to solve these simultaneous linear equations, we obtain the parabolic equation passing through the three points: $y = -0.1x^2 + 1.3x - 3.6$, and we find that the parabola is concave downwards.

By substituting the given values $(7.3, 0.56)$ into the above equation $(0.56 = -0.1 \cdot 7.3^2 + 1.3 \cdot 7.3 - 3.6)$ and verifying that both sides are equal, we can see that the parabola also passes through that point.

Using the formula to find the coordinates of the vertex V of a parabola can help to determine the pH that has the highest activity rate: at pH$=6.5$, the activity is the highest, *i.e.* $0.63\,\mu\text{M}\,\text{min}^{-1}$.

Then, in order to estimate the pH at which there is no possible enzymatic activity, we can solve the quadratic inequality $-0.1x^2 + 1.3x - 3.6 < 0$, and since a negative enzymatic activity is non-physical, the corresponding solution sets are $x_1 < 4$ and $x_2 > 9$.

Thus, it can be hypothesised that at pH< 4 and > 9 there might be no enzyme reactivity, but this conclusion would still need to be proved experimentally.

Lastly, we can plot the straight line $y = 0.3$ on the graph and calculate its intersection points with the parabola in order to find the pH values at which the $v_i = 0.3\,\mu\text{M min}^{-1}$.

Solving the following simultaneous equations using the substitution method

(i) $y = -0.1x^2 + 1.3x - 3.6$

(ii) $y = 0.3$

gives $\text{pH} = 4.7$ and $\text{pH} = 8.3$ as the solutions.

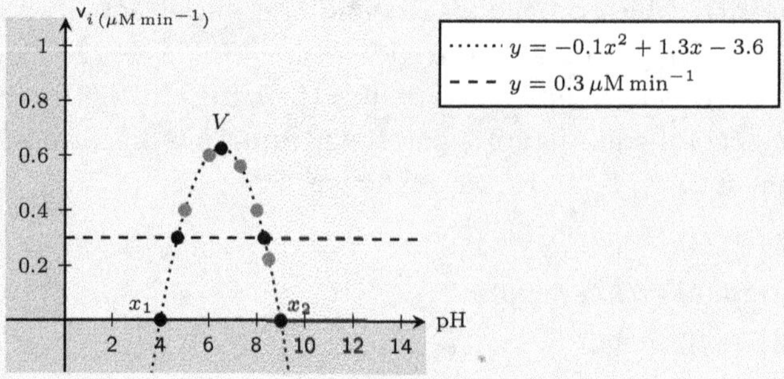

3.5 Exercises

(1) Calculate the length of the line segments AB for the pairs of A and B coordinates below.

(a) $A\,(2, 4)\ B\,(2, -1)$

(b) $A\,(1, 1)\ B\,(-1, 3)$

(c) $A\left(-\frac{1}{2}, -\frac{3}{2}\right)\ B\left(-\frac{3}{2}, -1\right)$

(d) $A\left(\frac{\sqrt{3}}{4}, \frac{1}{2}\right)\ B\left(-\frac{\sqrt{3}}{4}, \frac{3}{2}\right)$

(2) Solve the following problems.

(a) Given the points $A\,(7,0)$ $B\,(3,6)$ $C\,(-3,2)$ as the vertices of a triangle, calculate the length of the triangle perimeter.

(b) Calculate the coordinates of the point P which is on the y-axis and is at equal distances from the points $Q\,(3,2)$ and $T\,(-2,1)$.

(c) The parallelogram $ABCD$ has $A\,(2,1)$ $B\,(3,6)$ $C\,(5,3)$ as its first three vertices; find the coordinates of the fourth vertex D.

(d) Given the line segment AB with $A\,(-3,-1)$ and $B\,(0,4)$, find the coordinates of its midpoint M.

(3) Manipulate the following equations to make y the subject of the equation if possible; otherwise, make x the subject of the equation.
Then, plot the resulting straight line in the $x-y$ plane, if possible by identifying its slope m, and y- and x- intercepts.

(a) $2y+(3-5x)-3y+2(2-x)-[-(2-x)]=0$

(b) $(2x-1)(2x+1)-(6x^2y-3xy)\div 3xy-(12x^3y^2-6x^2y^3+8x^4y^2)\div 2x^2y^2=0$

(c) $4x-2y+\left(x^2-1\right)\left(x^2+1\right)+4\left(\frac{y}{2}-2\right)-\left(x^2+2x\right)\left(x^2-2x+4\right)=0$

(d) $2[(4y-1)(2y+x)-8y^2]-4y(2x-1)-3\left(\frac{1}{3}y+2x\right)=0$

(4) Find the equation of the straight line passing through each of the following pairs of points.

(a) $A\,(1,6)$ $B\,(-3,-2)$

(b) $A\,(-8,7)$ $B\,(4,1)$

(c) $A\left(\frac{1}{2},-\frac{1}{2}\right)$ $B(-2,2)$

(d) $A\left(\frac{7}{4},\frac{2}{5}\right)$ $B\left(-\frac{1}{3},\frac{2}{5}\right)$

(5) For each case, find the equation of the straight line that passes through each of the following points and has the stated m (slope) value.

(a) $A\,(3,-7)$, $m=-2$

(b) $A\,(-2,2)$, $m=-1$

(c) $A\left(\frac{1}{4},-2\right)$, $m=4$

(d) $A\left(\frac{5}{2},-3\right)$, $m=0$

(6) Calculate the distance between point A and the specified straight line for the following cases.

(a) $A\,(2,1)$, $y=\frac{3}{4}x+2$

(b) $A\,(0,3)$, $y=2x+1$

(c) $A\,(-1,-2)$, $y=-\frac{2}{3}x-\frac{5}{2}$

(d) $A\left(\frac{5}{2},\frac{4}{3}\right)$, $x=-3$

(7) Solve the following problems.

(a) Given the points $A(3,9)$, $B(1,1)$, and $C(5,7)$, calculate the length of the perimeter and the area of the triangle.

(b) Given the points $A(-1,1)$, $B(8,4)$, $C(4,6)$, and $D(3,-1)$, calculate the length of the perimeter and the area of the parallelogram.

(c) Given that the point $A(1,1)$ and the equation $y = kx + k$ representing an infinite set of lines depending on the value k, calculate the equation of the straight line which is a distance of 5 units from A.

(d) Given the straight line $y = 3x - 2$, find the points A and B that have the ordinate $y = -1$ and which are a distance of $\sqrt{10}$ units from that line.

(8) Find the equation of the straight line parallel to the specified straight line and which passes through the corresponding point A for the pairs given below.

(a) $y = -\frac{3}{2}x + \frac{1}{2}$, $A(2,1)$ (b) $y = \frac{1}{3}$, $A\left(-\frac{1}{2}, \frac{5}{4}\right)$

(c) $y = 4x - 2$, $A(-1,3)$ (d) $x = -3$, $A(4,-2)$

(9) Find the equation of the straight line perpendicular to the specified straight lines given in exercise **(8)** and passing through the specified points A in exercise **(8)**.

(10) Solve the following problems.

(a) Find the equations of the two straight lines which have a y-intercept $= 3$ and are perpendicular and parallel respectively to the bisector of quadrants I and III of the Cartesian plane.

(b) Given the points $A(-2,+3)$ and $B(4,-1)$, find the equation of the straight lines which pass through $C(-3,-5)$ and are perpendicular and parallel, respectively, to the line passing through A and B.

(c) First find the equation of the straight line forming a triangle of area $a = 8$ with the positive half-axes and passing through the point $A(1,4)$. Then, find the equation of the straight line perpendicular to that line and passing through the origin of the Cartesian axes.

(11) Find whether or not the following pairs of straight lines are parallel, and, if the lines intersect, find the coordinates of the intersection point A.

(a) $y = 4x - 2$, $y = -3x + 7$ (b) $6y = 3x - 2$, $8y = 4x + 1$

(c) $y = -\frac{x}{3} + \frac{2}{5}$, $y = \frac{x}{2} - \frac{3}{2}$ (d) $6y = 3x - 2$, $y + \frac{1}{3} = \frac{x}{2}$

(12) Solve the following problems.

(a) Given the lines $y = 2$ and $x = -3$, find the equation of the straight lines with slope $m = -2$ that form a triangle with an area of 8 units through intersection with the first two lines.

(b) Given the lines $y = x + 3$ and $y = -x - 1$, find the equation of the straight line with a y-intercept $c = 3$ that forms a triangle with an area of 16 units through intersection with the first two lines.

(c) Find the equation of the line that passes through the intersection point of the lines with equations $y = -3x + 10$ and $y = \frac{1}{2}x + \frac{13}{2}$ and is perpendicular to the line with equation $y = -\frac{1}{2}x + 1$.

(d) Given the lines with equations $y = x + 8$, $y = x - 3$, $y = -x - 6$, and $y = -x + 4$, find the type, the area, and the length of the perimeter of the polygon formed by intersection of the four lines, and also the equations of the diagonals and the coordinates of their corresponding intersection point A.

(13) Can x and y in the equations below be transformed into new variables X and Y to allow a suitable graph to be plotted if y is measured in an experiment as a function of x? If so, give the transformations and then the expressions for the gradient, X-intercept, and Y-intercept for each of your NEW equations.

(a) $y = \frac{a}{x} + b$ (b) $y = b - ax^2$ (c) $\frac{1}{y} = \frac{b}{1 - ax}$

(d) $y = a + b\sqrt{x}$ (e) $y = \frac{b}{x^2}$ (f) $\frac{1}{y} = \frac{a}{1 + bx}$

(14) The osmotic pressure of a solution of a protein is related to the concentration of that protein by the equation

$$Z = R \cdot T \cdot c$$

where Z is the osmotic pressure in kPa, T is the temperature in Kelvin, R is the gas constant ($R = 8.314\,\mathrm{kPa\,dm^3\,mol^{-1}K^{-1}}$), and c is the molarity of the protein in $\mathrm{mol\,dm^{-3}}$.

This equation becomes more accurate as the molarity of the solute approaches zero.

Plot a suitable graph to determine, as accurately as possible, the molecular mass (take care with units!) of the protein given the following data taken at room temperature (usually taken as $21\,^\circ\mathrm{C}$):

Protein concentration (in $\mathrm{g\,dm^{-3}}$)	7.3	18.4	27.6	42.1	57.4
Osmotic pressure (in kPa)	0.211	0.533	0.804	1.236	1.701

Hint: Compare the trend of the points with the equation of a straight line, $y = mx + c$, and think about the relationship between protein concentration, molar concentration, and molecular weight.

(15) Manipulate the following equations to make y the subject. Then, by finding its concavity, the equations of its asymptotes, the coordinates of its centre of symmetry, and possible intersections with both the x- and y-axes, plot the resulting hyperbola in a Cartesian plane.

(a) $\left(3xy+4x^2y^2\right)\div(2xy)-\left(x^3y^4+4x^2y^3\right)\div\left(x^2y^3\right)=0$

(b) $(7-3xy)-[-(5-2xy)]-(-6xy)=0$

(c) $-5xy-2x+1-y=0$

(d) $x(x-3y)-y(y-x)-3+y(y-4x)-x^2+y-x=0$

(16) Find the positions of the following pairs of hyperbolae and straight line, and, if they intersect, find the coordinates of the common points. Then, plot the hyperbola and the straight line, and mark any common points.

(a) $y=-\dfrac{2}{x},\ y=-x+3$ (b) $y=\dfrac{4}{x},\ y=-4x-8$

(c) $y=\dfrac{2x-3}{-x+2},\ y=y=-2x+1$ (d) $y=\dfrac{3-2x}{1+x},\ y=-\dfrac{x}{2}-3$

(17) The solubility equilibrium constant of $KClO_4$ (molecular weight: $138.55\,\mathrm{g\,mol^{-1}}$) is $K_{ps}=1.05\cdot10^{-2}\,\mathrm{M}^2$ at $25\,°\mathrm{C}$. Determine whether a $500\,\mathrm{mL}$ solution containing the following concentrations of K^+ and ClO_4^- ions is saturated or whether the ions get completely dissolved or precipitate:

(a) $[\mathrm{K^+}]=150\,\mathrm{mM},\ [\mathrm{ClO_4^-}]=70\,\mathrm{mM}$

(b) $[\mathrm{K^+}]=60\,\mathrm{mM},\ [\mathrm{ClO_4^-}]=120\,\mathrm{mM}$

(c) $[\mathrm{K^+}]=140\,\mathrm{mM},\ [\mathrm{ClO_4^-}]=110\,\mathrm{mM}$

If the solution is supersaturated, estimate the concentration of both the dissolved ions (in mM) and the mass of the precipitated salt (in mg).

(18) The rate at which a given enzyme catalyses a reaction is dependent upon the substrate concentration, S:

$$V=\frac{S}{m+cS}$$

where V is the rate of the reaction and m and c are constants.

(a) Rearrange the equation above to show that it corresponds to a rectangular hyperbola of the form $Y\cdot X=K$, where K is a constant, and describe its characteristics.

(b) What is the value of V when the substrate concentration $S=m$? What is the theoretical maximum attainable value of V, usually called V_{max}?

(c) At what substrate concentration is the reaction velocity V equal to one half of the maximum attainable velocity V_{max} for the reaction?

(d) How can we transform V and S to derive a straight line graph relating them? What will be the gradient, the x-intercept, and y-intercept of this graph?

(19) Manipulate the following equations to make y the subject. Then, for each equation, by finding its concavity, the equation of its symmetry axis, the coordinates of its vertex, and possible intersections with both the x- and y-axes, plot the resulting parabola in a Cartesian plane.

(a) $(x-3)(x+2)-(y+1)(x-2)+(x-2)(x+3)+xy=0$

(b) $(2x+1)\left(x^2-x+1\right)+\frac{4}{3}y-2\left(x^2-4\right)(x+3)+(-3x)^2-\frac{7}{3}y=0$

(c) $(x+y)(x-y)+(x-2y)^2+y(4x-3y)-y=0$

(d) $(x+1)(x-2)-\left(2-x^2\right)-(y-7)-x(x-3)+2y-3(1-x)=0$

(20) Find the equation of the parabola passing through each of the following sets of three points.

(a) $A(1,1)$, $B(3,6)$, $C(-2,4)$

(b) $A(-2,-4)$, $B(0,0)$, $C(3,-3)$

(c) $A(-3,-3)$, $B(0,-9)$, $C(6,-3)$

(d) $A(-3,-7)$, $B\left(0,-\frac{1}{4}\right)$, $C\left(2,-\frac{13}{4}\right)$

(21) Find the equation of the parabola having each of the following vertices and a values.

(a) $V(1,-1)$, $a=3$

(b) $V(4,-3)$, $a=-1$

(c) $V(2,-4)$, $a=-\frac{1}{2}$

(d) $V\left(-\frac{1}{4},-\frac{9}{8}\right)$, $a=2$

(22) Solve the following problems.

(a) Given the parabola with equation $y=-2x^2+bx+c$, find the values of the coefficients b and c so that the parabola passes through the point $P(1,-3)$ and its axis of symmetry is the straight line with equation $x=-1$.

(b) Find the equation of the parabola passing through the point $P(6,10)$ and having the vertex in $V(3,-8)$. Thus, calculate the distance between the intercepts of the parabola with the x-axis.

(c) Find the equation of the parabola that has the axis of symmetry with equation $x=-2$ and intercepts the y-axis at 2 and the x-axis at two different points which are 5 units apart from each other.

(d) Find the equation of the concave upward parabola passing through the points $A(1,6)$ and $B(4,3)$ and having its vertex on the straight line with equation $y=2x-4$.

(23) Find the positions of the following pairs of parabola and straight line, and, if they intersect, find the coordinates of the common points. Then, plot the parabola and the straight line, and mark any common points.

(a) $y = \frac{1}{2}x^2 - x + 1$, $y = 2x - 3$ (b) $y = -\frac{1}{2}x^2 + 2x - \frac{1}{2}$, $y = -x + 1$

(c) $y = -3x^2 - 2x - 4$, $y = 5x + 6$ (d) $y = 2x^2 - 6x + 4$, $y = 6x - 14$

(24) In an experiment on a sample of protein, the values of one of its properties, y, were observed for various values of its concentration, x, as follows:

x	1.6	2.4	3.2	4.1
y	2.55	2.39	3.51	6.3

It is thought that this result might obey a relationship of the form $y = x^2 + Ax + B$, where A and B are constants.

(a) Verify this by drawing an appropriate graph. Hence, estimate values for A and B. What is the value of y when $x = 0$?

(b) At what concentrations, x, is the property y the lowest?

(c) At what concentrations, x, does the property y equal 20 (where y is at its highest measurable value)?

(25) The growth rate R of a cell colony with N cells at time t can be represented by the equation $R = kN - bN^2$. This is called a logistic model. For this example, take the constants to be $k = 3.8\text{h}^{-1}$ and $b = 0.01\text{h}^{-1}$, respectively.

(a) What is the equilibrium size of the population?

(b) What is the size of the population when the rate of growth is maximum?

(c) Plot the graph of the parabola $R = f(N)$.

Answers

(1) (a) 5 (b) $2\sqrt{2}$ (c) $\frac{\sqrt{5}}{2}$ (d) $\frac{\sqrt{7}}{2}$

(2) (a) perimeter $= 4\sqrt{13} + 2\sqrt{26}$ (b) $P(0, 4)$
(c) $D(6, 8)$ (d) $M\left(-\frac{3}{2}, \frac{3}{2}\right)$

(3) (a) $y = 8x - 9$ (b) $y = \frac{8}{3}x$ (c) $x = -\frac{9}{4}$ (d) $y = -8x$

(4) (a) $y = 2x + 4$ (b) $y = -\frac{1}{2}x + 3$ (c) $y = -x$ (d) $y = \frac{2}{5}$

(5) (a) $y = -2x - 1$ (b) $y = -x$ (c) $y = 4x - 3$ (d) $y = -3$

(6) (a) 2 (b) $\frac{2}{5}\sqrt{5}$ (c) $\frac{11}{2}$ (d) $\frac{\sqrt{13}}{26}$

(7) (a) $\mathsf{p} = 3\sqrt{2} + 2\sqrt{29}$, $\mathsf{a} = \frac{21}{2}$ (b) $\mathsf{p} = 10\sqrt{2} + 4\sqrt{5}$, $\mathsf{a} = 6\sqrt{10}$
(c) $y = -2x - 2$ (d) $A\left(\frac{11}{3}, -1\right)$, $B(-3, -1)$

(8) (a) $y = -\frac{3}{2}x + 4$ (b) $y = \frac{5}{4}$ (c) $y = 4x + 1$ (d) $x = 4$

(9) (a) $y = \frac{2}{3}x - 1$ (b) $x = -\frac{1}{2}$ (c) $y = -\frac{1}{4}x - \frac{11}{4}$ (d) $y = -2$

(10) (a) $y = -x + 3$, $y = x + 3$ (b) $y = \frac{3}{2}x - \frac{1}{2}$, $y = -2x - 7$
(c) $y = -4x + 8$, $y = \frac{1}{4}x$

(11) (a) $A\left(\frac{9}{7}, \frac{22}{7}\right)$ (b) parallel lines (c) $A\left(\frac{57}{25}, -\frac{9}{25}\right)$ (d) coincident lines

(12) (a) $y = -2x$, $y = -2x - 8$ (b) $y = -3x + 3$ (c) $y = 2x + 5$
(d) rectangle with $a = 55$ and $p = 21\sqrt{2}$, diagonals: $y = -21x - 36$ and $y = -\frac{1}{21}x + \frac{2}{3}$, $A\left(-\frac{7}{4}, \frac{3}{4}\right)$

(13) (a) $Y = y,\ X = \frac{1}{x}$, equation: $Y = aX + b$, gradient $= a$, X-intercept $= b$, Y-intercept $= -\frac{b}{a}$
(b) $Y = y,\ X = x^2$, equation: $Y = b - aX$, gradient $= -a$, X-intercept $= b$, Y-intercept $= \frac{b}{a}$
(c) Take reciprocal of both sides: $y = \frac{1-ax}{b}$, $Y = y,\ X = x$, equation: $Y = \frac{1-aX}{b}$, gradient $= -\frac{a}{b}$, X-intercept $= \frac{1}{b}$, Y-intercept $= \frac{1}{a}$
(d) $Y = y,\ X = \sqrt{x}$, equation: $Y = a + bX$, gradient $= b$, X-intercept $= a$, Y-intercept $= -\frac{a}{b}$
(e) $Y = y,\ X = \frac{1}{x^2}$, equation: $Y = bX$, gradient $= b$, X-intercept $= 0$, Y-intercept $= 0$
(f) Rearrange as $\frac{1}{y} = \frac{1+bx}{a}$, so $Y = \frac{1}{y}\ X = x$, equation: $Y = \frac{1}{a} + \frac{bX}{a}$, gradient $= \frac{b}{a}$, X-intercept $= \frac{1}{a}$, Y-intercept $= -\frac{1}{b}$

(14) $y = 0.0296x - 0.0019$, protein molecular mass: $82{,}016\,\mathrm{g\,mol^{-1}}$

(15) (a) $y = \frac{5}{2x}$ (b) $y = -\frac{12}{x}$ (c) $y = \frac{1-2x}{1+5x}$ (d) $y = \frac{x+3}{1-6x}$

(16) (a) $A\left(\frac{3}{2} - \frac{\sqrt{17}}{2}, \frac{3}{2} + \frac{\sqrt{17}}{2}\right)$, $B\left(\frac{3}{2} + \frac{\sqrt{17}}{2}, \frac{3}{2} - \frac{\sqrt{17}}{2}\right)$ (b) $A(-1, -4)$
(c) $A(1, -1)$, $B\left(\frac{5}{2}, -4\right)$ (d) no common points

(17) (a) saturated solution (b) completely dissolved ions
(c) supersaturated solution, $1.52\,\mathrm{g}$ of precipitated salt

(18) (a) $Y = V - \frac{1}{c}$, $X = S + \frac{m}{c}$, $K = -\frac{m}{c^2}$, asymptote equations: $y = \frac{1}{c}$ and $x = -\frac{m}{c}$, symmetry centre: $C'\left(-\frac{m}{c}, \frac{1}{c}\right)$
(b) $V = \frac{1}{1+c}$, $V_{max} = \frac{1}{c}$ (c) $S = \frac{m}{2c}$
(d) $Y = \frac{1}{V}$, $X = \frac{1}{S}$, equation: $Y = mX + c$, gradient $= m$, X-intercept $= -\frac{c}{m}$, Y-intercept $= c$

(19) (a) $y = -x^2 + \frac{1}{2}x + 5$ (b) $y = 2x^2 + 9x + 25$
(c) $y = 2x^2$ (d) $y = -x^2 - 5x$

(20) (a) $y = \frac{7}{10}x^2 - \frac{3}{10}x + \frac{3}{5}$ (b) $y = -\frac{3}{5}x^2 + \frac{4}{5}x$
(c) $y = \frac{1}{3}x^2 - x - 9$ (d) $y = -\frac{3}{4}x^2 - \frac{1}{4}$

(21) (a) $y = 3x^2 - 6x + 2$ (b) $y = -x^2 + 8x - 19$
(c) $y = -\frac{1}{2}x^2 + 2x - 6$ (d) $y = 2x^2 + x - 1$

(22) (a) $-2x^2 - 4x + 3$ (b) $2x^2 - 12x + 10$, 4
(c) $-x^2 - 4x + 2$ (d) $x^2 - 6x + 11$

(23) (a) $A\,(4,5)$, $B\,(2,1)$ (b) $A\left(3+\sqrt{6},-2-\sqrt{6}\right)$, $B\left(3-\sqrt{6},-2+\sqrt{6}\right)$
(c) no common points (d) $A\,(3,4)$

(24) (a) $A=-4.20$, $B=6.71$, $y_0=6.71$ (b) $x_{min}=2.1$, $y_{min}=2.3$
(c) $x_{max}=6.3$

(25) (a) Equilibrium when $R=0, N=\dfrac{k}{b}=380$

(b) when $N=\dfrac{k}{2b}=190$, R is the parabola's vertex

(c)

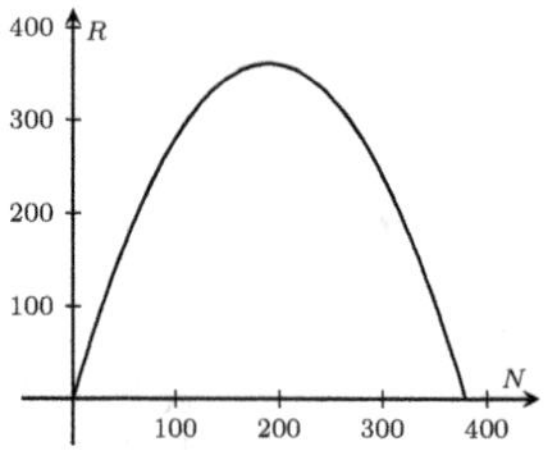

Chapter 4

Functions

Preamble

In this chapter, we consider how to link variables through the use of mathematical equations in which one physical quantity is a 'function' of another, *i.e.* a defined and predictable relationship exists between the variables. An example of this in Chemistry and the Biosciences is the relationship between the pH of a solution and its hydrogen ion (H^+) concentration.

First we must understand the language and nomenclature of functions. For instance, when we take a variable and carry out a mathematical operation (*e.g.* to square and to take the reciprocal) on it, we usually say that we have 'operated' on the variable.

4.1 Definition

4.1.1 Function

Given two sets of numbers A and B, a **function** from A to B is a rule which associates every element of the first set A to exactly one element of the second set B according to a pre-established mathematical expression.

The set A is called the **domain** of the function, whereas the set B is called the **co-domain** or the **target set** of the function.

The most usual notations used to express functions are

$$y = f(x) \quad \text{or} \quad f : x \to f(x)$$

where x is an element of the domain and is called the **independent variable** or the **argument** of the function, and y is called the **dependent variable** and is the element of the co-domain corresponding to x according to the relation f, which must be pre-defined.

Different letters of the alphabet can be also used in function notation: for instance, in the formula $s(t) = s_0 + v \cdot t$, t is the independent variable and s is the dependent variable.

According to the definition above, **straight lines, parabolae** and **rectangular hyperbolae** are the graphs of functions.

Important note: In this book, the terms *function* and *graph of function* are used interchangeably as synonyms.

4.1.2 Composite functions

Sometimes, given a function $f(x)$, the resulting dependent variable y can in turn be the independent variable of a second function, for instance, $g(x)$. This combination of both f and g functions gives a **composite function**, $g \circ f$.

The most common conventional notations used to express a composite function are:

$$y = g\left(f(x)\right) \quad \text{or} \quad g \circ f : x \to f(x) \to g\left(f(x)\right)$$

Warning! The composite functions $g \circ f$ and $f \circ g$ can give different results.

Example: Given $f(x) = x - 1$ and $g(x) = \dfrac{x}{1 + x^2}$, find the composite functions $g \circ f$ and $f \circ g$.

Solution: $g(f(x)) = \dfrac{f(x)}{1 + \left(f(x)\right)^2} = \dfrac{x - 1}{1 + (x-1)^2}$, which gives $g(f(x)) = \dfrac{x - 1}{x^2 - 2x + 2}$.

Alternatively, $f(g(x)) = g(x) - 1 = \dfrac{x}{1 + x^2} - 1 = \dfrac{x - 1 - x^2}{1 + x^2}$ which is completely different from $g(f(x))$.

4.2 Domain and Co-domain

The **domain** of a function can include one or more intervals of x and can be of different types: both-sides limited $(-3, 4]$, both-sides unlimited $(-\infty, +\infty)$, or just one-side unlimited $[5, +\infty)$.

Reminder: '[' denotes a closed limited interval (*e.g.* '4]' above), while ')' denotes an open limited interval (*e.g.* '(−3' above) or unlimited interval (*e.g.* '$(-\infty, +\infty)$') (see Section 2.5.2).

Also the **'co-domain'** or y **range** of a function can be limited to certain intervals of y, or be both-side or one-side unlimited.

If $x = 0, 1, -1$ are points in the domain, calculating $f(0)$, $f(1)$, $f(-1)$ can give **key points** which might be easy to plot, while calculating $f(x) = 0$ allows the intersection points between the graph of the function f and the x-axis to be found.

All this information can be helpful to start 'sketching' a function in a Cartesian plane.

Sometimes, a function can be defined using different mathematical expressions in different intervals of real numbers. This function is called a **piece-wise function**.

We can use • and ∘ on the graph of a piece-wise function to denote the closed and open points, respectively.

Example: Given the following piece-wise function,

$$f(x) = \begin{cases} \dfrac{1}{x} & x \leq -1 \\ -2x + 1 & -1 < x \leq 1 \\ x^2 - 3x & x > 1 \end{cases}$$

evaluate $f(-3)$, $f(0)$, and $f(2)$, plot the function, and define the range intervals of the function.

Solution: To evaluate $f(-3)$, we must use the function $y = \dfrac{1}{x}$ since $-3 < -1$. Thus, $f(-3) = -\dfrac{1}{3}$ and the point $\left(-3, -\dfrac{1}{3}\right)$ is labelled A in the plot below.

In the same way, $f(0) = -2 \cdot 0 + 1 = 1$ using the equation $y = -2x+1$ and $f(2) = 2^2 - 3 \cdot 2 = -2$ using the equation $y = x^2 - 3x$. Thus, points $(0, 1)$ and $(2, -2)$ are labelled B and C in the graph below.

We can use the knowledge acquired in Chapter 3 to plot the three functions making up $f(x)$, but only drawing the appropriate moiety of the three functions in their corresponding x interval.

For the $x > 1$ parabola $x^2 - 3x$, from the plot, the vertex $V\left(\frac{3}{2}, -\frac{9}{4}\right)$ seems to be the lowest value of the piece-wise function, whilst the highest value is $+\infty$.

Thus, the co-domain interval (*i.e.* the range of y-values) is $\left[-\frac{9}{4}, +\infty\right)$.

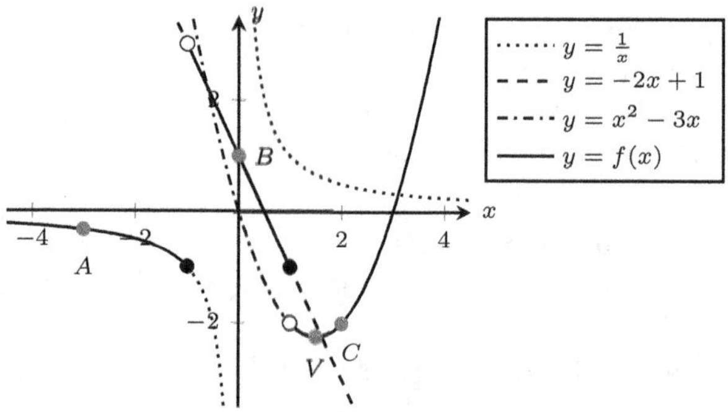

4.3 Symmetry and Sign

Studying the possible **symmetry** of a function can facilitate plotting it in the Cartesian plane.

The function $f(x)$ is called **even** if

$$f(-x) = f(x)$$

and thus is symmetrical across the y-axis. We need then to consider and plot the function only for positive x values and reflect the plot in the y-axis for negative x values.

The function $f(x)$ is called **odd** if

$$f(-x) = -f(x)$$

and thus is symmetrical around the point $O\,(0, 0)$, which is the origin of the Cartesian plane. We need then to consider and plot the function only for positive x values and then plot the function symmetrically with a $180°$ rotation around the point O for negative x values.

Example: Find whether the following functions are even, odd, or neither: $f_1(x) = \dfrac{4x^2-5}{3+x^2}$, $f_2(x) = \dfrac{x^2-2}{1-x^2}$, $f_3(x) = \dfrac{x}{9-x^2}$, and $f_4(x) = \dfrac{20x}{7x^2+4}$.

Solution: We first substitute $-x$ into each equation and then check whether $f(-x) = f(x)$, $= -f(x)$ or whether neither of these possibilities is true:

$$f_1(-x) = \frac{4(-x)^2-5}{3+(-x)^2} = \frac{4x^2-5}{3+x^2} = f_1(x)$$

$$f_2(-x) = \frac{(-x)^2-2}{1-(-x)^2} = \frac{x^2-2}{1-x^2} = f_2(x)$$

Thus, $f_1(x)$ and $f_2(x)$ are both even functions and their plots are symmetrical across the y-axis:

$$f_3(-x) = \frac{(-x)}{9-(-x)^2} = -\frac{x}{9-x^2} = -f_3(x)$$

$$f_4(-x) = \frac{20\cdot(-x)}{7(-x)^2+4} = -\frac{20x}{7x^2+4} = -f_4(x)$$

Thus, $f_3(x)$ and $f_4(x)$ are both odd functions and their plots are symmetrical round the origin point $O\,(0,0)$.

The following graphs show the particular symmetry of these four functions.

even functions

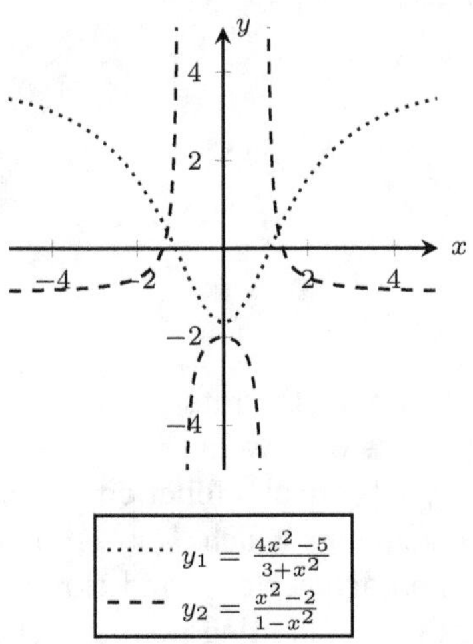

odd functions

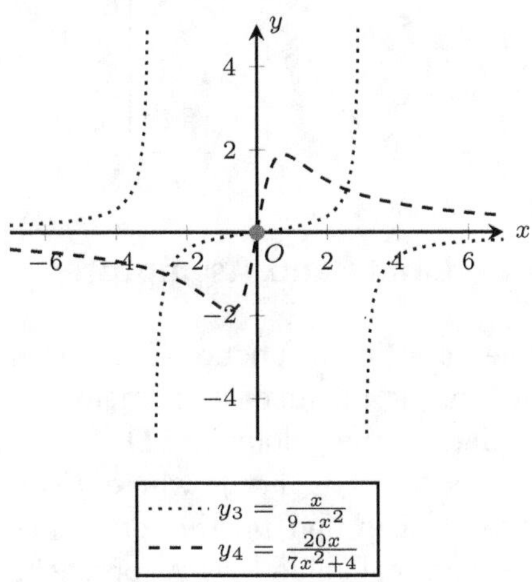

By studying the **sign** of a function, we can find over which intervals $f(x)$ is positive, using the inequality

$$f(x) \geq 0$$

Consequently, in the other intervals, the function $f(x)$ will be negative.

Thus, we can shade the y areas where the function should not be plotted in the Cartesian plane, simplifying the final drawing of the function.

Example: Investigate the **sign** of the function $f(x) = \dfrac{3x^2}{x^2 - 4}$ to find over which intervals $f(x)$ is positive or negative.

Solution: Solving $\dfrac{3x^2}{x^2 - 4} > 0$ gives $f(x) > 0$ if $x < -2$ OR $x > 2$, thus $f(x) < 0$ if $-2 < x < 2$. We can then shade negative y areas for $x < -2$ and $x > 2$, and a positive y area for $-2 < x < 2$. The final plot of the function occupies the white areas, as shown in the graph below.

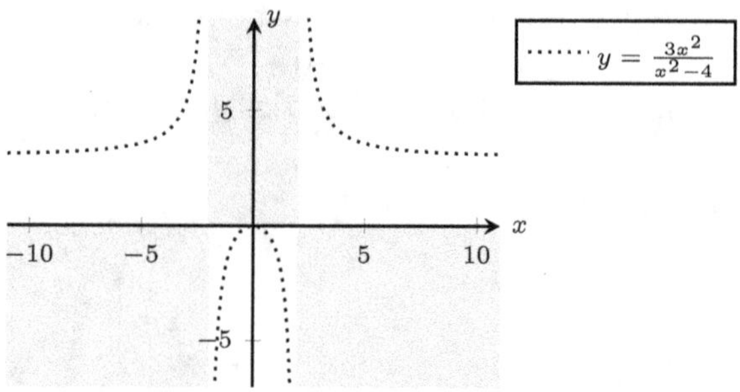

4.4 Limits and Asymptotes

When plotting a function, it is essential to understand its behaviour at the limiting points of its domain (*i.e.* its range of x values).

Given the domain $\mathbf{D}$ for a hypothetical function $f(x)$, $\mathbf{D}$:$(-\infty, a) \cup (a, +\infty)$, where a is the x value for which $f(x)$ is undefined, we can find whether or not $f(x)$ approaches closer and closer to a certain real value l, as x approaches one of the limiting points (x_0) of the function's domain:

- $-\infty$ (x is very large and negative);
- a^- (x approaches a from the left);
- a^+ (x approaches a from the right);
- $+\infty$ (x is very large and positive).

The mathematical notation to express the **limits** of the function is

$$\lim_{x \to x_0} f(x) = l$$

This method does not allow the calculation of the y value at x_0 but allows the definition of the function's behaviour around x_0.

If $x_0 = a^-$, we have a **left limit**, while if $x_0 = a^+$, we have a **right limit**.

When we write $f(\infty) = l$, we really mean $\lim_{x \to \infty} f(x) = l$, but 'infinity' ($\infty$) does not lie in the domain of the function.

If l is a real number, the limit is said to be **finite**.

However, as x approaches closer and closer to x_0, the function $f(x)$ can become extremely large, and can alternatively be said to be **infinite**, noted by the ∞ sign. The mathematical notation in this case is $\lim_{x \to x_0} f(x) = \infty$, which can be positive or negative.

4.4.1 Properties of limits

It can be proved that if the limit of a function exists, this limit is **unique**.

Additionally, if $\lim_{x \to x_0} f(x) = l$ and $\lim_{x \to x_0} g(x) = m$, with l and m being real numbers, the following properties can be used to simplify the calculation of the limits:

- $\lim_{x \to x_0} a \cdot f(x) = a \cdot \lim_{x \to x_0} f(x) = a \cdot l$, with $a \in \mathbb{R}$;
- $\lim_{x \to x_0} [f(x) \pm g(x)] = \lim_{x \to x_0} f(x) \pm \lim_{x \to x_0} g(x) = l \pm m$;
- $\lim_{x \to x_0} [f(x) \cdot g(x)] = \lim_{x \to x_0} f(x) \cdot \lim_{x \to x_0} g(x) = l \cdot m$;
- $\lim_{x \to x_0} \dfrac{f(x)}{g(x)} = \dfrac{\lim_{x \to x_0} f(x)}{\lim_{x \to x_0} g(x)} = \dfrac{l}{m}$, provided that $m \neq 0$;
- $\lim_{x \to x_0} [f(x)]^{g(x)} = \left[\lim_{x \to x_0} f(x)\right]^{\lim_{x \to x_0} g(x)} = l^m$, provided that $l > 0$.

4.4.2 Forms of limits

Particular attention should be paid during calculation of the limits involving infinite values (∞). The following forms of limit can be solved, with k being a real number:

- *Sum*:

  ◇ $+\infty + \infty = +\infty$ ◇ $-\infty - \infty = -\infty$
  ◇ $+\infty \pm k = +\infty$ ◇ $-\infty \pm k = -\infty$

- *Product* (sign according to the product law, $k \neq 0$):

  ◇ $\infty \cdot \infty = \infty$ ◇ $\infty \cdot k = \infty$

- *Division* (sign according to the product law):

  ◇ $\dfrac{0}{\infty} = 0$ ◇ $\dfrac{k}{\infty} = 0$ ◇ $\dfrac{\infty}{0} = \infty$ ◇ $\dfrac{\infty}{k} = \infty$

In order to **calculate the limit** of a function as x approaches a particular value x_0, x can be replaced with x_0 in the function and the expression rearranged to obtain the function limit value.

Important note: However, calculating some limits might involve infinite or null values, thus making the limit indeterminable, so further rearrangements are then needed to find them.

The following form of limits are **indeterminate** and cannot be solved without appropriate rearrangements:

$$\diamondsuit\ +\infty - \infty = ? \quad \diamondsuit\ \frac{\infty}{\infty} = ? \quad \diamondsuit\ 0 \cdot \infty = ? \quad \diamondsuit\ \frac{0}{0} = ?$$

Examples on how to solve some of these indeterminate forms are illustrated in the functions studied in Section 4.8.

4.4.3 Asymptotes of a function

We also need to check the curve behaviour around those values of x for which the function $f(x)$ is undefined, and also at $-\infty$ (x very large and negative) and $+\infty$ (x very large and positive), if $\pm\infty$ are included in the domain interval. The calculation of the limits of $f(x)$ at the limiting points of its domain is useful to find and mark the **asymptotes** of the function (*i.e.* the straight lines to which the function tends, but never actually reaches).

If $\lim_{x \to x_0} f(x) = \infty$, the straight line $\boxed{\mathbf{x = x_0}}$ is a **vertical asymptote** of $f(x)$ around x_0.

If $\lim_{x \to \pm\infty} f(x) = l$, the straight line $\boxed{\mathbf{y = l}}$ is a **horizontal asymptote** of $f(x)$ around $\pm\infty$.

If $\lim_{x \to \pm\infty} f(x) = \pm\infty$, a possible **slanting asymptote** of $f(x)$ should be sought in the form of a straight line $\boxed{\mathbf{y = mx + q}}$, where the constants $\mathbf{m} = \lim_{x \to \pm\infty} \frac{f(x)}{x}$ and $\mathbf{q} = \lim_{x \to \pm\infty} f(x) - mx$, if both m and q have finite limits.

Examples on how to find and mark the possible linear asymptotes of a function are illustrated in the cases studied in Section 4.8.

Examples from Biosciences:

(1) The binding of oxygen to haemoglobin is described by the Hill equation. Full occupancy of haemoglobin binding sites occurs as the oxygen concentration reaches very large values, which is the asymptotic limit of the process (example from Section 4.8.2).

(2) The height of a chromatographic column compared with the equivalent height of a theoretical plate under ideal conditions (HETP, height equivalent theoretical plate). The van Deemter equation relates the HETP to the column's flow speed by using a slightly complex function; when the flow speed is very high, the HETP asymptotically approaches linear behaviour (example addressed in Section 5.4).

4.5 Continuous Functions

A function $f(x)$ is said to be **continuous** at a specific point x_0 if the following three conditions hold:

(i) $f(x_0)$ exists and is a real number;

(ii) $\lim_{x \to x_0^-} f(x) = \lim_{x \to x_0^+} f(x) = l$ with l being a real number;

(iii) $\lim_{x \to x_0^\pm} f(x) = l = f(x_0)$;

i.e. the two-sided limit exists, is finite, and f assumes that value.

A function can also be continuous only from the left or from the right at a certain point x_0 if the corresponding left or right limit exists and equals $f(x_0)$.

Using the properties of limits (Section 4.4.1), it can be demonstrated that if two functions $f(x)$ and $g(x)$ are continuous at a specific point x_0, the following **combinations** of functions are **continuous** at x_0:

- $f(x) \pm g(x)$;
- $\dfrac{f(x)}{g(x)}$ provided that $g(x_0) \neq 0$ and f is continuous at $g(x_0)$;
- $f(x) \cdot g(x)$;
- $f(g(x))$.

If a function $f(x)$ is continuous at every point x in the open interval (a, b), $f(x)$ is said to be continuous in the open interval (a, b).

Similarly, if a function $f(x)$ is simultaneously continuous at every point x of the closed interval $[a, b]$, and continuous from the right starting at a and from the left finishing at b, $f(x)$ is said to be continuous in the closed interval $[a, b]$.

Practically, to say that a function is **continuous** in an interval means that the function can be drawn 'continuously' in that interval, *i.e.* 'without taking the pen off the paper.'

Examples of how to find and mark the intervals of x values where a function is continuous are illustrated in the cases presented in Section 4.8.

4.6 Discontinuous Functions

A function $f(x)$ is said to be **discontinuous** at a specific point x_0, if at least one of the three conditions given in Section 4.5 does not hold.

Different types of discontinuities exist and can be classified as jump, removable, infinite, endpoint, or mixed.

- **Jump discontinuity**: If $\lim_{x \to x_0^-} f(x) = L$ and $\lim_{x \to x_0^+} f(x) = M$, then property (ii) does not hold.

- **Removable discontinuity**: If $f(x_0)$ does not exist, but both the left limit $\lim_{x \to x_0^-} f(x)$ and the right limit $\lim_{x \to x_0^+} f(x)$ equal l, then property (i) does not hold.
- **Endpoint discontinuity**: If either the left limit $\lim_{x \to x_0^-} f(x)$ or the right limit $\lim_{x \to x_0^+} f(x)$ can not definitively be calculated (not even as infinite), while the other one exists as a real number, then property (ii) does not hold.
- **Infinite discontinuity**: If $\lim_{x \to x_0^{\pm}} f(x) = \infty$, with the sign of the ∞ being insignificant, then property (ii) does not hold.
- **Mixed discontinuity**: If $\lim_{x \to x_0^-} f(x)$ and $\lim_{x \to x_0^+} f(x)$ have different behaviours, either finite or infinite, then property (ii) does not hold.

Examples of how to find and mark the possible points of function discontinuity are illustrated in the cases presented in Section 4.8.

Note that discontinuous functions are very rare in the Biosciences as the relevant processes tend to be continuously dependent on their variables.

4.7 Geometric Transformations of $f(x)$

We often want to investigate the form of a function without plotting it accurately, so we can quickly 'sketch' it by using the graphs of some basic **elementary functions** which should become familiar to you and can be used as 'building blocks' (Figure 4.1).

The following rules should be used to sketch the form of a function, and which rule to use depends on the most appropriate geometric transformation. These are listed in the following subsections.

4.7.1 The shapes of $f(x) \pm a$ and $f(x \pm a)$

Rule A. If a function has equation $y = f(x) + a$ or $y = f(x) - a$ with $a > 0$, the following rules can be adopted to sketch it (Figure 4.2):

(i) to plot $y = f(x) + a$, the graph of $f(x)$ is translated up by a;

(ii) to plot $y = f(x) - a$, the graph of $f(x)$ is translated down by a.

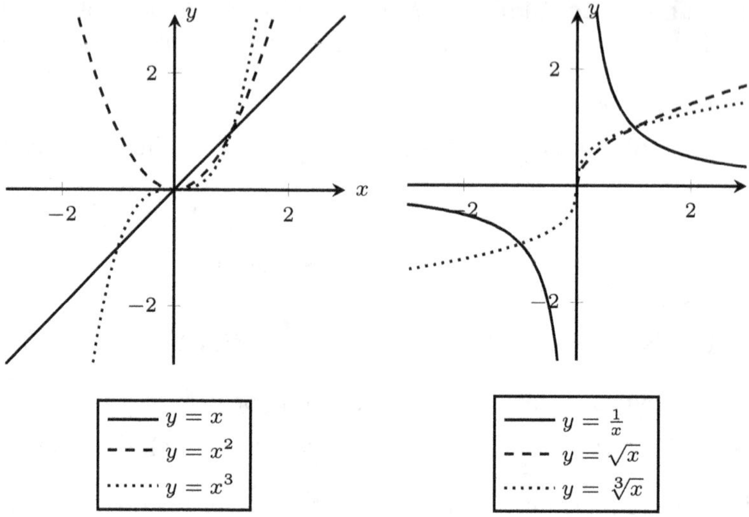

Figure 4.1: Graphs of some elementary functions

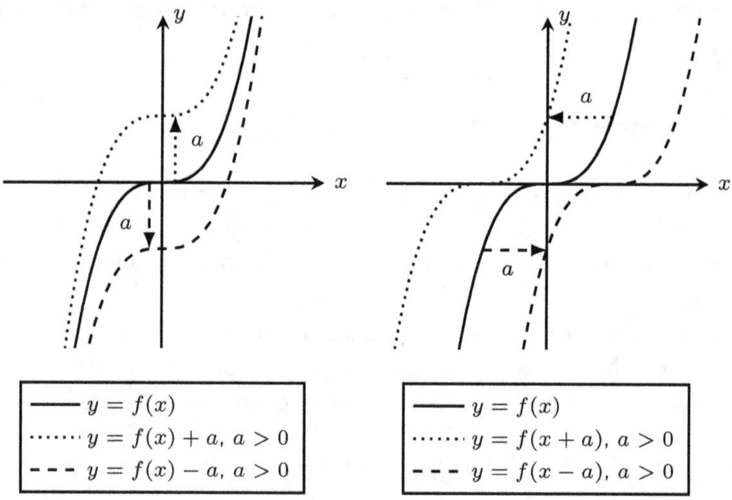

Figure 4.2: Graphs of $y = f(x) \pm a$ and $y = f(x \pm a)$

Rule B. If a function has equation $y = f(x + a)$ or $y = f(x - a)$ with $a > 0$, the following rules can be adopted to sketch it (Fig. 4.2):

(i) to plot $y = f(x + a)$, the graph of $f(x)$ is translated left by a;

(ii) to plot $y = f(x - a)$, the graph of $f(x)$ is translated right by a.

4.7.2 The shapes of $-f(x)$, $f(-x)$, $|f(x)|$, and $f(|x|)$

Rule C. If a function has equation $y = -f(x)$, we can reflect the whole plot of $f(x)$ in the x-axis to obtain its graph (Figure 4.3).

Rule D. If a function has equation $y = f(-x)$, we can reflect the whole plot of $f(x)$ in the y-axis to obtain its graph (Figure 4.3).

However, if a function has equation $y = f(a - x)$ or $y = f(-a - x)$ with $a > 0$, the following rules can be adopted to sketch it:

(i) to plot $y = f(a - x)$, the graph of $f(-x)$ is translated right by a;

(ii) to plot $y = f(-a - x)$, the graph of $f(-x)$ is translated left by a.

Rule E. If a function has equation $y = |f(x)|$, we can first sketch the function $f(x)$ without considering the modulus sign and by using one of our building blocks, and then reflect only the negative y values of this function in the x-axis to obtain the final graph (Figure 4.3).

Rule F. If a function has equation $y = f(|x|)$, we can initially sketch the elementary function $f(x)$ for only the positive x values without considering the modulus sign; then, we can complete the plot for negative x values by reflecting the first part of the curve in the y-axis to obtain the final graph (Figure 4.3).

4.7.3 The shapes of $a \cdot f(x)$ and $f(a \cdot x)$

Rule G. Let a be a positive real number > 1:

(i) If a function has equation $y = a \cdot f(x)$, we expand the graph vertically by a factor a.

(ii) If a function has equation $y = f(a \cdot x)$, we compress the graph horizontally by a factor of a (Figure 4.4).

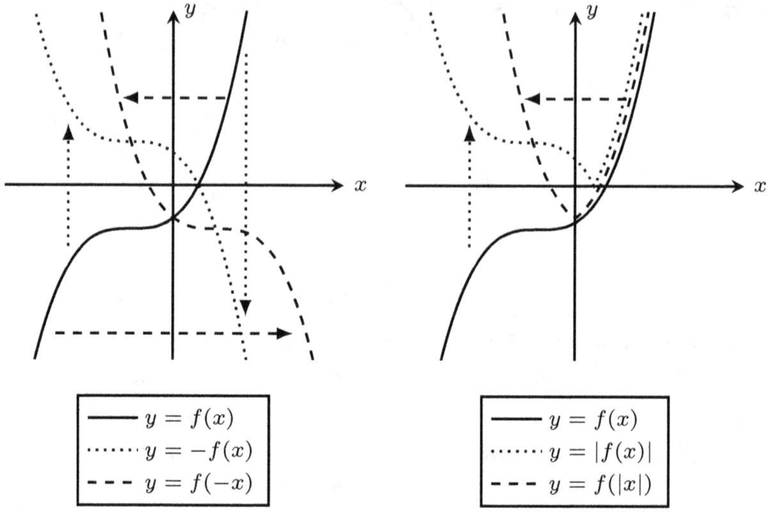

Figure 4.3: Graphs of $y = -f(x)$, $y = f(-x)$, $y = |f(x)|$, and $y = f(|x|)$

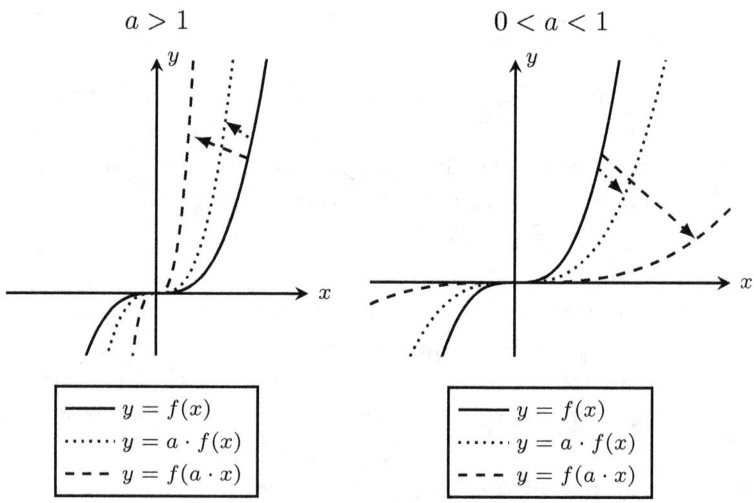

Figure 4.4: Graphs of $y = a \cdot f(x)$ and $y = f(a \cdot x)$

Rule H. Let a be a positive real number with $0 < a < 1$:

(i) If a function has equation $y = a \cdot f(x)$, we compress the graph vertically by a factor a.
(ii) If a function has equation $y = f(a \cdot x)$, we stretch the graph horizontally by a factor a (Figure 4.4).

Example: Sketch the functions $y = \dfrac{1}{1-x} - 3$ and $y = |4 - 2\sqrt{x+2}|$.

Solution: Initially, we can identify $y = \dfrac{1}{x}$ as the closest elementary function to $y = \dfrac{1}{1-x} - 3$, and start by plotting it in the Cartesian plane, as shown in the next figure.

Using Rule **D**, we can sketch $y = \dfrac{1}{-x}$ by reflecting the whole graph of $y = \dfrac{1}{x}$ in the y-axis.

Then, using Rule **D**(i), we can sketch $y = \dfrac{1}{1-x}$ by translating the graph of $y = \dfrac{1}{-x}$ by 1 unit to the right (positive x direction).

Finally, in order to sketch $y = \dfrac{1}{1-x} - 3$, we can use Rule **A**(ii) by translating the plot of $y = \dfrac{1}{1-x}$ down by 3 units.

The figure below summarises all the intermediate steps involved in sketching $y = \dfrac{1}{1-x} - 3$.

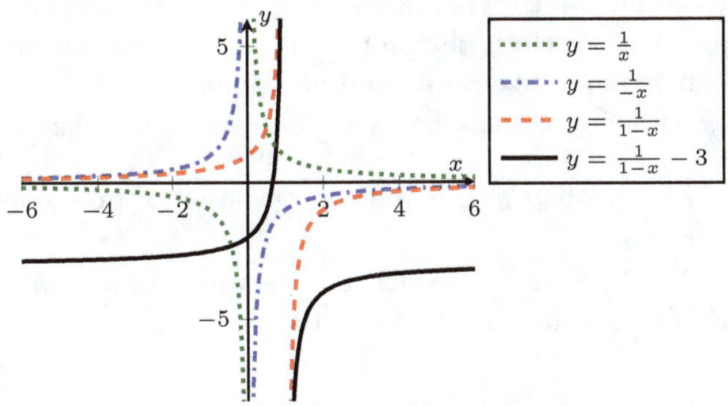

In order to sketch the function $y = |4 - 2\sqrt{x+2}|$, we can first identify $y = \sqrt{x}$ as its closest elementary function.

Using Rule **B**(i), we then sketch $y = \sqrt{x+2}$ by translating the graph of $y = \sqrt{x}$ by 2 to the left (negative x direction).

Then, using Rule **G**(i) for plotting $a \cdot f(x)$ with $a > 1$, we sketch $y = 2\sqrt{x+2}$ by compressing the whole graph of $y = \sqrt{x+2}$ towards the y-axis.

To sketch $y = -2\sqrt{x+2}$, we reflect the whole graph of $f(x)$ in the x-axis, using Rule **C**.

Next, to sketch $y = 4 - 2\sqrt{x+2}$, we translate the plot of $y = -2\sqrt{x+2}$ up by $+4$ according to Rule **A**(i) for functions, such as $y = f(x) + a$ with $a > 0$.

Finally, in order to sketch $y = |4 - 2\sqrt{x+2}|$, we can use Rule **E** to plot a function of the form $y = |f(x)|$ and only reflect the negative y values of $y = 4 - 2\sqrt{x+2}$ in the x-axis.

The figure below summarises all the possible intermediate steps involved in sketching $y = |4 - 2\sqrt{x+2}|$.

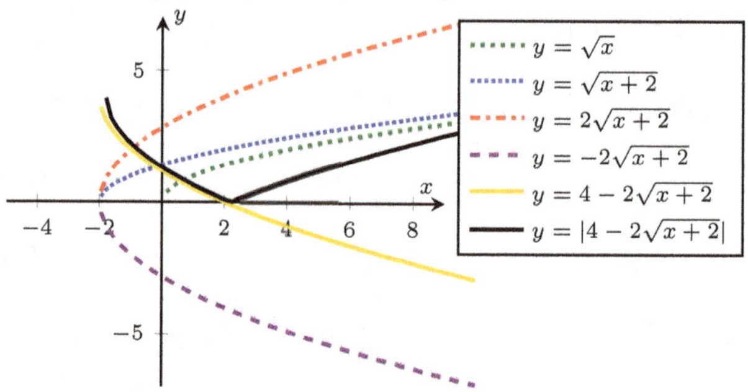

Example in Biochemistry: The fractional saturation θ of the myoglobin protein is the proportion of the total concentration which is made up of oxygen-bound myoglobin.

The variable θ can be also expressed by the equation $\theta = \dfrac{p_{O_2}}{p_{50} + p_{O_2}}$, where p_{O_2} is the concentration of free oxygen and p_{50} the partial pressure at which the fractional saturation is 50%. Myoglobin's p_{50} value for O_2 is 130 Pa.

Sketch the dependent variable θ as a function of the independent variable p_{O_2} by choosing the appropriate geometric transformations of a suitable elementary function.

Solution: First let us make the following substitutions into the equation:

$$\theta = \frac{p_{O_2}}{p_{50} + p_{O_2}} : \theta = y,\ p_{O_2} = x,\ p_{50} = 130$$

which gives the function $y = \dfrac{x}{130 + x}$ to be sketched.

We can then rearrange the myoglobin's saturation equation in order to use one of the cases of geometric transformations presented in this chapter:

$$\begin{aligned} y &= \frac{x}{130 + x} = \frac{x + 130 - 130}{130 + x} = \frac{x + 130}{130 + x} - \frac{130}{130 + x} \\ &= 1 - \frac{130}{130 + x} \end{aligned}$$

We can identify the equation of a hyperbola $y = \dfrac{130}{x}$ as the closest elementary function to $y = 1 - \dfrac{130}{130 + x}$, and start by plotting it only on the half-plane of positive x values, as shown in the figure below.

Using Rule **B**(i), we can sketch $y = \dfrac{130}{130 + x}$ by translating the hyperbolic graph by -130 to the left.

Then to sketch $y = -\dfrac{130}{130 + x}$, we can reflect the whole graph in the x-axis by using Rule **C**.

Finally, we can use Rule **A**(i) to sketch $y = 1 - \dfrac{130}{130 + x}$ by translating the the plot of $y = -\dfrac{130}{130 + x}$ up by $+1$.

This last sketch corresponds to that of the function $\theta = \dfrac{p_{O_2}}{p_{50} + p_{O_2}}$, as shown in the following figure.

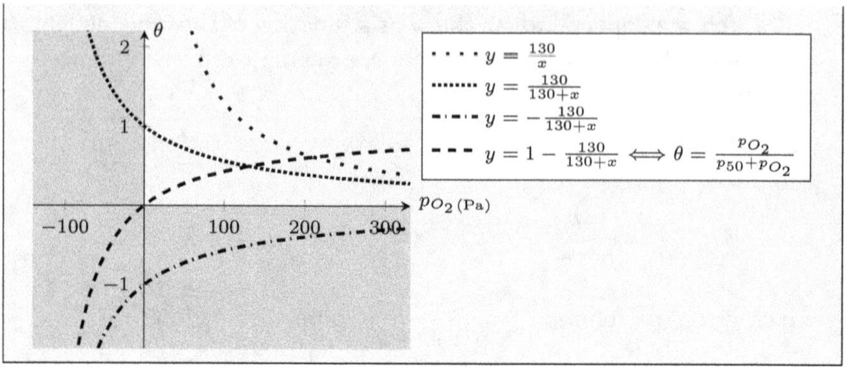

4.8 Curve Sketching

If the function to be sketched is more complex than the elementary functions shown in Section 4.7 and geometric transformations cannot be used, we can sketch it by using the information detailed in the above sections:

(i) **domain** through the interval notation, *e.g.* $(-\infty, a]$ (Section 4.2);

(ii) **key points** (Section 4.2);

(iii) possible **symmetry** (Section 4.3);

(iv) **sign** (Section 4.3);

(v) **limits** at the domain interval extremities and possible **asymptotes** (Section 4.4);

(vi) intervals of function **continuity** and possible points of **discontinuity** (Sections 4.5 and 4.6);

(vii) **co-domain** or range intervals of y values.

Important note: In most of the examples illustrated in this section, the guidelines presented above can only be used to 'sketch' a function in a Cartesian plane, and the sketches are shown using a **solid line** on the graph.

For clarity, in each example, the whole function plot is also displayed using a **dotted line**, but it can only be drawn after studying the differentiation methods explained in Chapter 5.

4.8.1 Polynomial functions

An expression containing the sum of different powers of x is called a **polynomial** in x. In general,

$$f(x) = a_n x^n + a_{n-1} x^{n-1} + a_{n-2} x^{n-2} + \cdots + a_1 x^1 + a_0 x^0$$

where $a_n, \ldots, a_0$ are constants and are called **coefficients**.

By convention, the expressions $a_1 x^1$ and $a_0 x^0$ are usually written as $a_1 x$ and a_0, respectively, and $a_n x^n$ is called the **leading term** of the polynomial.

The non-negative integer n is called the **degree** of the polynomial and the coefficients of a single polynomial of degree n are determined by at most $n+1$ points. Also, the graph of a polynomial of degree n has at most $n-1$ **bends** in it (*e.g.* if $n=3$, the graph will have two bends).

Thus, the graph of a polynomial of degree 1 is just a straight line, with no bends in it. If the degree is 0, 1, 2, or 3, we have **constant**, **linear**, **quadratic**, or **cubic** functions, respectively.

The key information useful to sketch a polynomial function is as follows:

(i) **Domain:** $\mathbb{R}$, *i.e.* $(-\infty, +\infty)$.

(ii) **Key points:** $f(0)$, $f(1)$ and $f(-1)$ can be calculated.

The equation $f(x) = 0$ can be solved by using some algebraic tools, such as the techniques of factorisation.

Example: Sketch the function $y = x^3 - 3x^2 - 9x + 27$.

Solution:

(i) **Domain:** $(-\infty, +\infty)$.

(ii) **Key points:** $f(0) = 27$, $f(1) = 16$, and $f(-1) = 32$, thus we can mark the points A $(0, 27)$, B $(1, 16)$, and C $(-1, 32)$ in the Cartesian plane.

In order to solve $f(x) = 0$, the polynomial function can be factorised as follows: $y = (x+3)(x-3)^2$. Thus, by equating each factor to 0 according to the zero-product property (see Sections 2.2.2 and 2.2.4), we have $x = -3$ and $x = 3$ as solutions of the equation $f(x) = 0$. So the intersection points between the function and the x-axis are D $(-3, 0)$ and E $(3, 0)$, which can be marked in the Cartesian plane.

(iii) **Symmetry**: We can seek a possible axis or point of symmetry to aid the function sketching.

(iv) **Sign**: Polynomial inequalities (Section 2.5.6) can be used to study the sign of a polynomial function.

Solution:

(iii) **Symmetry**: Let us start calculating $f(-x) = -x^3-3x^2+9x+27$ which does not equal either $f(x)$ or $-f(x)$. So, the function is neither even nor odd.

Alternatively, we can look for either an axis of symmetry of the type of $x = a$ or a point of symmetry $F(a, b)$.

Important note: In order to avoid elaborate calculations which do not result in useful conclusions, we can first study the sign of a polynomial function, as this can help suggest possible axial or point symmetry that might be present.

(iv) **Sign:** Let us solve the polynomial inequality $x^3 - 3x^2 - 9x + 27 > 0 \Longleftrightarrow (x+3)(x-3)^2 > 0$ (Section 2.5.6). We notice that $(x-3)^2$ is always > 0; thus, it is sufficient to solve the linear inequality $(x+3) > 0$ to know where the function is positive.

Since the function is positive for $x > -3$, we can shade the area below the x-axis for $x > -3$ and the area above the x-axis for $x < -3$. The function will not be plotted in the shaded areas in the Cartesian plane.

(v) **Limits and asymptotes:**

(a) If the function has an **even** polynomial degree,

- $\lim_{x\to\pm\infty} f(x) = +\infty$ if the coefficient of the leading term $a_n > 0$;
- $\lim_{x\to\pm\infty} f(x) = -\infty$ if $a_n < 0$.

(b) If the function has an **odd** polynomial degree,

- $\lim_{x\to-\infty} f(x) = -\infty$ and $\lim_{x\to+\infty} f(x) = +\infty$ if $a_n > 0$;
- $\lim_{x\to-\infty} f(x) = +\infty$ and $\lim_{x\to+\infty} f(x) = -\infty$ if $a_n < 0$.

Reminder: When we write $f(\pm\infty) = \pm\infty$, we really mean $\lim_{x\to\pm\infty} f(x) = \pm\infty$, but '$\pm$ infinity' ($\pm\infty$) does not lie in the domain of the function.

Generally, if the indeterminate form $\infty - \infty$ occurs as x approaches $\pm\infty$, the leading term x^n should be factorised to solve the indeterminate form.

Solution:

(v) **Limits:** As the function $y = x^3 - 3x^2 - 9x + 27$ has an odd polynomial degree, *i.e.* $n = 3$, and the coefficient of the leading term $a_n = 1 > 0$, we can state that $\lim_{x\to-\infty} f(x) = -\infty$ and $\lim_{x\to+\infty} f(x) = +\infty$.

Additionally, as $\lim_{x\to\pm\infty} f(x) = \pm\infty$, a possible slanting asymptote of $f(x)$ should be sought in the form of a straight line $y = mx + q$ (see Section 4.4.3).

However, $m = \lim_{x\to\pm\infty} \frac{f(x)}{x} = \lim_{x\to\pm\infty} \frac{x^3 - 3x^2 - 9x + 27}{x}$

$= \lim_{x\to\pm\infty} \left(x^2 - 3x - 9 + \frac{27}{x}\right) = +\infty$ which is not finite. Therefore, no slanting asymptote to the function can be drawn.

So, we can sketch solid dark arrows towards $-\infty$ and $+\infty$ as x approaches $-\infty$ and $+\infty$, respectively.

(vi) **Continuity:** Polynomial functions are continuous in every closed interval $[a, b]$ of their domain, which is $\mathbb{R}$.

(vii) **Co-domain:**

(a) If the function has an **odd** polynomial degree,
- the range of y values of a polynomial function is $\mathbb{R}$.

(b) If the function has an **even** polynomial degree,
- the co-domain is $(-\infty, max]$ if $a_n < 0$;
- the co-domain is $[min, +\infty)$ if $a_n > 0$.

min and *max* values will be estimated in Chapter 5, when we will study differentiation.

Solution:

(vi) **Continuity:** The polynomial function $y = x^3 - 3x^2 - 9x + 27$ is continuous in its domain $\mathbb{R}$.

(vii) **Co-domain:** As this function has an **odd** polynomial degree, the range of y values is $\mathbb{R}$.

Using all the information in the above sections about the function $y = x^3 - 3x^2 - 9x + 27$, we can now sketch it in a Cartesian plane, as shown in the graph below.

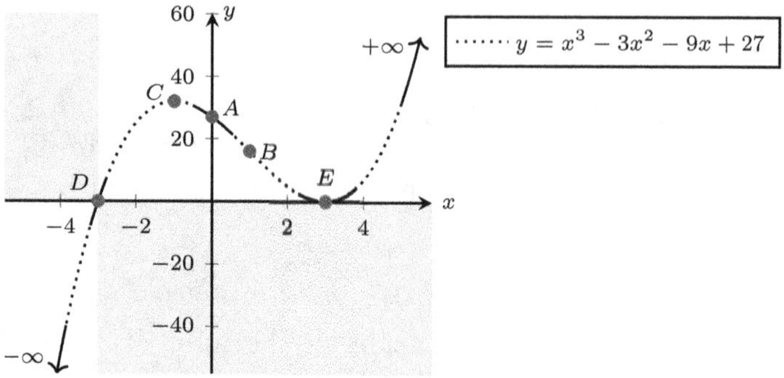

Example in Biology: To fight human vector-borne diseases, such as malaria and dengue, disease-refractory mosquitoes have been generated through transgenic techniques to replace wild types and thus stop transmission.

The synthetic gene drive mechanism to spread the required refractoriness-causing transgene (A) within future mosquito populations is based on the principle of underdominance, where the fitness of the heterozygous genotype is lower than that of the homozygous genotypes.

The 'allele frequencies' equilibrium can be obtained by solving the equation

$$f(p, q) = \frac{pq(s_2 q - s_1 p)}{1 + s_1 p^2 + s_2 q^2} = 0$$

where s_1 is the additional fitness value for the genotype AA, s_2 the additional fitness value for aa, p the frequency of dominant allele A, q the frequency of recessive allele a, and both p and q can only be positive numbers < 1.

The allele frequencies' equilibrium (p_e and q_e) are special values of p and q which satisfy $f(p, q) = 0$ and, once attained, this results in no change in allele frequency in future generations. This will help spread the required transgene through future populations.

As the denominator is always positive, it is sufficient to analyse the numerator to investigate the relationship between the change in frequency of recessive allele and q.

Using the *Hardy–Weinberg* equation $p+q=1$, make the appropriate substitution within the function equation in order to have only the variable q in the function f. Estimate the theoretical equilibrium allele frequency q_e for which $f(q)=0$ and, if we suppose that $s_1=0.2$ and $s_2=0.8$, discuss the evolution of $f(q)$ in relation to q.

Solution: Initially, by substituting $p=1-q$, we obtain

$$\begin{aligned} f(q) &= (1-q)q(s_2q - s_1(1-q)) \\ &= s_2q^2 - s_1q + s_1q^2 - q^3s_2 + q^2s_1 - s_1q^3 \\ &= -q^3\cdot(s_1+s_2) + q^2\cdot(2s_1+s_2) - q\cdot s_1 \end{aligned}$$

which is a cubic function in q.

To solve the equation $f(q)=0$, we can initially factorise $f(q)$ by the common factor $-q$:

$$f(q) = -q\cdot(q^2\cdot(s_1+s_2) - q^{\cdot}(2s_1+s_2) + s_1)$$

and then solve the quadratic $q^2\cdot(s_1+s_2) - q^{\cdot}(2s_1+s_2) + s_1 = 0$ using the quadratic formula (see Section 2.3.2):

$$q_e = \frac{2s_1+s_2 \pm \sqrt{4s_1^2+4s_1s_2+s_2^2-4s_1^2-4s_1s_2}}{2(s_1+s_2)} = \frac{2s_1+s_2 \pm s_2}{2(s_1+s_2)}$$

which gives (a) $q_{e1} = \dfrac{s_1}{s_1+s_2}$ and (b) $q_{e2}=1$, but (b) is not physically possible.

Using the values $s_1=0.2$ and $s_2=0.8$, we obtain

$$f(q) = -0.2q + 1.2q^2 - q^3$$

which can be factorised as follows $f(q) = -q\,(q-1)\,(q-0.2)$. From (a), the q_e value here is $0.2/(0.2+0.8)=0.2$.

In order to discuss the evolution of $f(q)$ in relation to q, we can sketch the function in a Cartesian plane and specifically focus on the interval of x values $[0,1]$:

(i) **Domain:** It is restricted to the interval $(0, 1)$, since the frequencies p and q can only have values from 0 to 1 and $p + q = 1$.

(ii) **Key points:** $f(0) = 0$, $f(1) = 0$; $f(q) = 0$ can be solved by factorising the polynomial $-q\,(q-1)\,(q-0.2) = 0$, which gives three solutions $x = 0$, $x = 1$, and $x = 0.2$. So, the function passes through the points $A\,(0, 0)$, $B\,(1, 0)$, and $C\,(0.2, 0)$, which corresponds to the allele frequencies' equilibrium point q_e.

(iv) **Sign:** $f(q) > 0$ is a polynomial inequality, which gives $0.2 < q < 1$ as the range of solutions. Thus, we can shade the area above the x-axis for $0 < x < 0.2$, while we can shade the area below the x-axis for $0.2 < x < 1$.

(vi) **Continuity:** $f(q)$ is continuous at every point in the closed interval $[0, 1]$.

(vii) **Co-domain:** As the graph of a polynomial of degree 3 must have at most two **bends** in it, the function $f(q)$ must have one negative bend in the interval $[0, 0.2]$ and one positive bend in $[0.2, 1]$. Also, there exists a point of absolute minimum min in the closed interval $[0, 0.2]$ and a point of absolute maximum max in the closed interval $[0.2, 1]$. So the range of y-values is $[min, max]$.

Other information, such as symmetry, limits, and asymptotes, are not essential to solve the problem.

Sketching the function $f(q) = -0.2q + 1.2q^2 - q^3$ using the information above gives the following graph.

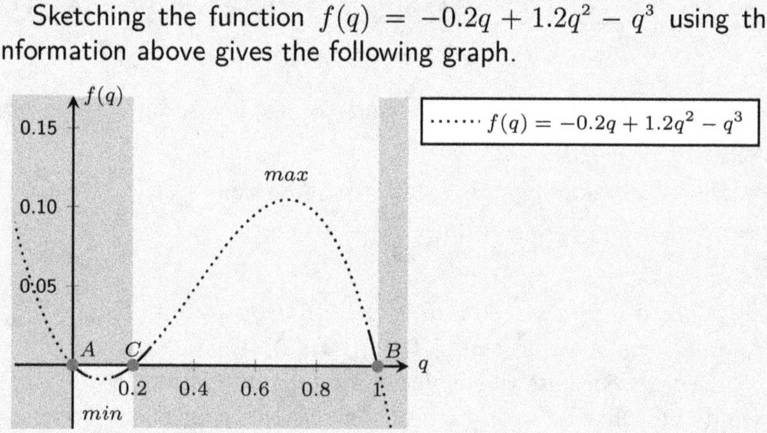

We can conclude that if $0 < q < q_e$, $f(q) < 0$, then the frequency of the recessive allele will decrease as equilibrium is approached, whereas if $q_e < q < 1$, $f(q) > 0$, the frequency of the recessive allele will increase as equilibrium is approached.

4.8.2 Rational functions

Functions containing the ratio of two expressions of x are called **rational functions**. The general form is

$$f(x) = \frac{N(x)}{D(x)}$$

where $N(x)$ and $D(x)$ can be any type of mathematical expression, **but** only polynomials are discussed in this chapter.

The key information useful to sketch a polynomial function are as follows:

(i) **Domain:** The condition $D(x) \neq 0$ is essential; all x values giving $D(x) = 0$ are excluded from the domain, thus avoiding $f(x)$ going to $\pm\infty$.

(ii) **Key points:** $f(0)$, $f(1)$, and $f(-1)$ can be calculated.

The equation $f(x) = 0$ can be solved by using rational equations.

Example: Sketch the function $y = \dfrac{1}{(x-1)(x-2)}$.

Solution:

(i) **Domain:** Let us solve $(x-1)(x-2) \neq 0$, which gives $x \neq 1$ and $x \neq 2$. Thus, the domain is $(-\infty, 1) \cup (1, 2) \cup (2 + \infty)$.

(ii) **Key points:** $f(0) = \dfrac{1}{2}$ and $f(-1) = \dfrac{1}{6}$, thus we can mark the points $A\left(0, \dfrac{1}{2}\right)$ and $B\left(-1, \dfrac{1}{6}\right)$ in the Cartesian plane.

In order to solve $f(x) = 0$, we must first exclude the x values giving $D(x) = 0$, which are $x = 1$ and $x = 2$, and then equate $N(x) = 0 \iff 1 = 0$, which is impossible. So, no intersection points exist between the function and the x-axis.

(iii) **Symmetry:** A possible axis or point of symmetry can be identified to aid the function sketching.

(iv) **Sign:** Rational inequalities (Section 2.5.6) can be used to study the sign of a rational function.

Solution:

(iii) **Symmetry:** Let us start by calculating $f(-x) = \dfrac{1}{(-x-1)(-x-2)}$, which does not equal either $f(x)$ or $-f(x)$. So, the function is neither even nor odd.

Alternatively, we can look for either an axis of symmetry of the type $x = a$ or a point of symmetry $F(a, b)$.

(iv) **Sign:** Let us solve the rational inequality $\dfrac{1}{(x-1)(x-2)} > 0$ (Section 2.5.6), which gives $x < 1$ and $x > 2$ as the ranges of solutions.

The function is positive for $x < 1$ and $x > 2$. So, we can shade the area below the x-axis for $x < 1$ and $x > 2$, and we can shade the area above the x-axis for $1 < x < 2$. The function will not be plotted in the shaded areas in the Cartesian plane.

(v) **Limits and asymptotes:**

(a) As x approaches all particular x values for which y is undefined (for instance a), $\lim_{x\to a} f(x)$ should be calculated by substituting a in the function equation. If $\lim_{x\to a} f(x) = \infty$, the sign of the ∞ is given by the function sign (point iv).

If the indeterminate form $\dfrac{0}{0}$ occurs, factorisation operations should be performed to simplify the fraction and calculate the limit.

(b) As x approaches ∞, if the indeterminate form $\dfrac{\infty}{\infty}$ occurs,

- $\lim_{x\to\infty} f(x) = \infty$ if the polynomial degree of $N(x)$ is **greater** than the polynomial degree of $D(x)$;
- $\lim_{x\to\infty} f(x) = 0$ if the polynomial degree of $N(x)$ is **lower** than the polynomial degree of $D(x)$;
- $\lim_{x\to\infty} f(x) = \dfrac{a_{n,N(x)}}{a_{n,D(x)}}$, where $a_{n,N(x)}$ and $a_{n,D(x)}$ are the coefficients of the leading terms of $N(x)$ and $D(x)$, respectively, if the polynomial degree of $N(x)$ **equals** the polynomial degree of $D(x)$.

If $\lim_{x\to\pm\infty} f(x) = \pm\infty$, possible slant asymptotes should be identified.

Reminder: When we write $f(\pm\infty) = \pm\infty$, we really mean $\lim_{x\to\pm\infty} f(x) = \pm\infty$, but '$\pm$ infinity' ($\pm\infty$) does not lie in the domain of the function.

Solution:

(v) **Limits:** As the function $y = \dfrac{1}{(x-1)(x-2)}$ is not defined for $x = 1$ and $x = 2$, we can calculate their corresponding limits by substituting 1 and 2 in the function equation, one at a time:

$$\lim_{x\to 1} \frac{1}{(1-1)(1-2)} = \frac{1}{0} = \infty$$
$$\lim_{x\to 2} \frac{1}{(2-1)(2-2)} = \frac{1}{0} = \infty$$

So, $y = 1$ and $y = 2$ are both vertical asymptotes of the function, and we can sketch solid dark arrows at $y = \infty$ close to $x = 1$ and $x = 2$ according to the function sign (point iii).

Then, we calculate $\lim_{x\to\pm\infty} f(x)$.

It is sufficient to calculate $\lim_{x\to+\infty} f(x)$, as $\lim_{x\to-\infty} f(x)$ has the same value as the axial symmetry of the function sign (point iii):

$$\lim_{x\to+\infty} \frac{1}{(\infty-1)(\infty-2)} = \frac{1}{\infty\cdot\infty} = \frac{1}{\infty} = 0$$

Thus, $y = 0$ is a horizontal asymptote for the function at $x = \pm\infty$, and we can sketch solid dark lines at $x = \pm\infty$ close to $y = 0$ (*i.e.* the x-axis), according to the function sign (point iii).

(vi) **Continuity:** Rational functions are continuous in any closed intervals $[a, b]$ of their corresponding domain.

All specific x values for which y is undefined are points of discontinuity to be determined.

(vii) **Co-domain:** The exact range of y values can be defined only after we have studied differentiation (see Chapter 5).

Solution:

(vi) **Continuity:** The rational function $y = \dfrac{1}{(x-1)(x-2)}$ is continuous in its domain $\mathbb{R}$. For $x = 1$ and $x = 2$, we have two points of infinite discontinuity.

Using all the information in the above sections regarding the function $y = \dfrac{1}{(x-1)(x-2)}$, we can now sketch it in a Cartesian plane, as shown in the graph below.

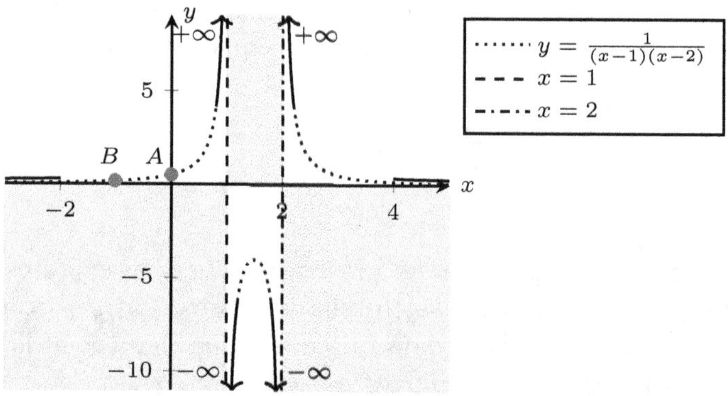

Example in Biochemistry: The *Hill equation* was initially applied to describe the binding of oxygen to haemoglobin, and subsequently has been widely used in Biosciences to analyse the binding equilibria in ligand–receptor interactions:

$$\theta = \frac{L^n}{L^n + K_d}$$

where θ is the percentage of binding sites occupied, $[L]$ the ligand concentration, K_d the dissociation constant, and n the degree of cooperativity.

- If $n = 1$, there is no cooperativity, and Micahelis–Menten kinetics are observed.
- If $n > 1$, there is positive cooperativity, which means that the binding of one ligand molecule increases the rate of binding of other ligands.

- If $n < 1$, there is negative cooperativity.

The scientist Archibald Hill initially supposed that $n = 4$ for oxygen-haemoglobin binding because of the four haemoglobin sub-units, giving $\theta = \dfrac{L^4}{L^4 + K_d}$.

Calculate the theoretical value of L to reach $\theta = 0.5$ and sketch the function $\theta = f(L)$.

Explain why positively cooperative kinetics with $n = 4$ are more favourable for haemoglobin towards oxygen than Michaelis–Menten kinetics.

Solution:

(i) **Domain:** Let us put the denominator $\neq 0$: $L^4 + K_d \neq 0$. Since $L^4 + K_d$ is always > 0, no real value will give the denominator $= 0$.

However, we must restrict the domain from $\mathbb{R}$ to $(0, +\infty)$, since the variable L is a physical entity and thus can only be positive.

(ii) **Key points:** $f(0) = 0$; $f(L) = 0$ can be solved by equating the numerator to 0: $L^4 = 0 \Longleftrightarrow L = 0$.

For $\theta = 0.5$, we can calculate $f(L) = \frac{1}{2}$: $\dfrac{L^4}{L^4 + K_d} = \dfrac{1}{2}$

$$\Longleftrightarrow L^4 = \frac{1}{2}(L^4 + K_d) \Longleftrightarrow \frac{1}{2}L^4 = \frac{1}{2}K_d \Longleftrightarrow L = \sqrt[4]{K_d}.$$

So, the function passes through the points $A(0, 0)$ and $B\left(\sqrt[4]{K_d}, \frac{1}{2}\right)$.

(iv) **Sign:** $f(L) > 0 \Longleftrightarrow \dfrac{L^4}{L^4 + K_d} > 0$. The function $f(L)$ is always positive in the domain range as it is the ratio of two positive quantities. This is consistent with θ being a physical quantity.

Thus, we can shade the area above the x-axis for $x > 0$.

(v) **Limits and asymptote:** Let us calculate $\lim\limits_{L \to +\infty} \dfrac{L^4}{L^4 + K_d} = 1$ as the polynomial degree of the numerator equals the polynomial degree of the denominator. So, $\theta = 1$ is a horizontal asymptote of $f(L)$ going to $+\infty$.

(vi) **Continuity:** $f(L)$ is continuous at every point in the domain.

Other information, such as the symmetry and co-domain, are not necessary to solve the problem.

Sketching the function $\theta = \dfrac{L^4}{L^4 + K_d}$ using the information above gives the following graph.

However, plotting the function $\theta = \dfrac{L}{L + K_d}$ using the information in Chapter 3, we notice that 50% of maximal binding-site occupancy is only reached at $L = K_d$ if haemoglobin follows Michaelis–Menten kinetics, but is reached at $L\sqrt[4]{K_d}$ with positively cooperative kinetics.

Thus, the positive cooperativity in oxygen-haemoglobin binding is a favourable biochemical way to reach higher occupancy at lower oxygen levels, which is obviously advantageous for all mammals, including humans at high altitudes.

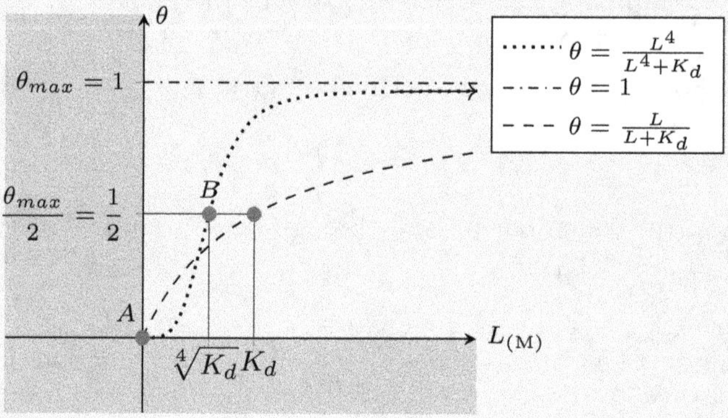

4.8.3 Irrational functions

A function containing at least one n^{th} root of x in its formula is called an **irrational function**. The general form of the irrational functions discussed in this book is

$$f(x) = \sqrt[n]{R(x)}$$

where $R(x)$ can be any mathematical expression. However, only the forms of polynomial and rational functions are discussed here.

Important note: Since a function relates only a single y value to each x value in the domain (see Section 4.1.1), the irrational function containing the **even root** of a mathematical expression **conventionally** relates only the **positive root** to every x value of the function's domain:

e.g. the function $y = \sqrt{x}$ associates $y = +3$ to $x = 9$

Algebraically, there are two **square roots** of a real number,

e.g. $\sqrt{9} = \pm 3$

The key information useful for sketching an irrational function is as follows:

(i) **Domain:**

(a) If n is **even**, the condition $R(x) \geq 0$ is necessary, since the nth root of a negative number does not exist in the set of the real numbers.

(b) If n is **odd**, the domain is $\mathbb{R}$, *i.e.* $(-\infty, +\infty)$.

(ii) **Key points:** $f(0)$, $f(1)$, and $f(-1)$ can be calculated, if this is possible.

The equation $f(x) = 0$ is solved by raising both sides of the equation to the nth power and then using appropriate techniques for solving polynomial or rational equations.

Warning! If n is even, then only the x values included in the function's domain can be considered as solutions $(R(x) > 0)$.

Example: Sketch the function $y = \sqrt{x^2 - x}$.

Solution:

(i) **Domain:** Let us solve the inequality $x^2 - x \geq 0$. We first factorise the polynomial as $x(x-1)$ and find its zeros: $x_1 = 0$ and $x_2 = 1$.

From Table 2.1, we can define the ranges of solution as $x \leq 0$ OR $x \geq 1$, which gives the domain $(-\infty, 0] \cup [1, +\infty)$. Thus, we can shade the region of the Cartesian plane between $x = 0$ and $x = 1$.

(ii) **Key points:** $f(0) = 0$, $f(1) = 0$, and $f(-1) = \sqrt{2}$, so we can mark the points $A\ (0,0)$, $B\ (1,0)$, and $C\ \left(-1, \sqrt{2}\right)$ in the Cartesian plane.

As shown earlier, $f(x) = 0$ gives $x_1 = 0$ and $x_2 = 1$. Thus, two intersection points exist between the function and the x-axis: A and B, which can be marked in the Cartesian plane.

(iii) **Symmetry:** A possible axis or point of symmetry can be sought to aid the function sketching.

(iv) **Sign:**

(a) If n is **even**, the irrational function is always **positive**.
(b) If n is odd, the inequality $R(x) > 0$ must be solved.

Solution:

(iii) **Symmetry:** Substituting $-x$ in the function equation gives $y = \sqrt{(-x)^2 - (-x)} = \sqrt{x^2 + x}$ which is $\neq f(x)$ and $\neq -f(x)$, thus the function is neither even nor odd. Alternatively, we can look for either an axis of symmetry of the type of $x = a$ or a point of symmetry $F(a, b)$.

(iv) **Sign:** We can shade the half-plane of negative y values, as the n of the root is even and the irrational function is thus always positive.

(v) **Limits and asymptotes:** Techniques for polynomial and rational functions should be used, bearing in mind that $\sqrt[n]{\infty} = \infty$.

If $\lim_{x\to\pm\infty} f(x) = \pm\infty$, possible slant asymptotes should be identified.

Warning! Sometimes, in order to calculate the limits of some irrational functions as x tends to $\pm\infty$, it can be helpful to simplify the root. In particular, if n is even, $\sqrt[n]{x^{m\cdot n}} = |x^m|$ which corresponds to $+x^m$ as x tends to $+\infty$, or $-x^m$ as x tends to $-\infty$, while if n is odd, $\sqrt[n]{x^{m\cdot n}} = x^m$.

Solution:

(v) **Limits:** First let us calculate $\lim_{x\to+\infty} \sqrt{x^2 - x}$:

$$\lim_{x\to+\infty} \sqrt{x^2 - x} = \sqrt{(+\infty)^2 - (+\infty)} = \sqrt{+\infty - \infty}$$

where $+\infty - \infty$ is an indeterminate form to be solved. We can then factorise the leading term of the polynomial to solve this:

$$\lim_{x\to+\infty} \sqrt{x^2\left(1 - \frac{1}{x}\right)} = \sqrt{+\infty \cdot (1 - \frac{1}{+\infty})} = \sqrt{+\infty} = +\infty$$

Thus, no horizontal asymptote for the function is possible.

We should look for a slanting asymptote of the type $y = mx + q$:

$$m = \lim_{x\to+\infty} \frac{\sqrt{x^2 - x}}{x} = \frac{+\infty}{+\infty}$$

which constitutes an indeterminate form to be solved. We can factorise the numerator and then simplify the fraction:

$$m = \lim_{x\to+\infty} \frac{\sqrt{x^2\left(1 - \frac{1}{x}\right)}}{x} = \lim_{x\to+\infty} \frac{|x| \cdot \sqrt{(1 - \frac{1}{x})}}{x}$$

$$= \lim_{x\to+\infty} \frac{x \cdot \sqrt{(1 - \frac{1}{x})}}{x}$$

as $|x| = +x$ if $x > 0$, as explained in the 'Warning!' above.

Finally, we simplify the fraction and get

$$m = \lim_{x\to+\infty} \sqrt{\left(1 - \frac{1}{x}\right)} = \sqrt{\left(1 - \frac{1}{+\infty}\right)} = 1$$

Now we calculate $q = \lim_{x\to+\infty} \sqrt{x^2 - x} - x = +\infty - \infty$ which is an indeterminate form to be solved.

We can thus multiply and divide the expression by $\left(\sqrt{x^2 - x} + x\right)$:

$$q = \lim_{x\to+\infty} \left(\sqrt{x^2 - x} - x\right) \cdot \frac{\sqrt{x^2 - x} + x}{\sqrt{x^2 - x} + x} = \lim_{x\to+\infty} \frac{x^2 - x - x^2}{\sqrt{x^2 - x} + x}$$

$$= \lim_{x\to+\infty} \frac{-x}{\sqrt{x^2 - x} + x}$$

We can then factorise the denominator, simplify, and substitute $x = +\infty$:

$$q = \lim_{x\to+\infty} \frac{-x}{\sqrt{x^2 \cdot (1 - \frac{1}{x})} + x} = \lim_{x\to+\infty} \frac{-x}{x \cdot \left(\sqrt{1 - \frac{1}{x}} + 1\right)}$$

$$= \frac{-1}{1+1} = -\frac{1}{2}$$

Thus, the slanting asymptote of the function as x approaches $+\infty$ has equation $y = x - \dfrac{1}{2}$.

In a similar way, it can be demonstrated that $y = -x + \dfrac{1}{2}$ is the slanting asymptote to the function as x approaches $-\infty$.

Reminder: When we write $f(\pm\infty) = \pm\infty$, we really mean $\lim_{x\to\pm\infty} f(x) = \pm\infty$, but '± infinity' ($\pm\infty$) does not lie in the domain of the function.

(vi) **Continuity:** Irrational functions are continuous in all closed intervals $[a, b]$ of their domain.

If the n root is even, an endpoint discontinuity can be observed for those x values giving $f(x) = 0$.

Generally, all specific x values for which the expression $R(x)$ within the irrational function $f(x)$ is undefined can be points of discontinuity to be determined.

(vii) **Co-domain:** The exact range of y values can be defined only after we have studied differentiation in Chapter 5.

Solution:

(vi) **Continuity:** The irrational function $y = \sqrt{x^2 - x}$ is continuous in all closed intervals $[a, b]$ of the function's domain $(-\infty, 0] \cup [1, +\infty)$.

For $x = 0$ and $x = 1$, we have two points of endpoint discontinuity.

Using all the information in the above sections about the function $y = \sqrt{x^2 - x}$, we can now sketch it in a Cartesian plane as shown in the graph below.

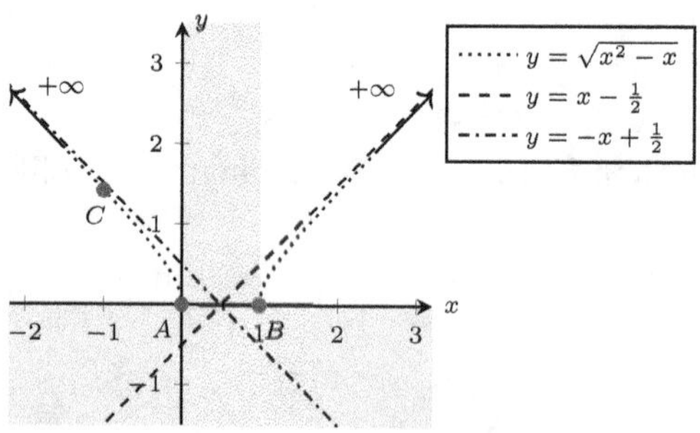

Example in Biochemistry: In order to find the three-dimensional structure of proteins and their complexes, single particles are cryocooled to $77\,\mathrm{K}$ and placed in an electron microscope, and the scattered electrons are detected to give images of the molecules.

The electrons are accelerated in the microscope to very high velocities, v, governed by a relativistically invariant form of Einstein's relation, where E is the energy of the electrons which are produced:

$$E = m_0 \cdot c^2 \sqrt{\frac{c^2}{c^2 - v^2}}$$

where m_0 is the rest or invariant mass (independent of the overall motion of the system) of the particle used, in this case an electron (*i.e.* $0.511\,\mathrm{MeV\,c^{-2}}$).

If the speed of light, c, is rounded to $300\,\mathrm{Mm\,s^{-1}}$, sketch the function $E = f(v)$.

Solution: After we have substituted the values of m_0 and c into the relativistically invariant form of Einstein's relation, we obtain the function

$$E = f(v) = 0.511\sqrt{\frac{300^2}{300^2 - v^2}}$$

(i) **Domain:** We can write the expression $\dfrac{300^2}{300^2 - v^2} \geq 0$.

The rational inequality gives the range of solutions as $-300 < v < 300$, which in turn gives the domain of the function: $[0, 300)$, as the physical variable v can only be ≥ 0.

(ii) **Key points:** $f(0) = 0.511$: the function passes through $A\,(0, 0.511)$.

To solve $f(v) = 0$, we can equate the numerator of the fraction: $300^2 = 0$ which is impossible. Thus, no intersection points exist between the function and the x-axis.

(iv) **Sign:** As n of the equation's root $= 2$, it is even, and the irrational function is always > 0. Thus, we can shade the area above the x-axis for $x > 0$.

(v) **Limits and asymptote:** Let us calculate the following limit:

$$\lim_{v \to 300^-} 0.511\sqrt{\frac{300^2}{300^2 - v^2}} = 0.511\sqrt{\frac{300^2}{300^2 - (300^-)^2}}$$
$$= 0.511\sqrt{\frac{300^2}{0^+}} = 0.511\sqrt{+\infty} = +\infty$$

thus $v = 300$ is a vertical asymptote to $f(v)$.

(vi) **Continuity:** $f(v)$ is continuous at every point in the domain.

(vii) **Co-domain:** The range of y values of $f(v)$ is therefore $[0, +\infty)$.

Other information, such as symmetry, are not needed to solve the problem.

Sketching the function $0.511\sqrt{\frac{300^2}{300^2 - v^2}}$ using the information above gives the following graph.

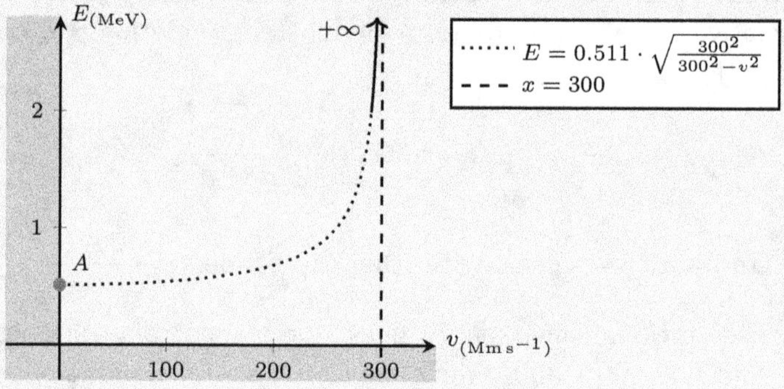

4.9 Exercises

(1) Given the following pairs of functions, $f(x)$ and $g(x)$, find the composite functions $g \circ f = g(f(x))$ and $f \circ g = f(g(x))$.

(a) $f(x) = \dfrac{1}{x^2 - x}$, $g(x) = \dfrac{1}{x - 3}$

(b) $f(x) = (2 - x)^2$, $g(x) = \sqrt{x + 3}$

(c) $f(x) = \dfrac{\sqrt{1 + 4x^2}}{2x - 3}$, $g(x) = \dfrac{7x}{3 - x}$

(2) Sketch the following functions using geometric transformations of appropriate elementary functions.

(a) $y = \dfrac{1}{x+3} - 2$ (b) $y = \sqrt[3]{x-4} + 5$ (c) $y = (1+x)^3 - 3$

(d) $y = \left|\dfrac{1}{-x} + 2\right|$ (e) $y = 1 - \sqrt{|3\,(x-3)|}$ (f) $y = -\,|x-2|^3 + 3$

(3) Sketch the following polynomial functions.

(a) $y = (x-4)\left(2+x^2\right)$ (b) $y = 3x\left(x^4 - 4\right)$

(c) $y = 16 - x^4$

(4) In a biochemical experiment on a sample of protein, the variations (y) of one of its properties compared to a standard value $y_0 = 0$ were measured for increasing values of the parameter x, as follows:

x	1.5	2.2	3.0	4.5
y	−13.125	9.240	75.000	365.625

It is thought that this result might obey a relationship of the form $y = Ax^3 + Bx$, where A and B are constants.

(a) Estimate values for A and B. For which values of x is $y = 0$?

(b) In what range of x values is y the lowest compared to the standard value y_0?

(c) For what range of x values is y always greater than y_0?

(d) What are the values of y when $x = 1$ and $x = 6$?

(e) What is the potential value of y if the parameter x is increased to a very large value?

(f) Sketch a graph showing y as a function of the variable x.

(5) Sketch the following rational functions.

(a) $y = \dfrac{x-4}{x^2+4x+3}$ (b) $y = \dfrac{x^3}{x^2+3}$ (c) $y = \dfrac{3}{x\,(x^2+2)}$

(d) $y = \dfrac{x^3+x^2}{9-x^2}$ (e) $y = \dfrac{6+5x+x^2}{x-6}$ (f) $y = \dfrac{\left(x^2-9\right)2x^2}{3-x}$

(6) The Lennard-Jones potential energy between two non-polar atoms may be simply given by the equation

$$V(R) = \frac{A}{R^{12}} - \frac{B}{R^6}$$

where A and B are constants, $V(R)$ is the potential energy, measured in Joules, and R is the internuclear distance measured in Å.

(a) For which values of R is $V(R)$ positive, negative, and zero?

(b) Sketch a graph showing the potential energy between the two atoms as a function of R (in Å) given that $A = 7$ and $B = 4$.

(c) What is the potential energy between the two atoms at infinite separation?

(d) What would happen to the two atoms if they were brought very close together?

(e) What is the physical interpretation of the sign of $V(R)$?

(f) What are the dimensions ([Length], [Mass], [Time], [Temperature], [Electric Charge]) and units of the constants A and B?

(7) Sketch the following irrational functions.

(a) $y = \sqrt{x^2 - 4x + 3}$ (b) $y = \sqrt[3]{x^2 - 4}$ (c) $y = \sqrt{-x^2 + 5x}$

(8) In an experiment on a chemical reaction, the values of its percentage yield (y in %) were observed for different values of pH, as follows:

pH	5.5	9	13.5
y in %	23.98	56.75	54.54

It is thought that this result might obey a relationship of the form $y(\text{pH}) = 10 \cdot \sqrt{(\text{pH} - A)(B - \text{pH})}$, where A and B are positive integer constants.

(a) Estimate values for A and B.

(b) What are the values of y when pH $= 7$ and pH $= 11$? At what pH value is the percentage yield $= 0$?

(c) Sketch a graph showing the percentage yield of the reaction, y, as a function of the pH, the possible range of which is $[0, 14]$.

(d) Is there any symmetry possible for the $y(\text{pH})$ function? Can you suggest what the maximum value of the percentage yield might be?

Answers

(1) (a) $f(g(x)) = \dfrac{-x^2 + 6x - 9}{x - 4}$; $g(f(x)) = \dfrac{x^2 - x}{3x^2 - 3x - 1}$

(b) $f(g(x)) = x + 7 - 4\sqrt{x + 3}$; $g(f(x)) = \sqrt{x^2 - 4x + 7}$

(c) $f(g(x)) = \dfrac{(3 - x)\sqrt{9 - 6x + 197x^2}}{17x|x - 3|}$; $g(f(x)) = \dfrac{7\sqrt{1 + 4x^2}}{6x - 9 - \sqrt{1 + 4x^2}}$

(2)

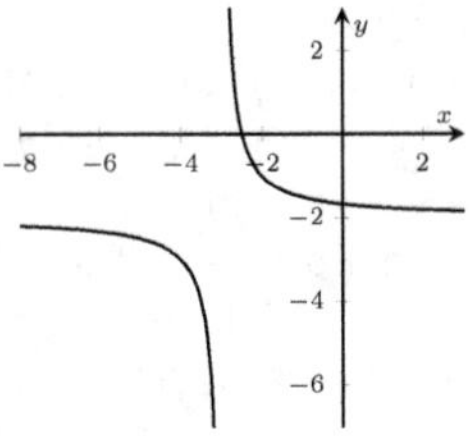

(a)

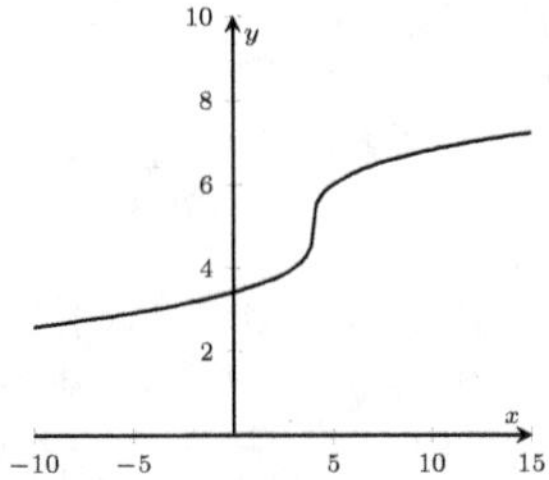

(b)

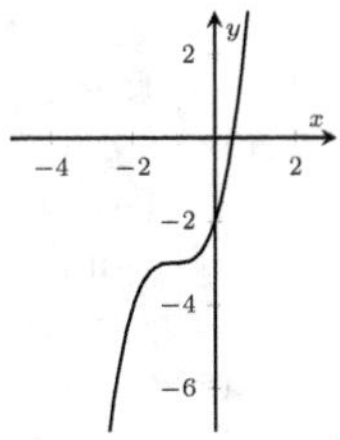

(c)

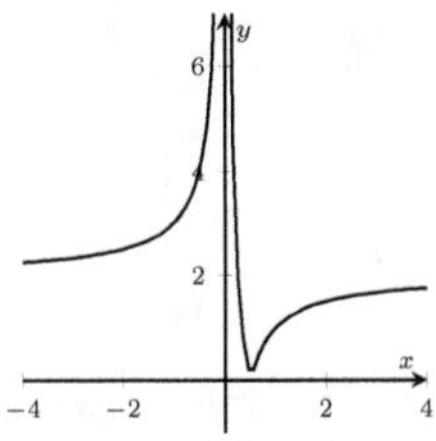

(d)

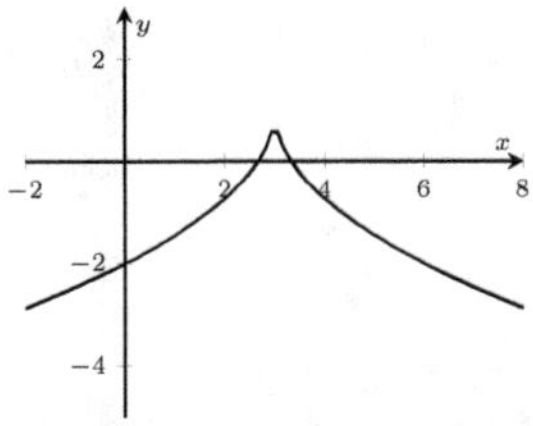

(e)

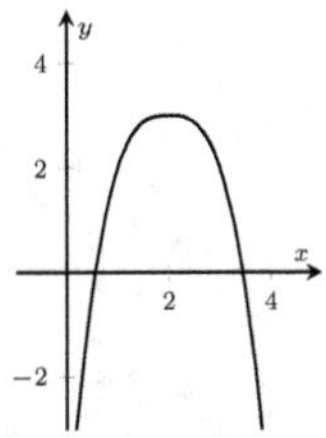

(f)

(3) *(a) Dom: $(-\infty,+\infty)$, Sym: $O'\left(\frac{4}{3},-\frac{272}{27}\right)$, Pos: $(4,+\infty)$, Asy: None, Dis: None

(b) Dom: $(-\infty,+\infty)$, Sym: axis $x=0$, Pos: $(-2,+2)$, Asy: None, Dis: None

(c) Dom: $(-\infty,+\infty)$, Sym: $O\,(0,0)$, Pos: $(-\sqrt{2},0)\cup(\sqrt{2},+\infty)$, Asy: None, Dis: None

(4) (a) $A=5$, $B=-20$, $x=0$, $x=2$ (b) $0<x<2$ (c) $x>2$

(d) $y=-15$, $y=960$ (e) $+\infty$

(f)

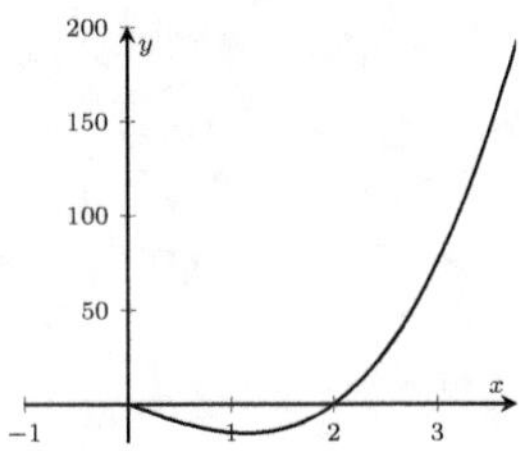

(5) *(a) Dom: $(-\infty,-3)\cup(-3,-1)\cup(-1,-\infty)$, Sym: None,
Pos: $(-3,-1)\cup(4,+\infty)$, Asy: $x=-3$, $x=-1$, $y=0$, Dis: $x=-3$, $x=-1$ infinite
(b) Dom: $(-\infty,+\infty)$, Sym: $O\,(0,0)$, Pos: $(0,+\infty)$, Asy: $y=x$, Dis: None
(c) Dom: $(-\infty,0)\cup(0,+\infty)$, Sym: $O\,(0,0)$, Pos: $(0,+\infty)$, Asy: axis $y=0$, Dis: $x=0$ infinite
(d) Dom: $(-\infty,-3)\cup(-3,3)\cup(3,+\infty)$, Sym: None, Pos: $(-\infty,-3)\cup(0,3)$, Asy: $x=-3$, $x=3$, $y=-x-1$, Dis:$x=-3$, $x=3$ infinite
(e) Dom: $(-\infty,6)\cup(6,+\infty)$, Sym: $O'\,(6,17)$, Pos: $(-3,-2)\cup(6,+\infty)$, Asy: $x=6$, $y=x+11$, Dis: $x=6$ infinite
(f) Dom: $(-\infty,3)\cup(3,+\infty)$, Sym: $O'\,(-1,-4)$, Pos: $(-\infty,3)$, Asy: None, Dis: $x=3$ removable

(6) (a) $V=0$ if $R=\sqrt[6]{\frac{A}{B}}$, $V>0$ if $R<\sqrt[6]{\frac{A}{B}}$, $V<0$ if $R>\sqrt[6]{\frac{A}{B}}$
(b)

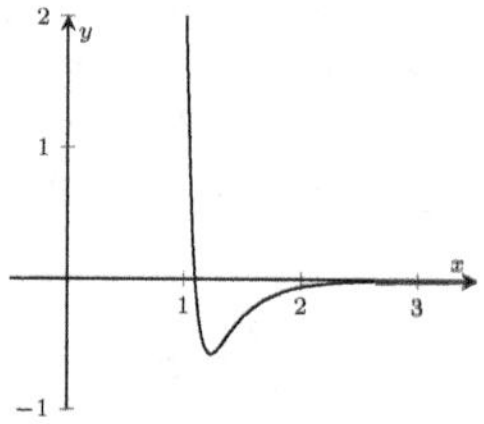

y-axis units are Joules, x axis units are Å

(c) $V=0$ (d) $V=+\infty$
(e) if $V>0$, repulsion; if $V<0$, attraction
(f) $[A]=[\mathrm{M}][\mathrm{L}]^{14}[\mathrm{T}]^{-2}$, $[B]=[\mathrm{M}][\mathrm{L}]^{8}[\mathrm{T}]^{-2}$

(7) *(a) Dom: $(-\infty,1]\cup[3,+\infty)$, Sym: axis $x=2$, Pos: $(-\infty,1)\cup(3,-\infty)$, Asy: $y=x-2$, $y=-x+2$, Dis: $x=+1,+3$ endpoint
(b) Dom: $(-\infty,+\infty)$, Sym: axis $x=0$, Pos: $(-\infty,-2)\cup(2,+\infty)$, Asy: None, Dis: None
(c) Dom: $[0,5]$, Sym: axis $y=\frac{5}{2}$, Pos: $(0,5)$, Asy: None, Dis: None

(8) (a) $A=5$, $B=17$ (b) $y=4.472$, $y=6$
(c)

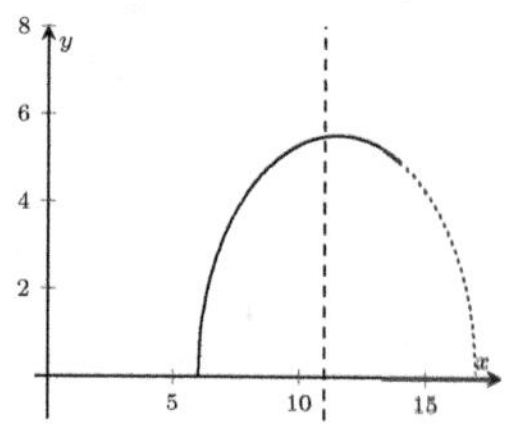

(d) $x=11$ axis of symmetry, y-maximum at $x=11$

*Abbreviations: Dom = domain; Sym = symmetry; Pos = function positive sign; Asy = asymptotes; Dis = discontinuity.

Chapter 5

Differentiation

Preamble

We often want to know about the rate at which one quantity, or one 'variable' as it is usually called in Mathematics, changes over time: the rate of disappearance of substrate over time in an enzyme reaction (*i.e.* how quickly is the substrate being converted into product?), the rate of decay of a radio-labelled substance, and the rate of bacterial growth over time.

To answer these questions, we need to introduce the concept of mathematical differentiation as applied to Biosciences. Knowing the rates of change allows us to calculate quantities of interest for these processes. For instance, for how long will my radioactive substance have an activity rate above a certain level, or when will the number of bacteria reach the required growth rate value for my experiment?

5.1 Gradient and Derivative

The gradient is known as the **rate of change** (of the variable y as a function of x) or the **derivative**, and the process of finding the gradient is known as **differentiation**.

The approximate gradient of the curve $y = f(x)$ at the point $P(x, y)$ is denoted in several different ways. The most common is $\dfrac{\Delta y}{\Delta x}$.

We begin by noting that the gradient between any two points $P_1(x_1, y_1)$ and $P_2(x_2, y_2)$ on the curve $y = f(x)$ is always given by

$$\text{gradient} = \frac{y_2 - y_1}{x_2 - x_1} = \frac{\Delta y}{\Delta x}$$

Given the function $y = f(x)$, we use $P(x, y) = P[x, f(x)]$ to represent the point on the curve $y = f(x)$ at which we wish to calculate the gradient, and we use $Q(x + \Delta x, y + \Delta y) = Q[x + h, f(x + h)]$ to represent a nearby point on this curve, where the value h represents a small change in the value of x.

If we draw a right angle triangle PQR, we obtain $PR = \Delta x = h$ and $RQ = \Delta y$ (Figure 5.1).

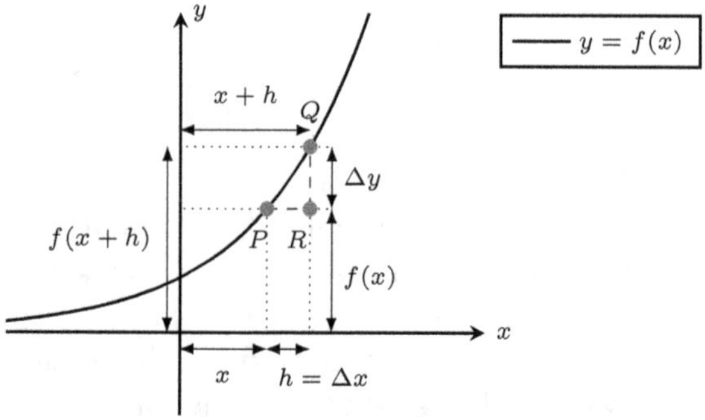

Figure 5.1: Gradient

Thus, in the triangle PQR, the gradient of the straight line passing through the points P and Q is $\dfrac{\Delta y}{\Delta x}$.

Let point Q approach P so that Δx and Δy get smaller. ΔPQR then shrinks to a point P. As the chord PQ gets reduced until the points P and Q coincide, the chord becomes a straight line which "just touches" the curve at point P. This type of straight line is called a **tangent**.

So the gradient of the tangent at P is the limit of $\dfrac{\Delta y}{\Delta x}$ as Δx and Δy approach zero. The gradient of the straight line passing through the points P and Q is thus given by

$$\text{gradient of } PQ = \frac{\Delta y}{\Delta x} = \frac{f(x + h) - f(x)}{h} = m$$

with the letter m also being used to identify the gradient of a straight line (see Section 3.2.1).

To obtain the **exact value** of the gradient at P, we have to evaluate the behaviour of this expression as the value h approaches 0.

We define the **derivative** $f'(x)$ at P as the **limit of the approximating gradient** of function $y = f(x)$:

$$f'(x) = \lim_{h \to 0} (\text{gradient of line } PQ_h) = \lim_{h \to 0} \frac{\Delta y}{\Delta x} = \lim_{h \to 0} \frac{f(x+h) - f(x)}{h}$$

A function $f(x)$ is defined as **differentiable** if this limit exists and is finite.

Note that not all functions are differentiable (*e.g.* $|x|$ at $x = 0$).

The derivative of a function $y = f(x)$ is denoted in several different ways. The three most common are as follows:

$$y' \ , \ f'(x) \text{ or } \frac{dy}{dx}$$

Important note: The prefix d **cannot** be 'cancelled', as $\frac{d}{dx}$ is called an **operator** since it acts on y, and is said verbally as 'd by dx', NOT as 'd divided by dx'.

Example: Calculate the approximate gradient of $y = f(x) = x^3 + 2$ at the point $P(x, y) = (x, x^3 + 2)$. Then compute the corresponding derivative with respect to x by taking the limit as $\lim\limits_{h \to 0}$.

Solution: We can initially use the definition of gradient given above:

$$\frac{\Delta y}{\Delta x} = \frac{f(x+h) - f(x)}{h} = \frac{[(x+h)^3 + 2] - [x^3 + 2]}{h}$$

Then, by expanding brackets, we obtain

$$\frac{\Delta y}{\Delta x} = \frac{[(x+h)(x^2 + h^2 + 2hx) + 2] - [x^3 + 2]}{h}$$
$$= \frac{[x^3 + 3x^2h + 3xh^2 + h^3 + 2] - [x^3 + 2]}{h} = \frac{3x^2h + 3xh^2 + h^3}{h}$$

Then, by dividing the top and bottom by h, we can remove the denominator $\frac{\Delta y}{\Delta x} = \frac{3x^2h + 3xh^2 + h^3}{h} = 3x^2 + 3xh + h^2$ to obtain the gradient.

Now let $h \to 0$ so that

$$\frac{dy}{dx} = \lim_{h\to 0}\frac{\Delta y}{\Delta x} = \lim_{h\to 0}(3x^2 + 3xh + h^2) = 3x^2 + 0 + 0 = 3x^2$$

Example in Chemistry: Photon energy is directly proportional to its frequency according to the well-known Planck–Einstein relation: $E = h_P \cdot f$, where E is energy in J, h_P is Planck's constant, *i.e.* $6.62606957 \cdot 10^{-34}\,\mathrm{m^2kg\,s^{-1}}$, and f is frequency in Hz.

Additionally, $E = \dfrac{h_P \cdot c}{\lambda}$, where λ is the photon's wavelength in m and c is the speed of light in vacuum, *i.e.* $3 \cdot 10^8\mathrm{m\,s^{-1}}$.

Calculate the approximating gradient of the second function at the point $P(\lambda, E)$ and then the corresponding derivative with respect to λ.

Solution: We can initially use the definition of gradient given above:

$$\frac{\Delta E}{\Delta\lambda} = \frac{E(\lambda + h) - E(\lambda)}{h} = \frac{\dfrac{h_P \cdot c}{\lambda + h} - \dfrac{h_P \cdot c}{\lambda}}{h}$$

Then, by simplifying the two terms in the numerator and rearranging the expression, we obtain

$$\frac{\Delta E}{\Delta\lambda} = \frac{h_P \cdot c \cdot \lambda - h_P \cdot c \cdot \lambda - h_P \cdot c \cdot h}{(\lambda + h)\lambda \cdot h} = -\frac{h_P \cdot c \cdot h}{(\lambda + h) \cdot \lambda \cdot h}$$

By dividing top and bottom by h, we obtain $\dfrac{\Delta E}{\Delta\lambda} = -\dfrac{h_P \cdot c}{\lambda^2 + h \cdot \lambda}$.

Now let $h \to 0$ so that $\lim\limits_{h\to 0}\dfrac{\Delta E}{\Delta\lambda} = \dfrac{dE}{d\lambda} = -\dfrac{h_P \cdot c}{\lambda^2}$.

5.2 Applications of Derivatives

5.2.1 Tangent line

The **derivative** of y at x, written by $\dfrac{dy}{dx} = f'(x)$, represents the **gradient** of the curve at (x, y) which is also the rate of change of y with respect to x at (x, y).

Thus, to find the equation of the **straight line tangent** to the curve of function $f(x)$ at point $P(x_P, y_P)$, we can use the equation $y - y_P = m \cdot (x - x_P)$ (see Section 3.2.2) and then since the gradient $m = \dfrac{y - y_P}{x - x_P} = f'(x_P)$ we get:

$$y - f(x_P) = f'(x_P) \cdot (x - x_P) \tag{5.1}$$

Example: Find the equation of the straight line tangent to the graph of the function $y = f(x) = x^3 + 2$ at the point P with $x_P = 1$. Then sketch the function and the line in a Cartesian plane.

Solution: We can use the derivative of $f(x)$ obtained in the previous example, $y' = 3x^2$, to calculate $f'(x_P) = f'(1) = 3$.

Then, after finding that $f(x_P) = f(1) = 3$, we can use equation (5.1) to define the equation of the straight line tangent to the function $y = f(x)$ at the point P:

$$y - f(x_P) = f'(x_P) \cdot (x - x_P) \Longleftrightarrow y - 3 = 3 \cdot (x - 1)$$

which gives $y = 3x$.

Using the rules given in Section 4.7, we can sketch both the function and its tangent line at the point $P(1, 3)$.

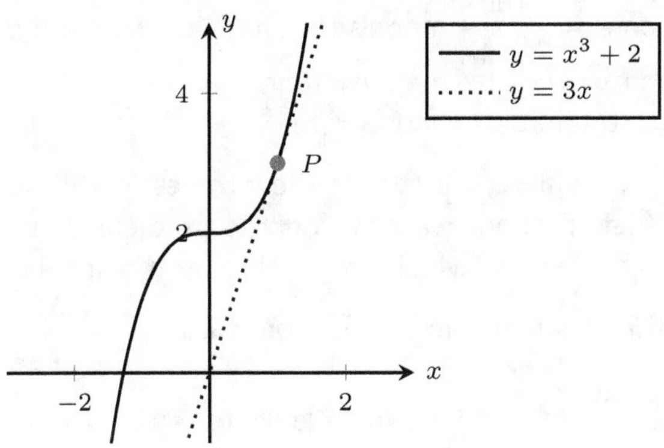

Example in Biochemistry: In a biochemical experiment, $100\,\mu\text{mol}$ of a chemical substrate, s, is metabolised by an enzyme over time. The amounts of the substrate remaining at specific times are reported in the following table:

$t_{(\text{min})}$	0.00	0.25	0.50	1.00	2.00	3.00	4.00
$s_{(\mu\text{mol})}$	100.0	66.7	50.0	33.3	20.0	14.3	11.1

The graph below shows the amount of substrate metabolised over time for the enzyme-controlled reaction:

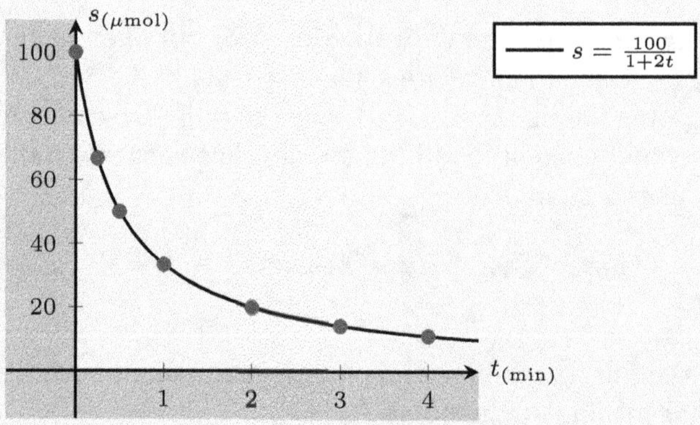

The function $s = \dfrac{100}{1+2t}$ models the shape of the trend of substrate consumption by the enzyme over time.

Estimate the initial reaction rate v_i.

Solution: Two different methods allow the estimation of v_i.

In the first method, we can find the gradient of the equation of the straight line, m, which passes through points $A\,(0, 100)$ and $B\,(0.25, 66.7)$, as explained in Section 3.2.1: $m = \dfrac{\Delta s}{\Delta t} = \dfrac{s_B - s_A}{t_B - t_A}$ $= \dfrac{66.7 - 100.0}{0.25} = -133.3$, which gives $v_i = 133.3\,\mu\text{mol}\,\text{min}^{-1}$.

This value is approximate as it is calculated using two distinct points on the graph.

Alternatively, we can obtain the exact value of the gradient of the function at point $x = 0$ by finding the derivative.

In order to calculate v_i, we differentiate the function $s = \dfrac{100}{1+2t}$ and then evaluate the derivative at $t = 0$:

$$\frac{\Delta s}{\Delta t} = \frac{\dfrac{100}{1+2(t+h)} - \dfrac{100}{1+2t}}{h} = \frac{-\dfrac{200h}{4t^2+4th+4t+2h+1}}{h}$$
$$= -\frac{200h}{(4t^2+4th+4t+2h+1)\cdot h}$$

The h can be cancelled from the top and bottom. Now, let $h \to 0$ so that

$$\lim_{h\to 0}\frac{\Delta s}{\Delta t} = \lim_{h\to 0} -\frac{200}{4t^2+4th+4t+2h+1} = -\frac{200}{4t^2+4t+1}$$
$$= -\frac{200}{(2t+1)^2} = \frac{ds}{dt} = f'(t)$$

Note also that we can use equation (5.1) to define the equation of the straight line tangent to the function $s = f(t)$ at any point, but specifically here at the point $t_P = 0$: $s - f(t_P) = f'(t_P)\cdot(t - t_P)$ $\iff s - 100 = -200\cdot(t-0)$, which gives $s = -200t + 100$, as shown in the graph below.

The gradient of the tangent gives a more accurate v_i than the value obtained by taking two separated points on the curve.

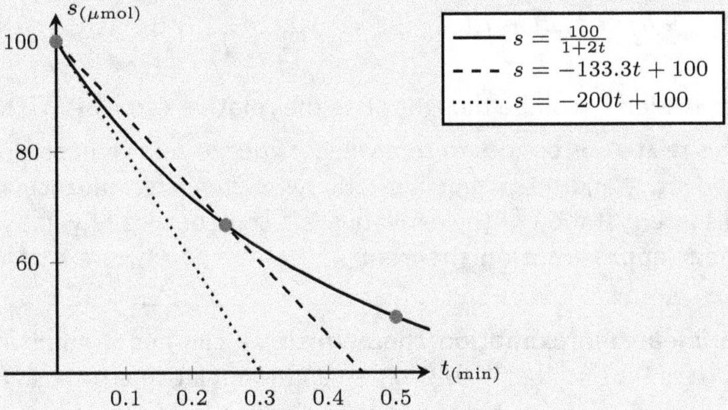

Thus, the exact value of the initial velocity of the enzyme v_i corresponds to $200\,\mu\text{mol}\,\text{min}^{-1}$, which is 50% higher than the earlier estimate for the value of v_i.

We can see from this example that 'differentiation' is a very useful tool in Biosciences.

5.2.2 Linear approximation

The **derivative** of y at x, written $dy/dx = f'(x)$, can represent different mathematical concepts:

- the **gradient** of the curve at (x, y);
- the **gradient** of the tangent line of the curve at (x, y);
- the **rate of change** of y with respect to x at (x, y).

We can therefore think of the derivative $f'(x)$ as the quantity by which we need to multiply a small change in x in order to find the corresponding change in y.

Using h to denote the small change in x, we can write it as

$$f(x+h) \approx f(x) + f'(x) \cdot h \tag{5.2}$$

This formalism accounts for nearly all of the change. The only error left, which we will call the **'infinitesimal' part**, is usually of a smaller order of magnitude than h itself.

The true value is

$$f(x+h) \approx f(x) + f'(x) \cdot h + \epsilon \quad \text{where } \frac{\epsilon}{h} \to 0 \quad \text{as} \quad h \to 0$$

The expression $\frac{\epsilon}{h}$ can be thought of as the **relative error in** h. The fact that this relative error goes to zero when h goes to zero characterises the first derivative and often simplifies otherwise cumbersome calculations.

This interpretation of the derivative is a consequence of what we call the **linear approximation theorem**.

The linear approximation theorem: If we can find a number M with $f(x+h) = f(x) + M \cdot h + \epsilon_h$ and the relative error $\frac{\epsilon_h}{h} \to 0$ as $h \to 0$, then M **must be** the **derivative** $f'(x)$.

Proof: If we have M and ϵ_h as given above, then we can write

$$\frac{f(x+h)-f(x)}{h} = \frac{f(x)+M\cdot h+\epsilon_h - f(x)}{h} = \frac{M\cdot h+\epsilon_h}{h} = M + \frac{\epsilon_h}{h}$$

Taking the limit of both sides as $h \to 0$ we see that the left-hand side converges to $f'(x)$ while the right-hand side converges to $M+0 = M$. These two are therefore the same. That is, $M = f'(x)$.

Example: Estimate $\sqrt{5218}$ using the linear approximation theorem.

Solution: First we look for a number which, when squared, gives a convenient number nearby, such as $4900 = 70^2$, and work from there. The function involved is $f(x) = \sqrt{x}$, and we can calculate its derivative by using the definition in Section 5.1:

$$\frac{\Delta y}{\Delta x} = \frac{f(x+h)-f(x)}{h} = \frac{\sqrt{x+h}-\sqrt{x}}{h}$$

In order to calculate the limit of $\dfrac{\Delta y}{\Delta x}$ as $h \to 0$, we can multiply both the numerator and denominator by $\sqrt{x+h}+\sqrt{x}$, thereby obtaining

$$\frac{\Delta y}{\Delta x} = \frac{\sqrt{x+h}-\sqrt{x}}{h}\cdot\frac{\sqrt{x+h}+\sqrt{x}}{\sqrt{x+h}+\sqrt{x}} = \frac{x+h-x}{h\cdot(\sqrt{x+h}+\sqrt{x})}$$

x cancels out in the numerator and h cancels top and bottom.

Now let $h \to 0$ so that $\lim\limits_{h\to 0}\dfrac{\Delta y}{\Delta x} = \lim\limits_{h\to 0}\dfrac{1}{\sqrt{x+h}+\sqrt{x}} = \dfrac{1}{2\sqrt{x}} = \dfrac{dy}{dx}$.

Since $5218 = 4900 + 318$, we can let $x = 4900$ and $h = 318$. Now we can use equation 5.2 to see that:

$$\sqrt{5218} = \sqrt{4900+318} \approx \sqrt{4900} + \frac{1}{2\sqrt{4900}}\cdot 318 = 70 + \frac{318}{2\times 70}$$

$$= 70 + \frac{318}{140} = 70 + 2.27 = 72.27$$

Observation: This technique will give an over-estimate or an under-estimate, according to whether the graph of $y = f(x)$ is downwardly or upwardly concave, respectively (see Section 4.7). The graph of $y = \sqrt{x}$

is downwardly concave (see Section 4.7), so the estimate 72.27 is just a little bigger than the true value of $\sqrt{5218}$ which is 72.24.

Disclaimer: As mentioned above, this technique accounts for nearly all of the change. In our example, h was about 3×10^2, while the remaining error in the approximation was 3×10^{-2}.

5.3 Calculation of Derivatives

5.3.1 Derivatives of power-of-x functions

By using the techniques such as those applied in the two examples in Section 5.1, the connection between powers of x and their corresponding derivatives can be generalised as follows:

$$\text{if } y = a \cdot x^n \text{ with } a \in \mathbb{R}, \text{ then } \frac{dy}{dx} = a \cdot n \cdot x^{n-1} \tag{5.3}$$

Important note:

If $y = a = a \cdot x^0$, then $\dfrac{dy}{dx} = 0$.

Proof: Given the function $y = a \cdot x^n$ with $n \in \mathbb{N}$, we use the limit of the approximating gradient of the function $y = f(x)$:

$$\frac{dy}{dx} = \lim_{h \to 0} \frac{f(x+h) - f(x)}{h} = \lim_{h \to 0} \frac{a \cdot (x+h)^n - a \cdot x^n}{h}$$

We can initially factorise the scalar constant a in the numerator and put it before the limit notation according to the properties of limits (Section 4.4.1):

$$\frac{dy}{dx} = a \cdot \lim_{h \to 0} \frac{(x+h)^n - x^n}{h}$$

We can then use the Binomial Expansion (Section 2.1.1) on the numerator:

$$(x+n)^n - x^x = x^n + nx^{n-1}h + \frac{n(n-1)}{2!}x^{n-2}h^2 + \frac{n(n-1)(n-2)}{3!}x^{n-3}h^3 + \cdots + h^n - x^n$$

The first and last terms cancel, so we obtain

$$\frac{dy}{dx} = a \cdot \lim_{h \to 0} \frac{nhx^{n-1} + \frac{n(n-1)}{2!}x^{n-2}h^2 + \frac{n(n-1)(n-2)}{3!}x^{n-3}h^3 + \cdots + h^n}{h}$$

We can then factorise h in the numerator, so h cancels top and bottom:

$$\frac{dy}{dx} = a \cdot \lim_{h \to 0} nx^{n-1} + \frac{n(n-1)}{2!}x^{n-2}h + \frac{n(n-1)(n-2)}{3!}x^{n-2}h^2 + \cdots + h^{n-1}$$

which gives $\frac{dy}{dx} = anx^{n-1}$ upon calculation of the limit as $h \to 0$.

Example: Find the derivative of $y = x^3 + 2x^2 + 4$.

Solution: According to the formula $\frac{\Delta y}{\Delta x} = \frac{f(x+h) - f(x)}{h}$, we can take the point $(x, y) = (x, x^3 + 2x^2 + 4)$ and calculate $\frac{\Delta y}{\Delta x}$:

$$\begin{aligned}\frac{\Delta y}{\Delta x} &= \frac{[(x+h)^3 + 2(x+h)^2 + 4] - [x^3 + 2x^2 + 4]}{h} \\ &= \frac{[x^3 + 3x^2h + 3xh^2 + h^3 + 2x^2 + 4xh + 2h^2 + 4] - [x^3 + 2x^2 + 4]}{h} \\ &= \frac{3x^2h + 3xh^2 + h^3 + 4xh + 2h^2}{h}\end{aligned}$$

h in the numerator and denominator then cancels, giving

$$3x^2 + 3xh + 4x + 2h + h^2$$

Now let $h \to 0$ so that $\lim_{h \to 0} \frac{\Delta y}{\Delta x} = 3x^2 + 4x = \frac{dy}{dx}$, which corresponds to the formula $\frac{dy}{dx} = anx^{n-1}$ applied to each term of the polynomial function.

The formula $\frac{dy}{dx} = anx^{n-1}$ can be generalised to differentiate $y = a \cdot x^n$ with all $n \in \mathbb{Q}$, *i.e.* all rational numbers.

Example: Using equation (5.3), find the derivative of $y = \dfrac{2}{\sqrt{x}}$.

Solution: $y = \dfrac{2}{\sqrt{x}}$ can also be expressed as $y = 2 \cdot \dfrac{1}{\sqrt{x}} = 2 \cdot x^{-\frac{1}{2}}$, with $a = 2$ and $n = -\dfrac{1}{2}$.

Therefore, $\dfrac{dy}{dx} = anx^{n-1} = 2 \cdot \left(-\dfrac{1}{2}\right) \cdot x^{-\frac{3}{2}} = -\dfrac{1}{x^{\frac{3}{2}}} = -\dfrac{1}{\sqrt{x^3}}$.

Example in Medicine: In order to maintain an adequate interstitial homeostasis (*i.e.* the proper nutritional environment surrounding all cells in your body), it is necessary that blood flows almost continuously through each of the millions of capillaries in the body.

The French scientist J. L. Poiseuille (1799–1869), in an attempt to understand the flow of blood which is often a turbulent fluid, derived an equation describing how the resistance R to flow through a cylindrical tube or vessel depends on several factors: $R = \dfrac{8\eta l}{\pi r^4}$, where r is the vessel radius, l is its length, and η is the viscosity of the fluid (blood).

Differentiate Poiseuille's equation with respect to the variable r to find the rate of change of R with r.

Solution: Poiseuille's equation can be expressed as $R = \dfrac{8\eta l}{\pi} \cdot r^{-4}$, where $\dfrac{8\eta l}{\pi}$ is a constant and r^{-4} is a 'power-of-x' function.

Thus, using equation (5.3):

$$\frac{dR}{dr} = \frac{8\eta l}{\pi} \cdot (-4) \cdot r^{-4-1} = -\frac{32\eta l}{\pi} \cdot r^{-5}$$

5.3.2 Rules for differentiating functions

For the derivative of $u = f(x)$, we use several different notations interchangeably: the first uses the notation $f(x)$ and $g(x)$ and can be traced back to *Newton*, and the second uses the notations u and v, and is due to *Leibniz*. For consistency, we will use Newton's notation.

Rule A. Sums or Differences: $\boxed{(\mathbf{f(x) \pm g(x)})' = \mathbf{f'(x) \pm g'(x)}}$

Example: Differentiate the following functions and their sum: $y_1 = 5x^4 + x^{\frac{1}{2}}$ and $y_2 = x - x^2$.

Solution: $\dfrac{dy_1}{dx} = 20x^3 + \dfrac{1}{2}x^{-\frac{1}{2}}$ and $\dfrac{dy_2}{dx} = 1 - 2x$.

For $\dfrac{d(y_1 + y_2)}{dx}$, we can add these derivatives together:

$$\frac{d(y_1 + y_2)}{dx} = 20x^3 + \frac{1}{2}x^{-\frac{1}{2}} + 1 - 2x$$

Rule B. Scalar Multiple: $\boxed{(\mathbf{a \cdot f(x)})' = \mathbf{a \cdot f'(x)}}$

Example: Differentiate the following function: $y = 4 \cdot \sqrt{x^3}$.

Solution: Since $y = 4 \cdot \sqrt{x^3} = 4 \cdot x^{\frac{3}{2}}$, $\dfrac{dy}{dx} = 4 \cdot \dfrac{3}{2}x^{\frac{1}{2}} = 6\sqrt{x}$.

Example in Chemistry: The Lennard-Jones potential energy V between two non-polar atoms with an internuclear distance r measured in Å ($10^{-10}\,\mathrm{m}$) is given by $V(r) = 4\epsilon\left[\left(\dfrac{\sigma}{r}\right)^{12} - \left(\dfrac{\sigma}{r}\right)^{6}\right]$, where ϵ is the depth of the potential well (usually known as the 'dispersion energy') and σ is the distance at which the particle–particle potential energy V is zero (often known as the 'size of the particle').

Find the derivative of V with respect to r.

Solution: We can initially write the equation as follows:

$$V(R) = 4\epsilon\left(\sigma^{12} \cdot r^{-12} - \sigma^{6} \cdot r^{-6}\right)$$

Using Rules **B** and **A** sequentially, we obtain

$$\frac{dV}{dr} = 4\epsilon\left[\sigma^{12} \cdot (-12) \cdot r^{-13} - \sigma^{6} \cdot (-6) \cdot r^{-7}\right] = 4\epsilon\left(-12\frac{\sigma^{12}}{r^{13}} + 6\frac{\sigma^{6}}{r^{7}}\right)$$

Rule C. Products: $\boxed{(\mathbf{f}(\mathbf{x}) \cdot \mathbf{g}(\mathbf{x}))' = \mathbf{f}'(\mathbf{x}) \cdot \mathbf{g}(\mathbf{x}) + \mathbf{f}(\mathbf{x}) \cdot \mathbf{g}'(\mathbf{x})}$

Proof: Use the linear approximation theorem: the idea here is to change the value of x by a small quantity h, and then to see by what we must multiply h to account for all but the 'infinitesimal' parts of the change in y: ϵ_1 and ϵ_2 for $f(x)$ and $g(x)$, respectively.

From equation (5.2), we know that $f(x+h) = f(x)+h\cdot f'(x)+\epsilon_1$ and $g(x+h) = g(x) + h \cdot g'(x) + \epsilon_2$ so

$$\begin{aligned} f(x+h)\cdot g(x+h) &= (f(x) + f'(x)\cdot h + \epsilon_1)\cdot(g(x) + g'(x)\cdot h + \epsilon_2) \\ &= f(x)\cdot g(x) + f'(x)\cdot h\cdot g(x) + f(x)\cdot g'(x)\cdot h + K + \epsilon_3 \\ &= f(x)\cdot g(x) + (f'(x)\cdot g(x) + f(x)\cdot g'(x))\cdot h + K + \epsilon_3 \\ &= f(x)\cdot g(x) + M\cdot h + K + \epsilon_3 \end{aligned}$$

where $M = f'(x)\cdot g(x) + f(x)\cdot g'(x)$, $K = f'(x)\cdot g'(x)\cdot h^2$, and

$$\epsilon_3 = (f(x) + f'(x)h)\epsilon_2 + (g(x) + g'(x)h)\epsilon_1 + \epsilon_1\epsilon_2$$

Now recalling that $\dfrac{dy}{dx} = \lim\limits_{h\to 0}\dfrac{f(x+h) - f(x)}{h}$, for $y = f(x)g(x)$, we have, from above:

$$\frac{dy}{dx} = \lim_{h\to 0}\frac{f(x)g(x) + M\cdot h + f'(x)g'(x)h^2 + \epsilon_3 - f(x)\,g(x)}{h}$$

First and last terms go, h cancels top and bottom, and using the fact that the relative errors $\dfrac{\epsilon_1}{h}$ and $\dfrac{\epsilon_2}{h}$ both go to 0 as $h \to 0$, it is clear that this also holds for the relative error $\dfrac{\epsilon_3}{h}$ giving

$$\frac{dy}{dx} = \lim_{h\to 0}\left(f'(x)\;g(x) + g'(x)\;f(x) + hf'(x)\;g'(x)\right)$$

which gives the **Product Rule** $\dfrac{dy}{dx} = f'(x)\cdot g(x) + g'(x)\cdot f(x)$ when we calculate the limit as h approaches 0.

Example: Differentiate the function $y = (6x^3 - 1)(3x^2 + 4)$.

Solution: We could multiply this out and then differentiate each term using the rule for sums:

$$y = 18x^5 + 24x^3 - 3x^2 - 4 \Rightarrow \frac{dy}{dx} = 90x^4 + 72x^2 - 6x$$

Alternatively we could use the Product Rule:

$$y = (6x^3 - 1)(3x^2 + 4) = f(x) \cdot g(x)$$

with $f(x) = (6x^3 - 1)$ and $g(x) = (3x^2 + 4)$.

Thus, $\frac{dy}{dx} = f'(x) \cdot g(x) + g'(x) \cdot f(x) = 18x^2 \cdot (3x^2 + 4) + 6x \cdot (6x^3 - 1)$

$$= 54x^4 + 72x^2 + 36x^4 - 6x = 90x^4 + 72x^2 - 6x.$$

This agrees with what we had when we multiplied out the brackets and differentiated every term separately.

In combination with the Chain Rule (Rule **D**), the Product Rule (Rule **C**) will enable us to differentiate functions that cannot be multiplied out.

Rule D. Composite Functions: $\boxed{(\mathbf{f}(\mathbf{g}(\mathbf{x})))' = \mathbf{f}'(\mathbf{g}(\mathbf{x})) \cdot \mathbf{g}'(\mathbf{x})}$

Proof: First, we change x by a small amount h. This will change the intermediate function $g(x)$ by a small amount k according to equation (5.2):

$$g(x + h) = g(x) + g'(x)h + \epsilon_1 = g(x) + k$$

where $k = g'(x)h + \epsilon_1$.

We can use this to see by what quantity we must multiply h in order to find all but the infinitesimal bit of change in y:

$$\begin{aligned} f(g(x + h)) &= f(g(x) + k) = f(g(x)) + f'(g(x)) \cdot k + \epsilon_2 \\ &= f(g(x)) + f'(g(x))(g'(x) \cdot h + \epsilon_1) + \epsilon_2 \\ &= f(g(x)) + f'(g(x)) \cdot g'(x) \cdot h + f'(g(x)) \cdot \epsilon_1 + \epsilon_2 \end{aligned}$$

The coefficient of $f'(g(x))\cdot g'(x)$ is h and the last two terms with ϵ_1 and ϵ_2 represent the **infinitesimal error** for the linear approximation theorem and tend to 0 as $h \to 0$. Thus,

$$\begin{aligned}\frac{dy}{dx} &= \lim_{h\to 0}\frac{f(g(x+h)) - f(g(x))}{h}\\ &= \lim_{h\to 0}\frac{f(g(x)) + f'(g(x))\,g'(x)h - f(g(x))}{h}\end{aligned}$$

First and last terms go and h cancels top and bottom, thereby obtaining

$$\frac{dy}{dx} = f'(g(x))\,g'(x)$$

which proves the **Chain Rule**.

Example: Differentiate $y = (5x^2 + 2)^4$.

Solution: Here we have $y = f(g(x))$, where $g(x) = 5x^2 + 2$ and $f(g(x)) = (g(x))^4$. We could expand this:

$$y = (5x^2 + 2)^4 = 625x^8 + 1000x^6 + 600x^4 + 160x^2 + 16$$

and then differentiate each term:

$$\frac{dy}{dx} = 5000x^7 + 6000x^5 + 2400x^4 + 320x$$

But recognising that $y = (5x^2 + 2)^4 = f(g(x))$ is a composite function and using the Chain Rule we get

$$\begin{aligned}\frac{dy}{dx} &= f'(g(x))\cdot g'(x) = \underbrace{4(g(x))^3}_{\text{equation (5.3)}}\cdot g'(x) = 4(5x^2+2)^3\cdot \underbrace{10x}_{\text{Rule }\mathbf{A}\text{, equation (5.3)}}\\ &= 40x\cdot(5x^2+2)^3\end{aligned}$$

If we expand out the brackets, we obtain

$$\begin{aligned}\frac{dy}{dx} &= 40x\cdot(5x^2+2)^3 = 40x\cdot(125x^6 + 150x^4 + 60x^2 + 8)\\ &= 5000x^7 + 6000x^5 + 2400x^4 + 320x\end{aligned}$$

which agrees with the result above.

Additionally, in combination with the the Product Rule (Rule **C**), the Chain Rule (Rule **D**) enables us to differentiate functions that cannot be multiplied out, *e.g.* $y = (3x+4)^{\frac{1}{2}}(x+1)$.

Example in Biology: The *Brière equation* is a nonlinear model which describes the temperature (T)-dependent developmental rates r of insects:

$$r = aT(T - T_1)(T_2 - T)^{\frac{1}{b}}$$

where a and b are constants, and T_1 and T_2 represent the 'conceptual' lower and upper developmental temperature thresholds, respectively, at which developmental rates are 0.

Differentiate the function $r(T)$ with respect to T.

Solution: We can initially rearrange the equation to facilitate the subsequent differentiation, as follows: $r = (aT^2 - T_1T)(T_2 - T)^{\frac{1}{b}}$.

Here we have $r = f(T) \cdot g(T)$ with $f(T) = (aT^2 - T_1T)$ and $g(T) = (T_2 - T)^{\frac{1}{b}}$.

Thus, according to Rule **C**, $\dfrac{dr}{dT} = f'(T) \cdot g(T) + f(T) \cdot g'(T)$.

Using Rules **B** and **A**, we can calculate $f'(T) = 2aT - T_1$, whereas we have to use Rule **D** to calculate $g'(T)$, since $g(T)$ is a composite function $g\left(v(T)\right)$ with $v(T) = T_2 - T$:

$$g'(T) = g'\left(v(T)\right) \cdot v'(T) = \underbrace{\frac{1}{b}\left(v(T)\right)^{\frac{1}{b}-1}}_{\text{eq. (5.3)}} \cdot v'(T)$$

$$= \frac{1}{b}(T_2 - T)^{\frac{1}{b}-1} \cdot \underbrace{(-1)}_{\text{Rule A, eq. (5.3)}} = -\frac{1}{b}(T_2 - T)^{\frac{1}{b}-1}$$

Finally, $\dfrac{dr}{dT} = f'(T) \cdot g(T) + f(T) \cdot g'(T)$
$= (2aT - T_1) \cdot (T_2 - T)^{\frac{1}{b}} + (aT^2 - T_1T) \cdot \left(-\frac{1}{b}(T_2 - T)^{\frac{1}{b}-1}\right)$ which can be re-arranged by factorisation, as follows:

$$\frac{dr}{dT} = (T_2 - T)^{\frac{1}{b}} \cdot \left[2aT - T_1 - \frac{aT^2 - T_1T}{b(T_2 - T)}\right]$$

Rule E. Quotients: $$\left(\frac{f(x)}{g(x)}\right)' = \frac{f'(x) \cdot g(x) - f(x) \cdot g'(x)}{(g(x))^2}$$

Using the Chain Rule, it can be demonstrated that the Quotient and Product Rules are equivalent.

Example: Differentiate the function $y = \dfrac{3x-1}{4x+2}$ by using both the Quotient and Product Rules.

Solution: We can put $f(x) = 3x - 1$ and $g(x) = 4x + 2$, so their corresponding derivatives are $f'(x) = 3$ and $g'(x) = 4$.

By applying the Quotient Rule, we obtain

$$\frac{dy}{dx} = \frac{f'(x) \cdot g(x) - f(x) \cdot g'(x)}{(g(x))^2} = \frac{(4x+2) \cdot 3 - (3x-1) \cdot 4}{(4x+2)^2}$$
$$= \frac{12x + 6 - 12x + 4}{(4x+2)^2} = \frac{10}{(4x+2)^2}$$

Alternatively, we can write $y = \dfrac{3x-1}{4x+2} = (3x-1)(4x+2)^{-1}$ and differentiate by using both the Product and Chain Rules.

We can initially put $f(x) = 3x - 1$ and $g(x) = (4x+2)^{-1}$ and then calculate $\dfrac{dy}{dx} = 3 \cdot (4x+2)^{-1} + (-1) \cdot 4 \cdot (4x+2)^{-2} \cdot (3x-1)$

$$= \frac{3 \cdot (4x+2)}{(4x+2)^2} - \frac{4 \cdot (3x-1)}{(4x+2)^2} = \frac{12x + 6 - 12x + 4}{(4x+2)^2} = \frac{10}{(4x+2)^2}$$

which agrees with the answer from the quotient formula.

Example in Biochemistry: *Cooperativity* constitutes a phenomenon in which the shape of one subunit of an enzyme, which is formed by several subunits, is modified by the substrate so as to change the shape of the neighbouring subunits and favour the binding of the substrate to a second subunit. This results in the progressive increase in the velocity of substrate binding to the subunits upon gradual occupancy of the subunits.

The *Hill–Langmuir equation* reflects the substrate occupancy of a macromolecule, *i.e.* the fraction that is saturated or bound by the ligand:

$$\theta([L]) = \frac{[L]^n}{K_d + [L]^n}$$

where θ is the fraction of the receptor protein concentration that is bound to the ligand, $[L]$ is the total ligand concentration, K_d is the apparent dissociation constant, and n is the Hill coefficient.

Differentiate the function $\theta([L])$ with respect to $[L]$.

Solution: As the Hill–Langmuir equation is a rational function, we can use the Quotient Rule (Rule **E**) to differentiate it.

We can put $f([L]) = [L]^n$ and $g([L]) = K_d + [L]^n$, so their corresponding derivatives are $f'([L]) = n \cdot [L]^{n-1}$ and $g'([L]) = 0 + n \cdot [L]^{n-1}$.

By applying the Quotient Rule, we obtain:

$$\frac{d\theta}{d[L]} = \frac{f'([L]) \cdot g([L]) - f([L]) \cdot g'([L])}{(g([L]))^2}$$

$$= \frac{n \cdot [L]^{n-1} \cdot (K_d + [L]^n) - [L]^n \cdot n \cdot [L]^{n-1}}{(K_d + [L]^n)^2} = \frac{n \cdot [L]^{n-1} \cdot K_d}{(K_d + [L]^n)^2}$$

5.3.3 Differentiability and continuity

It can be demonstrated that if a function $f(x)$ is **differentiable** at a point x_0, then $f(x)$ is **continuous** at x_0.

It follows that if a function $f(x)$ is **not continuous** at a point x_0, it is also **not differentiable** at x_0.

However, it is possible that a function $f(x)$ is continuous at a point x_0, but it is not differentiable at the same point.

These peculiar points often occur when piece-wise functions, functions with moduli, or $n\text{th}$-root functions are studied.

When an irrational function of the type $f(x) = \sqrt[n]{R(x)} = (R(x))^{\frac{1}{n}}$ with $n < 1$ and $\in \mathbb{N}$ is differentiated, we obtain

$$f'(x) = (R(x))^{\frac{1}{n}-1} \cdot R'(x) = (R(x))^{\frac{1-n}{n}} \cdot R'(x) = \frac{R'(x)}{\sqrt[n]{(R(x))^{n-1}}}$$

While x values making $R(x) = 0$ are included in the domain of the original function $f(x) = \sqrt[n]{R(x)}$, the derivative of the irrational function

$f(x)$ **cannot** be calculated at these x values, as it would give **infinite values**. Thus, the irrational function $f(x)$ is **NOT differentiable** at these points.

However, when $f'(x) = \pm\infty$ at a particular point $P(x_P, y_P)$, the tangent line to the function has equation of the type $x = x_P$, its graph is parallel to the y-axis.

Example: Find whether the function $f(x) = \sqrt[3]{4 - x^2}$ is differentiable over all of its domain.

Solution: The domain of the function $f(x) = \sqrt[3]{4 - x^2}$ is the whole set of real numbers $\mathbb{R}$, as the root index is odd, *i.e.* 3.

We can now differentiate $f(x)$ as $f(x) = (4 - x^2)^{\frac{1}{3}}$:

$f'(x) = (4-x^2)^{\frac{1}{3}-1}\cdot(-2x) = \dfrac{-2x}{\sqrt[3]{(4-x^2)^2}}$ which **cannot** be calculated at points which make the denominator equal to 0, that is, at $x = -2$ and $x = 2$.

Therefore, the function $f(x)$ is not differentiable at either points $x = -2$ or $x = 2$.

We can calculate now $\lim_{x\to\pm2} f'(x) = \lim_{x\to\pm2} \dfrac{-2x}{\sqrt[3]{(4-x^2)^2}} = \mp\infty$, which means that the tangent lines at points $x = -2$ and $x = 2$ to the function are parallel to the y-axis and have equation $x = -2$ and $x = 2$, respectively, as shown in the graph below.

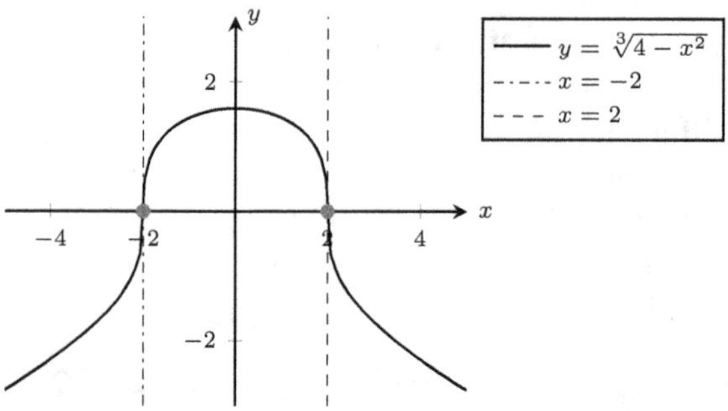

5.3.4 Higher derivatives

If $f(x)$ is differentiable, then $f'(x)$ is called the **first derivative** of $f(x)$.

If $f'(x)$ is also differentiable, its derivative is called the **second derivative** of $f(x)$. The three most commonly used notations are

$$y'' \quad , \quad f''(x) \quad \text{or} \quad \frac{d^2y}{dx^2}$$

If $f''(x)$ is also differentiable, its derivative is called the **third derivative** of $f(x)$. The corresponding ways of writing this are

$$y''' \quad , \quad f'''(x) \quad \text{or} \quad \frac{d^3y}{dx^3}$$

Generally, if $f^{n-1}(x)$ is also differentiable, its derivative is called the ***n*th derivative** of $f(x)$. The most common unambiguous notation for this is

$$\frac{d^ny}{dx^n}$$

Example: Calculate the third derivative of $f(x) = \frac{5}{3}x^3 - 2x$.

Solution: Using the appropriate rules from Section 5.3.2, we can sequentially calculate

$$f'(x) = \frac{5}{3} \cdot 3x^2 - 2 \cdot x^0 = 5x^2 - 2$$
$$f''(x) = 5 \cdot 2x - 0 = 10x$$
$$f'''(x) = 10$$

5.4 Plotting Functions

5.4.1 Information from derivatives

Derivatives provide useful pieces of information about a function $f(x)$.

The sign of the first derivative $y' = f'(x) = \frac{dy}{dx}$ tells us how $f(x)$ is increasing or decreasing.

A function $f(x)$ is said to be **increasing** in an interval I from its domain set, if, given any x_1 and x_2 from the interval I with $x_1 < x_2$, $f(x_1) < f(x_2)$.

Conversely, a function $f(x)$ is said to be **decreasing** in an interval I from its domain set, if, given any x_1 and x_2 from the interval I with $x_1 < x_2$, $f(x_1) > f(x_2)$.

Thus, from the information given by the derivatives, we can distinguish three different cases:

(1) If $y' = f'(x) > 0$ in an interval I of the domain (*i.e.* the gradient is positive), then y is **increasing** in I.
(2) If $y' = f'(x) < 0$ in an interval I of the domain (*i.e.* the gradient is negative), then y is **decreasing** in I.
(3) Although an unusual case, if $y' = f'(x) = 0$ in an interval I of the domain (*i.e.* the gradient is null), then y is **flat** in I.

The sign of the second derivative $y'' = f''(x) = \frac{d^2y}{dx^2}$ tells us about a function's curvature.

We can distinguish three different cases:

(1) If $y'' = f''(x) > 0$ in an interval I of the domain, then the function's graph is **concave up** in I.
(2) If $y'' = f''(x) < 0$ in an interval I of the domain, then the function's graph is **concave down** in I.
(3) If $y'' = f''(x) = 0$ in an interval I of the domain, then the function's graph is a **straight line** in I.

5.4.2 Critical points of a function

A **critical point** of a continuous function $f(x)$ within its domain is a point at which the first **and/or** second derivatives are **zero or undefined**.

Graphically, critical points are the points on the function's graph where the function changes either from increasing to decreasing, in concavity, or in some unpredictable fashion.

Functions can have different types of possible critical points, and they are described below.

Turning points occur at those values of x at which y changes from increasing to decreasing (*i.e.* **local maximum**, max) or decreasing to increasing (*i.e.* **local minimum,** min) (Figure 5.2).

These are the places where the first derivative $f'(x) = \frac{dy}{dx} = 0$, and the gradient of the tangent is flat $\longrightarrow$, according to Fermat's Theorem (see below and Figure 5.2).

The x value satisfying this equation, along with its corresponding y value, is known as a **stationary point**. Note that not all the stationary points are turning points, since the term stationary points includes saddle points too (see case (i) Figure 5.4).

Fermat's theorem: If a real-valued function $f(x)$ is differentiable on an open interval (a, b) with $a < x < b$, and $f(x)$ has a *local maximum* or *local minimum* at point $c \in (a, b)$, then $f'(c) = 0$.

Points of inflection, F, on the graph of $y = f(x)$ occur at those values of x for which the curve changes concavity (Figure 5.2). These will be the places where the second derivative $f''(x) = \frac{d^2y}{dx^2} = 0$ or does not exist (*e.g.* the second derivative of $y = x^{\frac{1}{3}}$ does not exist at $x = 0$ but there is an inflection point at $x = 0$).

Note that the first derivative is not necessarily zero for a point of infection.

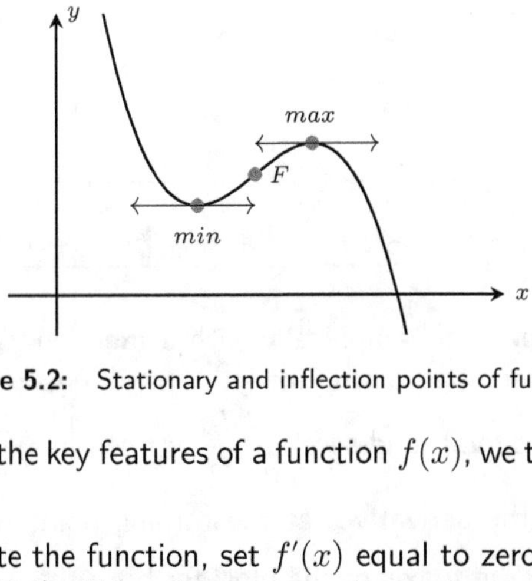

Figure 5.2: Stationary and inflection points of function

Thus to find the key features of a function $f(x)$, we take the following steps:

(i) Differentiate the function, set $f'(x)$ equal to zero, and solve for x (see Table 5.1). This tells us where the stationary points occur but not whether they are maxima, minima, or points of inflection.

(ii) Differentiate a second time, substitute the maxima, minima, and any points of inflection found in step (i) into $f''(x)$ (see Table 5.1):

(a) if the result is positive, then the point is a **local minimum**: $min\,(x_{min}, y_{min})$;

(b) if it is negative, then the point is a **local maximum**: $max\,(x_{max}, y_{max})$;

(c) if it is zero or it can not be calculated, carry out further investigation of the **sign** of $f'(x)$ on each side of the values found in step (i) in order to find in which intervals the function is

increasing $\nearrow$ or **decreasing** $\searrow$. This tells us where **turning points**, such as maxima and minima, occur.

(iii) Set $f''(x)$ equal to zero and solve for x. Then, carry out further investigation of the **sign** of $f''(x)$ on each side of the values of x giving $f''(x) = 0$ in order to find in which intervals the function is **concave up** $\smile$ or **concave down** $\frown$ and where the function's curvature changes (see Table 5.1). This tells us where **points of inflection**, $F\,(x_F, y_F)$, occur.

Lastly, calculating the equation of the tangent line at the inflection point F helps show that the curvature of the plot of the function lies on opposite sides of this tangent line.

Table 5.1: Summary of differentiable key points

Function $f(x)$		max		F_1		min		F_2	
First derivative $f'(x)$	+ve	0	−ve		−ve	0	+ve	0	+ve
Function's trend	$\nearrow$	$\longrightarrow$	$\searrow$		$\searrow$	$\longrightarrow$	$\nearrow$		$\nearrow$
Second derivative $f''(x)$	−ve	−ve	−ve	0	+ve	+ve	+ve	0	−ve
Function's curvature	$\frown$	$\frown$	$\frown$		$\smile$	$\smile$	$\smile$		$\frown$

Figure 5.3 shows this graphically with a function $f(x)$ (blue curve) which has a minimum, a point of inflection, and a maximum, along with its first, $\frac{dy}{dx}$ (red curve), and second, $\frac{d^2y}{dx^2}$ (green line), derivatives.

The value of the derivatives at the turning points (max and min) can be read from the y-axis of the plot: for both these points, $\frac{dy}{dx} = 0$, whereas the $\frac{d^2y}{dx^2} > 0$ for min and $\frac{d^2y}{dx^2} < 0$ for max.

For the inflection point F, $\frac{d^2y}{dx^2} = 0$.

To summarise:

- for a **local minimum**: $\frac{dy}{dx} = 0$ and $\frac{d^2y}{dx^2} > 0$;
- for a **local maximum**: $\frac{dy}{dx} = 0$ and $\frac{d^2y}{dx^2} < 0$;

- for an **inflection point**: $\frac{d^2y}{dx^2} = 0$ and concavity of $\frac{d^2y}{dx^2}$ changes sign at this point, or is undefined.

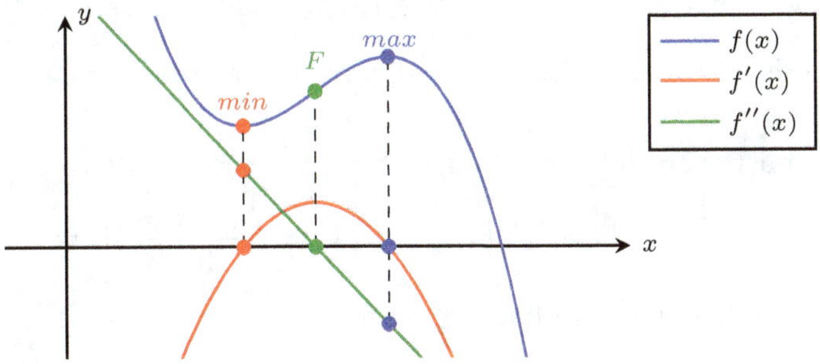

Figure 5.3: Plots of f, f' and f''

Sometimes, it is useful to draw the tangent to the curve at the point of inflection in order to better highlight the change of curvature.

In particular,

(i) If a point of inflection x_F also has $f'(x_F) = 0$, then a horizontal tangent of equation $y = f(x_F)$ occurs at x_F and the point is defined to be a **saddle point**.

(ii) If a point of inflection x_F has $f'(x_F) \neq 0$, then a slanting tangent to the curve occurs at x_F and can be calculated using the formula given in Section 5.2.1.

(iii) If a point of inflection x_F has $f'(x_F) \to \infty$, a vertical tangent of equation $x = x_F$ occurs at x_F.

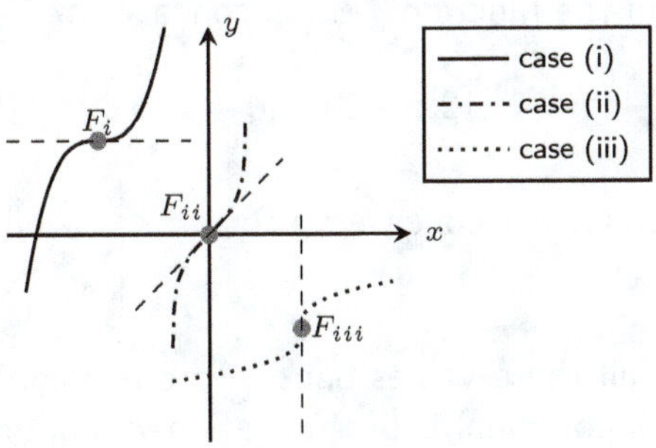

Figure 5.4: Tangents at inflection points

Example: Find the stationary and inflection points of the function $f(x) = 2x^3 - 5x^2 - 4x$ and define the domain intervals where $f(x)$ is increasing or decreasing and is concave up or down.

Solution: First we differentiate $f(x)$: $f'(x) = 6x^2 - 10x - 4$, set the derivative equal to zero $6x^2 - 10x - 4 = 0$, and solve for x, in this case by factorising $(3x+1)(2x-4) = 0$, which gives the roots as $x = -\frac{1}{3}$ or $x = 2$.

We must then differentiate again: $f''(x) = 12x - 10$.

At $x = -\frac{1}{3}$, we have $f''\left(-\frac{1}{3}\right) = 12 \times \left(-\frac{1}{3}\right) - 10 = -14 < 0$. This is negative, so $f(x)$ has a local maximum at $x = -\frac{1}{3}$.

At $x = 2$, we have $f''(2) = 12 \times (2) - 10 = +14 > 0$; this is positive, so $f(x)$ has a local minimum at $x = 2$.

Solving the inequality $f'(x) > 0$ allows us to define in what intervals of the domain the function $f(x)$ is increasing: $6x^2 - 10x - 4 > 0 \iff (3x+1)(2x-4) > 0$. Now we can use the method for solving polynomial inequalities (Section 2.5.6): $x < -\frac{1}{3}$ or $x > 2$. Therefore, $f(x)$ is increasing in the intervals $x < -\frac{1}{3}$ or $x > 2$, while $f(x)$ is decreasing in the interval $-\frac{1}{3} < x < 2$.

To look for points of inflection, we set $f''(x) = 0$ and solve $12x - 10 = 0$ to get $x = \frac{10}{12} = \frac{5}{6}$.

Solving the inequality $f''(x) > 0$ allows us to define in what intervals of the domain the function $f(x)$ is concave up:

$$12x - 10 > 0 \iff x > \frac{5}{6}$$

Therefore, $f(x)$ is concave up for $x > \frac{5}{6}$, while it is concave down for $x < \frac{5}{6}$.

If we put all these values back into our original function $f(x)$, we can find the y coordinates of the stationary and inflection points.

We find a maximum at $max\left(-\frac{1}{3}, \frac{19}{27}\right)$, a minimum at $min\,(2, -12)$, and an inflection point at $F\left(\frac{5}{6}, -\frac{305}{54}\right)$.

By using equation (5.1), we can calculate the equation of the tangent line at the inflection point F to show how the curvature of the function lies on either side with respect to the tangent line to F:

$$y - f\left(\frac{5}{6}\right) = f'\left(\frac{5}{6}\right) \cdot \left(x - \frac{5}{6}\right)$$

$$\iff y - \left(-\frac{305}{54}\right) = \left(-\frac{49}{6}\right) \cdot \left(x - \frac{5}{6}\right) \iff y = -\frac{49}{6}x + \frac{125}{108}$$

We can show all these findings on the graph of the function.

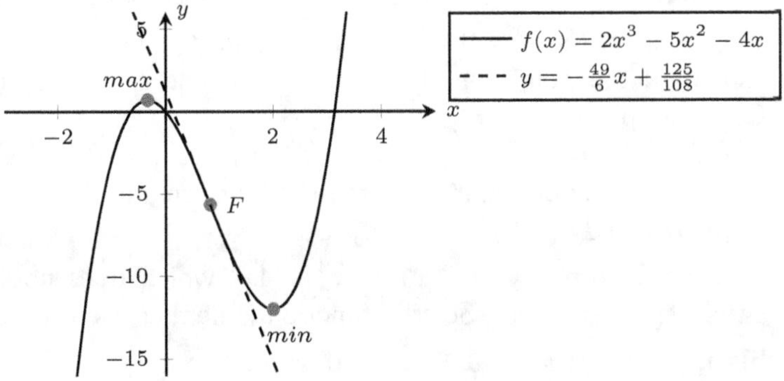

5.4.3 Detailed plotting of functions

Now we can move from simply **sketching** to completely **plotting** a function $f(x)$ by using all the various pieces of information obtained in Chapters 4 and 5, as detailed below:

(i) **domain** through the interval notation, *e.g.* $(-\infty, a]$ (Section 4.2);

(ii) **key points** (Section 4.2);

(iii) possible **symmetry** (Section 4.3);

(iv) **sign** (Section 4.3);

(v) **limits** at the domain interval extremities and possible **asymptotes** (Section 4.4);

(vi) intervals of function **continuity** and possible points of **discontinuity** (Sections 4.5 and 4.6);

(vii) **stationary and inflection points** (Section 5.4.2);

(viii) **co-domain** or range intervals of y values.

Note that there are now many software packages available online for characterising and plotting functions, so you can easily check your results.

Example: Investigate the function $f(x) = x^4 - 4x^3$ and plot it in a Cartesian plane.

Solution: Let us follow the list of steps suggested in Section 5.4.3:

(i) **Domain:** $\mathbb{R}$, *i.e.* $(-\infty, +\infty)$, as $f(x)$ is a polynomial function.

(ii) **Key points:** $f(0) = 0$, $f(1) = -3$, and $f(-1) = 5$, thus we can mark the points $A\ (0, 0)$, $B\ (1, -3)$, and $C\ (-1, 5)$ in the Cartesian plane.

$f(x) = 0 \Longleftrightarrow x^3\,(x - 4) = 0$, which gives $x = 0$ and $x = 4$. So the intersection points between the function and the x-axis are $A\ (0, 0)$ and $D\ (4, 0)$.

(iii) Possible **Symmetry**: $f(-x) = x^4 + 4x^3$ which does not equal either $f(x)$ or $-f(x)$. So, the function is neither even nor odd.

(iv) **Sign:** $f(x) > 0 \Longleftrightarrow x^4 - 4x^3 > 0 \Longleftrightarrow x^3\,(x - 4) > 0$.

Using the method for solving linear inequalities (Section 2.5.3), we obtain $f(x) > 0$ when $x < 0$ or $x > 4$.

Thus, the function is positive for $x < 0$ or $x > 4$.

So, we can shade the area below the x-axis for $x < 0$ or $x > 4$, and the area above the x-axis for $0 < x < 4$.

The function will not be plotted in these shaded areas in the Cartesian plane.

(v) **Limits** and **asymptotes**: As $f(x)$ has an even polynomial degree and the coefficient of the leading term $a_n = 1 > 0$, then $\lim_{x \to \pm\infty} f(x) = +\infty$.

Reminder: When we write $f(\pm\infty) = \pm\infty$, we really mean $\lim_{x \to \pm\infty} f(x) = \pm\infty$, but '$\pm$ infinity' ($\pm\infty$) does not lie in the domain of the function.

Additionally, as $\lim_{x\to\pm\infty} f(x) = \infty$, a possible slanting asymptote of $f(x)$ should be sought in the form of a straight line $y = mx + q$ (see Section 3.2.1).

However, $m = \lim_{x\to\pm\infty} \frac{f(x)}{x} = \lim_{x\to\pm\infty} (x^3 - 4x^2) = \pm\infty$. Therefore, no slanting asymptote to the function can be drawn as x approaches $\pm\infty$.

(vi) **Continuity:** Polynomial functions are continuous at every point in $\mathbb{R}$.

(vii) **Stationary and inflection points**: We differentiate $f(x)$: $f'(x) = 4x^3 - 12x^2$, and then set it equal to zero:

$$4x^3 - 12x^2 = 0 \Longleftrightarrow x^2(4x - 12) = 0$$

giving roots of $x = 0$ or $x = 3$.

We differentiate again: $f''(x) = 12x^2 - 24x$; at $x = 0$, we have $f''(0) = 0$, so we cannot make any assumption about the function at $x = 0$; at $x = 3$, we have $f''(3) = 60 > 0$, which is positive, so $f(x)$ has a local minimum at $x = 3$.

We solve the inequality $f'(x) = 4x^3 - 12x^2 > 0$ to find in which intervals of the domain $f(x)$ is increasing: by using the polynomial inequality, $x^2(4x - 12) > 0$ we get the range of solutions $x > \frac{12}{4} \Longleftrightarrow x > 3$. Thus, $f(x)$ is increasing for $x > 3$ and conversely decreasing for $x < 3$.

For points of inflection, we set $f''(x) = 12x^2 - 24x = x(12x - 24) = 0$ and solve it to get $x = 0$ and $x = \frac{24}{12} = 2$.

By solving the inequality $f''(x) > 0 \Longleftrightarrow x(12x - 24) > 0$, we get the intervals of solutions $x < 0$ or $x > 2$; thus, $f(x)$ is concave up for $x < 0$ or $x > 2$ and it is concave down for $0 < x < 2$.

If we put all these values back into our original function $f(x)$, we can find the y coordinates of the stationary values and inflection points.

Thus, there is a minimum at $min\,(3, -27)$ and two inflection points at $F_1\,(0, 0)$ and $F_2\,(2, -16)$.

Since $f'(0) = 0$, the equation of the tangent line at the inflection point F_1 is $y = f(0) = 0$, which is the x-axis. However, by using equation 5.1, we can calculate the equation of the tangent line at the inflection point F_2 to show that the curvature of the function lies on opposite sides of the tangent line to F_2: $y - f(2) = f'(2) \cdot (x-2) \iff y - (-16) = (-16) \cdot (x-2) \iff y = -16x + 16$.

(viii) **Co-domain:** y values go from the minimum $f(3) = -27$ to $+\infty$: $[3, +\infty)$.

We can summarise all the properties obtained in the list above in a table and plot the graph of the function in a Cartesian plane.

x	$-\infty$	0		2	3	$+\infty$
$f(x)$	$+\infty$	0		-16	-27	$+\infty$
$f'(x)$	$-$ve	0	$-$ve	$-$ve	0	$+$ve
$f(x)$'s trend	↘	?	↘	↘	*min*	↗
$f''(x)$	$+$ve	0	$-$ve	0	$+$ve	$+$ve
$f(x)$'s curvature	⌣	F_1	⌢	F_2	⌣	⌣

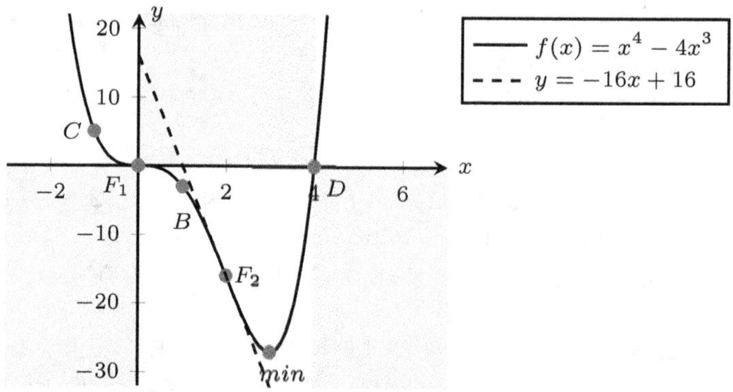

Example in Biochemistry: The height of a chromatographic column can be compared to the equivalent height of a theoretical plate under ideal conditions (HETP, height equivalent theoretical plate). The *van Deemter* equation relates the HETP to the flow speed of the eluant.

The equation is HETP $= f(u) = A + \dfrac{B}{u} + C \cdot u$, where HETP is a measure of the resolving power of the column (units of m), u is the flow speed of the chromatography column ($\mathrm{m\,s^{-1}}$), and A, B, and C are positive constants characteristic of the chromatography column, with $B > C$ in this specific case.

What is the graph of the van Deemter equation?

Solution: In order to draw a general plot of the van Deemter relationship, we can follow the key steps defined in Section 5.4.3.

(i) **Domain:** $(0, +\infty)$, u can only have positive physical values (> 0).

(ii) **Key points:** $f(1) = A + B + C$, we can mark the point $(1, A + B + C)$ in the Cartesian plane.

Since A, B, and C are positive constants and the variable u can only be physically positive, the function can only be positive in its domain and so is never equal to 0.

(iii) Possible **symmetry:** From (ii), we see that u can only be positive throughout its domain, so there is no symmetry to consider.

(iv) **Limits** and **asymptotes:**

$\lim_{u \to 0} f(u) = A + \dfrac{B}{0} + C \cdot 0 = A + 0 + \infty = +\infty$, thus $u = 0$ is a vertical asymptote of the function.

$\lim_{u \to +\infty} f(u) = A + \dfrac{B}{\infty} + C \cdot \infty = A + 0 + \infty = +\infty$, thus there is no horizontal asymptote of the function.

However, as $\lim_{u \to +\infty} f(x) = +\infty$, a possible slanting asymptote of $f(u)$ should be sought in the form of a straight line $y = mu + q$ (see Section 3.2.1).

Now, $m = \lim_{u \to +\infty} \frac{f(u)}{u} = \lim_{u \to +\infty} \left(\frac{A}{u} + \frac{B}{u^2} + C \right)$
$= \frac{A}{+\infty} + \frac{B}{(+\infty)^2} + C = 0 + 0 + C$, thus $m = C$.
Then,

$$q = \lim_{u \to +\infty} f(u) - m \cdot u = \lim_{u \to +\infty} \left(A + \frac{B}{u} + Cu - Cu \right)$$
$$= \lim_{u \to +\infty} A + \frac{B}{u} = A + 0, \text{ thus } q = A$$

Lastly, the equation of the slanting asymptote to $f(u)$ as x approaches $+\infty$ is $y = Cu + A$. Note that for large u, $\frac{B}{u}$ is nearly 0, so $f(u)$ is nearly $A + Cu$.

(v) **Continuity:** $f(u)$ is a rational function, so it is continuous at every point in its domain, which is $u > 0$.

(vi) **Stationary and inflection points:** We differentiate $f(u)$: $f'(u) = 0 - \frac{B}{u^2} + C$. Then we set it equal to zero: $-\frac{B}{u^2} + C = 0 \Longleftrightarrow u^2C - B = 0$ giving roots of $u = +\sqrt{\frac{B}{C}}$, which is physically acceptable and > 1, or $u = -\sqrt{\frac{B}{C}}$, which is physically meaningless.

We differentiate again: $f''(u) = \frac{2B}{u^3}$ which can never equal zero for any value of u, thus the function has no inflection points.

At $u = \sqrt{\frac{B}{C}}$, we have $f''\left(\sqrt{\frac{B}{C}}\right) = 2B \cdot \left(\sqrt{\frac{C}{B}}\right)^3$, which is positive, so $f(u)$ has a local minimum min at $u = \sqrt{\frac{B}{C}}$.

HETP achieves a minimum value at a particular flow velocity, and at this flow rate, the resolving power of the column is maximized, although in practice the elution time is likely to be impractically long.

(vii) **Co-domain** $f(u)$ goes from $f\left(\sqrt{\frac{B}{C}}\right) = A + 2\sqrt{BC}$ to $+\infty$.

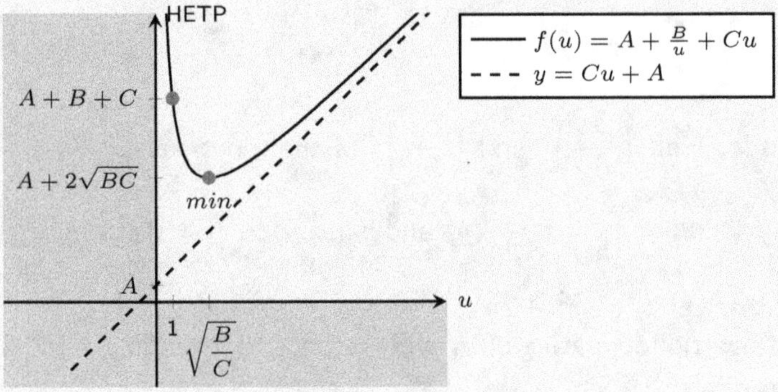

5.5 Partial Differentiation

5.5.1 First-order partial derivatives

Recall that if $y = f(x)$, then $y' = \dfrac{dy}{dx} = f'(x)$ gives the **rate** at which the dependent variable y is changing with respect to the independent variable x.

This works well for quantities such as 'distance travelled' that depend on a single variable such as 'time elapsed', but there are lots of quantities that depend on **several variables**. We can analyse rates of change in this **multivariable** context by means of a process called **partial differentiation**.

Given a function $z = f(x, y)$ of two variables x and y, the **partial derivative of z with respect to x** is the function obtained by differentiating $f(x, y)$ with respect to x, holding y constant. We denote this using ∂ (the 'curly' delta, usually pronounced 'del') as shown here:

$$\left(\frac{\partial z}{\partial x}\right)_y = \left(\frac{\partial f(x,y)}{\partial x}\right)_y = f_x(x,y)$$

Note that the subscript y on the first two expressions denotes the variable to be held constant, while the subscript x on the third expression denotes the variable for the differentiation.

In an entirely analogous manner, we can define the **partial derivative of z with respect to y** to be the function obtained by differentiating $f(x, y)$ with respect to y, holding x constant:

$$\left(\frac{\partial z}{\partial y}\right)_x = \left(\frac{\partial f(x, y)}{\partial y}\right)_x = f_y(x, y)$$

Example: Find $\left(\frac{\partial z}{\partial x}\right)_y$ and $\left(\frac{\partial z}{\partial y}\right)_x$ of the functions

$$f_1(x, y) = z_1 = x^2 - 2y^2 \text{ and } f_2(x, y) = z_2 = 3x^2y + 5xy^2$$

Solution: Let $z_1 = x^2 - 2y^2$.

The partial derivative of z_1 with respect to x holding y fixed is $f_x = \left(\frac{\partial z_1}{\partial x}\right)_y = \frac{\partial (x^2)}{\partial x} = 2x$, since the $2y^2$ term is constant with respect to x and the derivative of a constant is 0.

Similarly, the partial derivative of z_1 with respect to y holding x fixed is $f_y = \left(\frac{\partial z_1}{\partial y}\right)_x = \frac{\partial (x^2 - 2y^2)}{\partial y} = -4y$, where the x^2-term is constant with respect to y and its derivative is thus 0.

Let $z_2 = 3x^2y + 5xy^2$.

Then the partial of z_2 with respect to x holding y fixed is

$$\left(\frac{\partial z_2}{\partial x}\right)_y = \left(\frac{\partial (3x^2y + 5xy^2)}{\partial x}\right)_y = 3y \cdot 2x + 5y^2 \cdot 1 = 6xy + 5y^2$$

while the partial of z_2 with respect to y holding x fixed is:

$$\left(\frac{\partial z_2}{\partial y}\right)_x = \left(\frac{\partial (3x^2y + 5xy^2)}{\partial y}\right)_x = 3x^2 \cdot 1 + 5x \cdot 2y = 3x + 10xy$$

Example in Biochemistry: Consider the familiar *ideal gas law* that relates pressure P to volume V and temperature T *via* the formula $P \cdot V = n \cdot R \cdot T$, where n and R are constants.

Calculate the rate at which pressure P changes with respect to volume V when the temperature T is held fixed, and also find how pressure changes with respect to temperature when volume is held constant.

Solution: First, by rearranging this formula, we can see how pressure depends on temperature and volume: $P = \dfrac{nRT}{V}$.

To calculate the rate at which pressure P changes with respect to volume V when the temperature T is held fixed, we take the partial derivative of the quantity P with respect to the variable V and treat the quantity T as a constant:

$$\left(\frac{\partial P}{\partial V}\right)_T = \left(\frac{\partial\left(\frac{nRT}{V}\right)}{\partial V}\right)_T = nRT \cdot (-V^{-2}) = -\frac{nRT}{V^2}$$

The fact that this final expression is always negative implies that when the temperature is held constant, the pressure falls with a rise in volume.

Let us find out how pressure changes with respect to temperature when the volume is held constant:

$$\left(\frac{\partial P}{\partial T}\right)_V = \left(\frac{\partial\left(\frac{nRT}{V}\right)}{\partial T}\right)_V = \left(\frac{nR}{V}\right) \cdot 1 = \frac{nR}{V}$$

By looking at the equation, we see that if V is fixed then P must rise with T. The partial derivatives confirm this, as the final expression in the calculation above is always positive.

We can relate the above results to the behaviour of a balloon: if we heat it up and constrain its volume, the pressure rise will probably burst it!

5.5.2 Higher-order partial derivatives

Given the function $z = f(x, y)$, we can have two possible first-order partial derivatives, as shown in Section 5.5.1.

Now, if we differentiate each of them twice, first with respect to x holding y constant, and second with respect to y holding x constant, we then obtain four distinct possibilities for the second-order partial derivatives, as shown below:

(i) $\dfrac{\partial}{\partial x}\left(\dfrac{\partial z}{\partial x}\right) = \dfrac{\partial}{\partial x}\left(\dfrac{\partial f(x,y)}{\partial x}\right) = f_{xx}(x,y)$

(ii) $\dfrac{\partial}{\partial y}\left(\dfrac{\partial z}{\partial x}\right) = \dfrac{\partial}{\partial y}\left(\dfrac{\partial f(x,y)}{\partial x}\right) = f_{xy}(x,y)$

(iii) $\dfrac{\partial}{\partial y}\left(\dfrac{\partial z}{\partial y}\right) = \dfrac{\partial}{\partial y}\left(\dfrac{\partial f(x,y)}{\partial y}\right) = f_{yy}(x,y)$

(iv) $\dfrac{\partial}{\partial x}\left(\dfrac{\partial z}{\partial y}\right) = \dfrac{\partial}{\partial x}\left(\dfrac{\partial f(x,y)}{\partial y}\right) = f_{yx}(x,y)$

Important note: If all the partials of $f(x,y)$ exist and are continuous, then $f_{xy} = f_{yx}$ for all (x,y).

In this case, the operators $\dfrac{\partial}{\partial x}$ and $\dfrac{\partial}{\partial y}$ are **commutative** (*i.e.* the order of the operators does not change the result):

$$\boxed{\frac{\partial}{\partial \mathbf{x}}\left(\frac{\partial \mathbf{z}}{\partial \mathbf{y}}\right) = \frac{\partial}{\partial \mathbf{y}}\left(\frac{\partial \mathbf{z}}{\partial \mathbf{x}}\right)}.$$

Example: Consider the function $z = 2x^3y^2 - xy^3$ and calculate the four second-order partial derivatives.

Solution: In order to calculate the four second-order partial derivatives of $z = f(x,y)$, let us differentiate as follows with respect to:

(i) x twice: $\dfrac{\partial}{\partial x}\left(\dfrac{\partial z}{\partial x}\right) = \dfrac{\partial}{\partial x}(6x^2y^2 - y^3) = 12xy^2$

(ii) x first, then y: $\dfrac{\partial}{\partial y}\left(\dfrac{\partial z}{\partial x}\right) = \dfrac{\partial}{\partial y}(6x^2y^2 - y^3) = 12x^2y - 3y^2$

(iii) y twice: $\dfrac{\partial}{\partial y}\left(\dfrac{\partial z}{\partial y}\right) = \dfrac{\partial}{\partial y}(4yx^3 - 3xy^2) = 4x^3 - 6xy$

(iv) y first, then x: $\dfrac{\partial}{\partial x}\left(\dfrac{\partial z}{\partial y}\right) = \dfrac{\partial}{\partial x}(4yx^3 - 3xy^2) = 12x^2y - 3y^2$

Note that the results of items (ii) and (iv) are the same, validating the equality we expect from the commutative property of mixed partials.

Example in Chemistry: Let us consider Laplace's equation for the equilibrium temperature distribution on a copper plate.

If we take a large sheet of copper, hold a steady flame under it at one point and place a chunk of dry ice at another point, and then wait until each point on the plate has reached its own equilibrium temperature, we find that points near the flame are hotter than those near the dry ice.

Let $T(x, y)$ give the temperature at the point (x, y).

According to a result of the French mathematician Pierre Laplace (1749–1827), at every point (x, y) the second-order partials of T must satisfy the equation: $T_{xx} + T_{yy} = 0$.

Verify that the function $T(x, y) = y^2 - x^2$ satisfies Laplace's equation.

Solution: We first differentiate $T(x, y)$ twice with respect to x:

$$T_{xx} = \frac{\partial}{\partial x}\left(\frac{\partial T}{\partial x}\right) = \frac{\partial}{\partial x}(0 - 2x) = -2$$

We can then differentiate $T(x, y)$ twice with respect to y:

$$T_{yy} = \frac{\partial}{\partial y}\left(\frac{\partial T}{\partial y}\right) = \frac{\partial}{\partial y}(2y - 0) = 2$$

Thus, $T_{xx}(x, y) + T_{yy}(x, y) = 2 + (-2) = 0$, which verifies that $T(x, y) = y^2 - x^2$ satisfies Laplace's equation.

5.6 Exercises

(1) Using the definition $f'(x) = \lim_{h \to 0} \dfrac{f(x+h) - f(x)}{h}$, calculate the derivatives of the following functions.

(a) $y = 5x^2 + 2$ (b) $y = \dfrac{x}{2} + 3x^3$ (c) $y = 2x^3 + 4x^2 + 5$

(d) $y = 6x - 2$ (e) $y = \dfrac{1}{x^2}$ (f) $y = \dfrac{2x^2}{3} + \dfrac{x}{5}$

(2) Sketch the graphs of the following functions and draw the tangent at the given points in a Cartesian plane.

(a) $y = 2x^2 + 4$ when $x = -1$ (b) $y = 5x^3 - 10$ when $x = 2$

(c) $y = x^3 - 2x + 1$ when $x = 4$ (d) $y = \dfrac{2}{x^2}$ when $x = -3$

(3) In a biochemical experiment, $200\,\mu\text{mol}$ of a chemical substrate is metabolised by an enzyme over time.

The amounts of the product generated at specific times are shown in the following table:

$t_{(\text{sec})}$	0	10	20	30	50	80	120	240	300
$s_{(\mu\text{mol})}$	0	75	109	129	150	166	176	187	189

(a) Plot a graph showing the amount of product generated over time for the enzyme-controlled reaction.

(b) Given that the function $s = \dfrac{600t}{50+3t}$ models the shape of the trend of product formation by the enzyme over time, estimate the initial reaction rate v_i using the two different methods explained in Section 5.2.1.

(4) Estimate the following values using the linear approximation method:

(a) $\sqrt{6513}$ (b) $\dfrac{1}{\sqrt{171}}$ (c) $\sqrt[3]{130}$ (d) $\dfrac{1}{\sqrt[4]{250}}$

(5) Using the fact that if $y = ax^n$, $y' = anx^{n-1}$, differentiate the following functions using the sum, difference, and scalar multiple rules, as appropriate.

(a) $y = 5x^3 - 8x^2 + 1$

(b) $y = 8x^4 - 7 + 6x^5$

(c) $y = \dfrac{32}{x^2} + 16x - 1$

(d) $y = \dfrac{1}{x^3} + \dfrac{2}{x^4} + \dfrac{5}{x^2}$

(e) $y = \sqrt{x} + 9x^{15}$

(f) $y = -\sqrt[3]{x} + \sqrt[4]{x}$

(g) $y = \dfrac{1}{\sqrt{x^3}} + x$

(h) $y = \dfrac{9}{x^5} + 17x^{-3} - \sqrt[3]{x^2} + 2x - 4$

(6) Differentiate the following functions using the Product Rule.

(a) $y = (5x-1)(4x^2+3)$

(b) $y = \sqrt{x}(2x-4)$

(c) $y = \dfrac{1}{x^2}(x^3 + x^2)$

(d) $y = \dfrac{2}{\sqrt{x}} \cdot 6\sqrt{x}$

(e) $y = \sqrt{x^3} \cdot \dfrac{1}{x^3}$

(f) $y = 3x \cdot (x^4 - 3x^3)$

(g) $y = \dfrac{1}{x}(\dfrac{7}{x^2} - 1)$

(h) $y = 10x^2(5x-3)$

(i) $y = (13x - x^3)(5x - x^2)$

(j) $y = (8\sqrt{x} + x^2)(2x+4)$

(7) Differentiate the following functions using the Chain Rule.

(a) $y = (x+1)^2$ (b) $y = \sqrt{3x^2 - 2}$ (c) $y = \dfrac{1}{(x-3)^2}$

(d) $y = (\sqrt{x} + 3x)^2$ (e) $y = \dfrac{1}{\sqrt{2x-1}}$ (f) $y = (3x^3 + x)^3$

(g) $y = \dfrac{1}{9x - 3x^4}$ (h) $y = 5\sqrt{x^2 - x}$ (i) $y = \dfrac{8}{(5x^3 - 20)^2}$

(8) Differentiate the following functions using the Quotient Rule.

(a) $y = \dfrac{2x}{3x-1}$ (b) $y = \dfrac{4x^2 - 4}{x^3 - 6}$ (c) $y = \dfrac{x^2 - 5x}{\sqrt{x}}$

(d) $y = \dfrac{x + 3x^2}{2x}$ (e) $y = \dfrac{x^3}{3\sqrt{x}}$ (f) $y = \dfrac{x - x^2}{x^3 - x^4}$

(g) $y = \dfrac{\frac{1}{x} - \frac{1}{x^2}}{\frac{1}{x^3} + \frac{1}{x^4}}$ (h) $y = \dfrac{\frac{2}{x^3}}{\frac{3}{x}}$ (i) $y = \dfrac{5x^4}{2x^3 + x}$

(9) Differentiate the following functions using the Product, Chain, and Quotient Rules, as appropriate.

(a) $y = \dfrac{(x+5)(2x+2)}{x^2}$ (b) $y = \dfrac{3x^2 + 4}{x^2(x-2)}$

(c) $y = \dfrac{5x(x^2+1)}{3x}$ (d) $y = \dfrac{\sqrt{x^3} \cdot 5x}{\sqrt{x}}$

(e) $y = \dfrac{(8x+3)(2x^3 - 5x)}{\frac{1}{x}}$ (f) $y = \dfrac{(3x^2 + x)(x-1)}{x^3}$

(g) $y = \dfrac{(5x-1)^2}{x}$ (h) $y = 13x^2(x+3)$

(i) $y = 2x^3(2x-3)^{\frac{1}{2}}$ (j) $y = \dfrac{\sqrt{x^2 - 4}}{2x^3}$

(k) $y = \dfrac{(\sqrt{3x+5})^3}{x}$ (l) $y = \dfrac{(x-5)}{(2-2x)^2}$

(m) $y = 9x^2(x+5)^2$ (n) $y = x^8\left(\dfrac{1}{x^3} + 3\right)$

(10) Find whether the following functions are differentiable and if so, for which x values within their corresponding domains.

(a) $f(x) = \sqrt{x+2}$ (b) $f(x) = \sqrt[3]{x^2 - 9} + 2$

(c) $f(x) = \sqrt[5]{1-x}$

(11) Differentiate the following functions which are used in Biosciences, with respect to the argument in the brackets (...) in each case, using the appropriate differentiation rules. All the other parameters in the formulae should be considered as constants.

(a) $\Delta N(N) = r[1-N]N$

(b) $f(p) = ap^3 - bp^2 + cp$

(c) $\Lambda_m(c) = \Lambda_m^\circ - k\sqrt{c}$

(d) $\phi(r) = \dfrac{1}{r^2}$

(e) $p(v) = \dfrac{nRT}{v}$

(f) $p(\zeta) = \dfrac{1-\zeta}{1+\zeta}$

(g) $r([Y]) = \dfrac{kK_E[Y][M]}{1+K_E[Y]}$

(h) $i(t) = \dfrac{nFAc_j^\circ\sqrt{D_j}}{\sqrt{\pi t}}$

(i) $C_p(M) = \sqrt{\dfrac{2RT}{M}}$

(12) For the following functions, identify the coordinates of the stationary points and determine whether they are maxima, minima, or points of inflection.

(a) $y = x^3 - 6x^2 - 7$

(b) $y = \dfrac{1}{3}x - \sqrt[3]{x}$

(c) $y = 3x^2(1-x)^2$

(d) $y = \dfrac{3x}{x^2+1}$

(e) $y = \dfrac{x}{\sqrt{x^2+1}}$

(f) $y = \sqrt[3]{x^2-4}$

(13) For the following functions, determine whether, at the given point, y is increasing $\nearrow$, decreasing $\searrow$, or flat $\longrightarrow$, and state the concavity (upwards $\smile$ or downwards $\frown$) of the graph.

(a) $y = 9x - 10x^2$ at $x = 2$

(b) $y = 4x^4 + x^2 - 7x^3$ at $x = -1$

(c) $y = \sqrt{x^3} + x^3$ at $x = 4$

(d) $y = \dfrac{2}{x^3} - 3x$ at $x = 1$

(e) $y = x^2 + \sqrt{x}$ at $x = 2$

(f) $y = \dfrac{1}{x^2-4}$ at $x = -1$

(14) Plot graphs of the following functions by following the steps listed in Section 5.4.3.

(a) $y = 3x^3 + 5$

(b) $y = x^3 - 5x$

(c) $y = x - \dfrac{x^2}{2}$

(d) $y = -x^3 - 5x^2$

(e) $y = \dfrac{6x}{x^2+1}$

(f) $y = \dfrac{x^2-1}{x+2}$

(g) $y = \sqrt{x} - x$

(h) $y = \sqrt{\dfrac{x-1}{x+3}}$

(i) $y = \sqrt{x^2-5x+6}$

(15) Plot graphs of the following functions by using the rules listed in Section 4.7.

(a) $y = -(x-1)^3 + 2$ (b) $y = (x+2)^2 + 3$ (c) $y = \left|x^3 - 5\right| - 2$

(d) $y = \dfrac{1}{x-4} + 3$ (e) $y = \dfrac{2}{5-x}$ (f) $y = \left|1 - \dfrac{2}{x-1}\right|$

(16) The growth rate R of a cell colony with N cells at time t can be represented by the equation

$$R = kN^2 - bN^3$$

This is called a logistic model.
For this example, take the constants to be $k = 7.5\,\mathrm{h}^{-1}$ and $b = 0.05\,\mathrm{h}^{-1}$.

(a) What is the equilibrium size ($R = 0$) of the population?

(b) What is the number of cells N when the rate of growth is maximum?

(c) Plot a graph of the function $R = f(N)$.

(d) What happens if the population size N ever exceeds the equilibrium value $\dfrac{k}{b}$?

(e) At what population size is the curvature of the rate changing? What is the physical meaning of this point?

(17) In a biochemical experiment on a sample of cells, the variation, y, of one of its properties compared to a standard value were measured for increasing values of the parameter x.

It is thought that the results obtained might obey a relationship of the form

$$y = A^2x + \frac{3B}{x^2} + \frac{C}{2}$$

where A, B, and C are positive constants with $\dfrac{C}{B} > 6$ in this case.

(a) What is the graph of the function?

(b) In order to check your findings, investigate and then plot the function $y = f(x)$ when $A = 3$, $B = 1$, and $C = 7$.

(18) The power, P, that a certain machine develops is given by the formula

$$P = AI - BI^2$$

where I is the current and A and B are positive constants.

Find the maximum value of P as I varies.

(19) The formula for the Lennard-Jones potential energy, $V(r)$, between two non-polar atoms is

$$V(r) = \frac{A}{r^{12}} - \frac{B}{r^6}$$

where r is the internuclear distance and A and B are positive constants.

(a) Find the value of $V(r)$ at the stationary point of this function and determine if this point is a maximum, minimum, or point of inflection.

(b) Sketch the function and label the stationary point.

(c) Calculate the potential energy between two neon atoms $0.4\,\mathrm{nm}$ apart if $A = 4 \cdot 10^{-28}\,\mathrm{J\,nm}^{12}$ and $B = 1 \cdot 10^{-24}\,\mathrm{J\,nm}^6$.

(20) The focal length of the lens of the eye, f, can be controlled so that an object at distance u in front of the eye can be brought to perfect focus on the retina at $v = 1.8\,\mathrm{cm}$ behind the lens (assume v is fixed).

A fly is moving towards the eye at a speed of $0.7\,\mathrm{m\,s}^{-1}$. Assuming that the optics of the eye lens obeys the thin lens formula

$$\frac{1}{f} = \frac{1}{u} + \frac{1}{v}$$

find the rate of change of focal length required to keep the fly in perfect focus at a distance of $3\,\mathrm{m}$.

(21) Find $\left(\frac{\partial z}{\partial x}\right)_y$ and $\left(\frac{\partial z}{\partial y}\right)_x$, using the Product, Quotient and Chain Rules as appropriate:

(a) $z = xy + x^2y^2$ (b) $z = x^2 + 3y^3$ (c) $z = \sqrt{x} - \sqrt{y}$

(d) $z = y^2x - 4y$ (e) $z = y\sqrt{x} + x^2y$ (f) $z = xy - yx^3 + 2x^2$

(g) $z = \frac{x}{y} + 3y^2$ (h) $z = \frac{3x + 2y}{x^2}$ (i) $z = \sqrt{4x^2 - y}$

(22) If $G = H - TS$, where G is the Gibbs free energy, H is the enthalpy, T is the temperature, P is the pressure, and S is the entropy, show that the Gibbs equation $\left[\frac{\partial}{\partial T}\left(\frac{G}{T}\right)\right]_P = -\frac{H}{T^2}$ implies $\left(\frac{\partial G}{\partial T}\right)_P = -S$.

Hint: Start by using the Product Rule on the function $\frac{G}{T}$.

(23) Given the ideal gas law: $PV = nRT$, where P is the pressure, V is the volume, n is the number of moles, R is the universal gas constant, and T is the temperature, show that $\frac{\partial V}{\partial T} \times \frac{\partial T}{\partial P} \times \frac{\partial P}{\partial V} = -1$.

(24) Find $\left(\frac{\partial z}{\partial x}\right)_y$, $\left(\frac{\partial z}{\partial y}\right)_x$, $\frac{\partial^2 z}{\partial x^2}$, and $\frac{\partial^2 z}{\partial y^2}$ for the following functions.

Then prove for each one that $\frac{\partial^2 z}{\partial x \partial y} = \frac{\partial^2 z}{\partial y \partial x}$.

(a) $z = 2y^3 + 9x^2 y$ (b) $z = \sqrt[4]{x} - x^2\sqrt{y}$ (c) $z = (2x + 3y)^2$

(d) $z = \frac{x^2 - y^2}{x}$ (e) $z = \frac{x + 2y}{y}$ (f) $z = \frac{x}{y^2} - \frac{y^2}{x^3}$

(25) Given that $z = 9x^2 + 3y^3$, show that $\frac{\partial z}{\partial x} + \frac{\partial^2 z}{\partial y^2} = 18(x + y)$.

(26) Verify whether the function $T(x, y) = x^2 y - xy^2$ satisfies Laplace's equation $T_{xx} + T_{yy} = 0$ and thus whether it can be used to describe thermal equilibrium.

(27) The body mass index, B, is used as a parameter to classify people as underweight, normal, overweight, or obese. It is defined as their weight in kg, w, divided by the square of their height in m, h.

(a) Plot a graph of B against w for a person who is $1.7\,\text{m}$ tall.

(b) Find the rate of change of B with weight of this person.

(c) Plot a graph of B against h for a child whose weight is constant at $35\,\text{kg}$.

(d) Find the rate of change of B with height h of this child.

(e) Show that $\frac{\partial^2 B}{\partial h\, \partial w} = \frac{\partial^2 B}{\partial w\, \partial h}$.

Answers

(1) (a) $10x$ (b) $\frac{1}{2} + 9x^2$ (c) $6x^2 + 8x$ (d) 6 (e) $-\frac{2}{x^3}$ (f) $\frac{4x}{3} + \frac{1}{5}$

(2) (a) $y' = 4x$, $y = -4x + 2$ (b) $y' = 15x^2$, $y = 60x - 30$
(c) $y' = 3x^2 - 2$, $y = 46x - 127$ (d) $y' = -\frac{4}{x^3}$, $y = \frac{4}{27}x + \frac{2}{3}$

(3) (a)

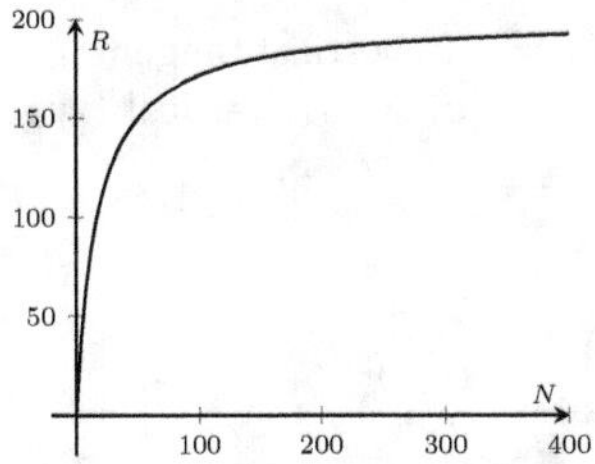

(b) $y_1 = \frac{15}{2}x$, $y_2 = 12x$

(4) (a) $\sqrt{6513} = \sqrt{80^2 + 113} = 80.71$ (b) $\frac{1}{\sqrt{171}} = \frac{1}{\sqrt{13^2+2}} = 0.077$
(c) $\sqrt[3]{130} = \sqrt[3]{5^3 + 5} = 5.067$ (d) $y = \frac{1}{\sqrt[4]{250}} = \frac{1}{\sqrt[4]{4^4-6}} = 0.2515$

(5) (a) $15x^2 - 16x$ (b) $32x^3 + 30x^4$ (c) $-\frac{64}{x^3} + 16$ (d) $-\frac{3}{x^4} - \frac{8}{x^5} - \frac{10}{x^3}$
(e) $\frac{1}{2\sqrt{x}} + 135x^{14}$ (f) $\frac{1}{3x^{\frac{2}{3}}} + \frac{1}{4x^{\frac{3}{4}}}$ (g) $-\frac{3}{2x^{\frac{5}{2}}} + 1$ (h) $-\frac{45}{x^6} - 51x^{-4} - \frac{2}{3x^{\frac{1}{3}}} + 2$

(6) (a) $y' = 60x^2 - 8x + 15$ (b) $y' = \frac{3x-2}{\sqrt{x}}$ (c) $y' = 1$ (d) $y' = 0$
(e) $y' = -\frac{3}{2x^{\frac{5}{2}}}$ (f) $y' = 3\left(5x^4 - 12x^3\right)$ (g) $y' = \frac{x^2-21}{x^4}$
(h) $y' = 10\left(15x^2 - 6x\right)$ (i) $y' = 5x^4 - 20x^3 - 39x^2 + 130x$
(j) $y' = 6x^2 + 8x + 24\sqrt{x} + \frac{16}{\sqrt{x}}$

(7) (a) $y' = 2(x+1)$ (b) $y' = \frac{3x}{\sqrt{3x^2-2}}$ (c) $y' = -\frac{2}{(x-3)^3}$
(d) $y' = 1 + 9\sqrt{x} + 18x$ (e) $y' = -\frac{1}{(2x-1)^{\frac{3}{2}}}$
(f) $y' = 3\left(3x^3 + x\right)^2\left(9x^2 + 1\right)$ (g) $y' = \frac{4x^3-3}{3x^2\left(x^3-3\right)^2}$
(h) $y' = \frac{5(2x-1)}{2\sqrt{x^2-x}}$ (i) $y' = -\frac{48x^2}{25\left(x^3-4\right)^3}$

(8) (a) $y' = -\frac{2}{(3x-1)^2}$ (b) $y' = \frac{-4x^4+12x^2-48x}{\left(x^3-6\right)^2}$ (c) $y' = \frac{3x-5}{2\sqrt{x}}$
(d) $y' = \frac{3}{2}$ (e) $y' = \frac{5x\sqrt{x}}{6}$ (f) $y' = -\frac{2}{x^3}$
(g) $y' = \frac{x\left(2x^2+2x-2\right)}{(x+1)^2}$ (h) $y' = -\frac{4}{3x^3}$ (i) $y' = \frac{5x^2\left(2x^2+3\right)}{\left(2x^2+1\right)^2}$

(9) (a) $y' = \frac{4(-3x-5)}{x^3}$ (b) $y' = -\frac{3x^3+12x-16}{x^3(x-2)^2}$
(c) $y' = \frac{10x}{3}$ (d) $y' = 10x$
(e) $y' = 80x^4 + 24x^3 - 120x^2 - 30x$ (f) $y' = \frac{2(x+1)}{x^3}$
(g) $y' = \frac{25x^2-1}{x^2}$ (h) $y' = 13\left(3x^2 + 6x\right)$
(i) $y' = \frac{2\left(7x^3-9x^2\right)}{(2x-3)^{\frac{1}{2}}}$ (j) $y' = \frac{-x^2+6}{x^4\sqrt{x^2-4}}$
(k) $y' = \frac{9x\sqrt{3x+5}-2(3x+5)^{\frac{3}{2}}}{2x^2}$ (l) $y' = -\frac{x-9}{4(x-1)^3}$
(m) $y' = 9\left(4x^3 + 30x^2 + 50x\right)$ (n) $y' = 24x^7 + 5x^4$

(10) Functions are not differentiable at (a) $x = -2$, vertical tangent $x = -2$
(b) $x = \pm 3$, vertical tangents $x = \pm 3$ (c) $x = 1$, vertical tangent $x = 1$

(11) (a) $\Delta'(N) = r(-2N + 1)$ (b) $f'(p) == 3s_1p^2 - 2s_2p + s_3$
(c) $\Lambda_m'(c) = -\frac{k}{2\sqrt{c}}$ (d) $\phi'(r) = -\frac{2}{r^3}$
(e) $p'(v) = -\frac{nRT}{v^2}$ (f) $p'(\zeta) = -\frac{2}{(1+\zeta)^2}$
(g) $r'([Y]) = \frac{kMk_E}{(1+[Y]k_E)^2}$ (h) $i'(t) = -\frac{nFAc_j^\circ\sqrt{D_j}}{2t\sqrt{\pi t}}$
(i) $C_p'(M) = -\frac{1}{2M}\sqrt{\frac{2RT}{M}}$

(12) (a) $max(0, -7)$, $min(4, -39)$, $F(2, -23)$
(b) $max\left(-1, \frac{2}{3}\right)$, $min\left(1, -\frac{2}{3}\right)$, $F(0, 0)$
(c) $max(\frac{1}{2}, \frac{3}{13})$, $min_1(0, 0)$, $min_2(1, 0)$, $F_1\left(\frac{3-\sqrt{3}}{6}, \frac{1}{12}\right)$, $F_2\left(\frac{3+\sqrt{3}}{6}, \frac{1}{12}\right)$

(d) $max\left(1, \frac{3}{2}\right)$, $min\left(-1, -\frac{3}{2}\right)$ (e) $F(0,0)$
(f) $min\left(0, -\sqrt[3]{4}\right)$, $F_1(-2,0)$, $F_2(2,0)$

(13) (a) $\searrow, \frown$ (b) $\searrow, \smile$ (c) $\nearrow, \smile$ (d) $\searrow, \smile$ (e) $\nearrow, \smile$ (f) $\nearrow, \frown$

(14) *(a) Dom: $(-\infty, +\infty)$; F: $(0, 5)$.
(b) Dom: $(-\infty, +\infty)$; max: $(-1.3, 4.3)$; min: $(1.3, -4.3)$; F: $(0, 0)$.
(c) Dom: $(-\infty, +\infty)$; max: $(1.2, 0.8)$; min: $(-1.2, -0.8)$; F: $(0, 0)$.
(d) Dom: $(-\infty, +\infty)$; max: $(0, 0)$; min: $(-3.3, -18.5)$; F: $(-1.7, -9.2)$.
(e) Dom: $(-\infty, +\infty)$; max: $(1, 3)$; min: $(-1, -3)$; F: $(-\sqrt{3}, -2.6), (0, 0)$, $(\sqrt{3}, 2.6)$.
(f) Dom: $(-\infty, -2) \cup (-2, +\infty)$; max: $(-3.7, -7.5)$; min: $(-2.3, -0.5)$.
(g) Dom: $[0, -\infty)$; max: $(0.25, 0.25)$.
(h) Dom: $(-\infty, -3) \cup [1, +\infty)$; min: $(1, 0)$.
(i) Dom: $(-\infty, 2] \cup [3, +\infty)$; min: $(2, 0), (3, 0)$

(15)

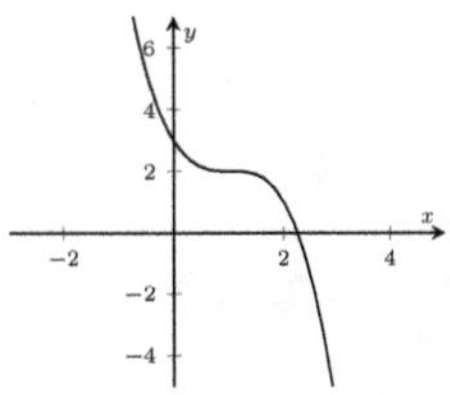

(a)

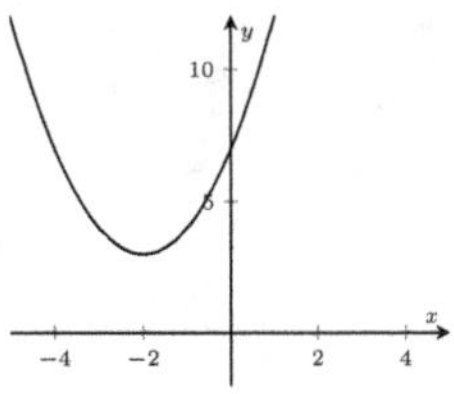

(b)

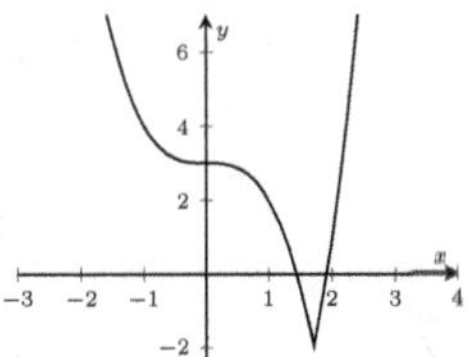

(c)

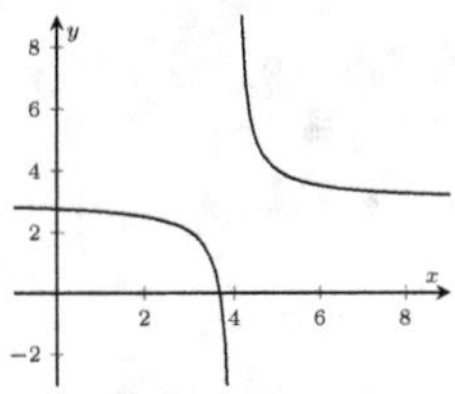

(d)

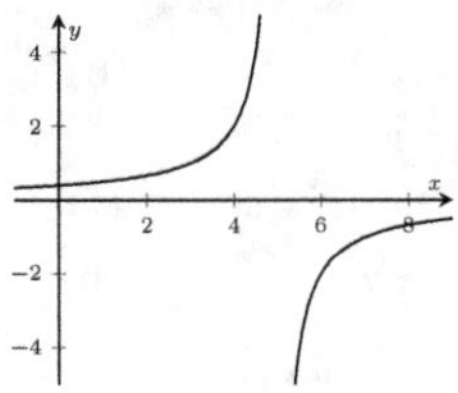

(e)

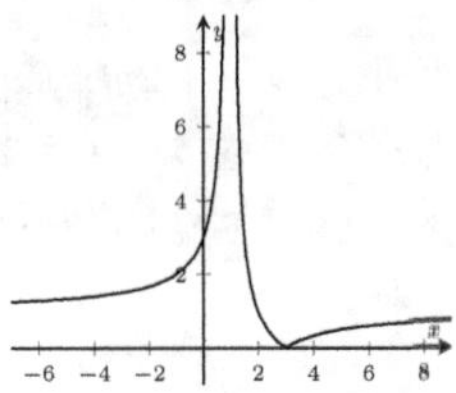

(f)

(16) (a) Equilibrium size when $R = 0$ at $N = 150$
(b) R is maximum when $N = \frac{2k}{3b} = 100$
(c)

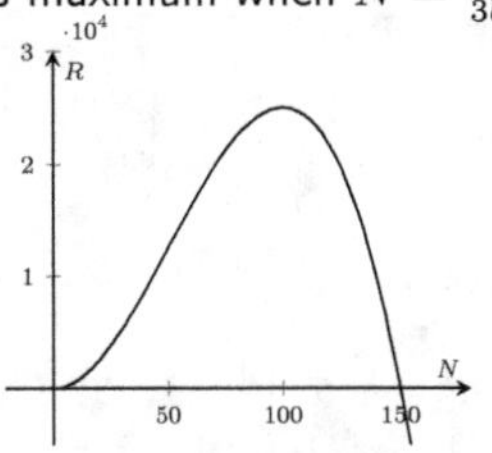

(d) The growth rate R will be negative, forcing N to decrease back to the equilibrium value
(e) When $R'(N) = 0$, that is, at $N = \frac{2k}{3b} = 100$; when $N < 100$, the rate R is growing faster then when $N > 100$.

(17) (a) Dom: $(0, -\infty)$; asymptote: $y = A^2x + \frac{C}{2}$ min at $x = \sqrt[3]{\frac{6B}{A^2}}$

(b)

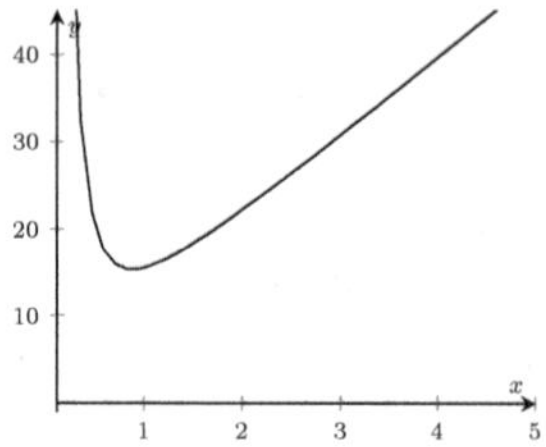

(18) P is maximum at $I = \dfrac{A}{2B}$

(19) (a) $\frac{dV(r)}{dr} = \frac{-12A}{r^{13}} - \frac{-6B}{r^7}$; $\frac{d^2V(r)}{dr^2} = \frac{156A}{r^{14}} - \frac{42B}{r^8}$; $r_{min} = \left(\frac{2A}{B}\right)^{\frac{1}{6}}$

(b)

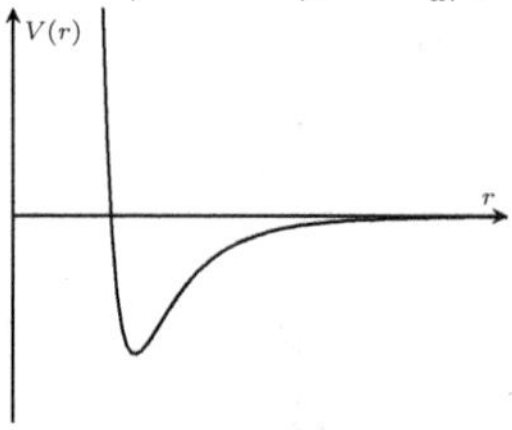

with $V(r_{min}) = -\frac{B^2}{4A}$

(c) $V(r) = -2.2 \cdot 10^{-22}$ J

(20) Rearrange equation to make f the subject $f = uv(u+v)^{-1}$.
Differentiate f with respect to t: $\frac{df}{dt} = v(u+v)^{-1} \cdot \frac{du}{dt} - uv(u+v)^{-2} \cdot \frac{du}{dt}$;
$\frac{df}{dt} = 0.028\,\text{mm}\,\text{s}^{-1}$.

(21) (a) $z_x = y + 2y^2x$, $z_y = x + 2x^2y$ (b) $z_x = 2x$, $z_y = 9y^2$
(c) $z_x = \frac{1}{2\sqrt{x}}$, $z_y = -\frac{1}{2\sqrt{y}}$ (d) $z_x = y^2$, $z_y = 2xy - 4$
(e) $z_x = \frac{y}{2\sqrt{x}} + 2yx$, $z_y = \sqrt{x} + x^2$ (f) $z_x = y - 3yx^2 + 4x$, $z_y = -x^3 + x$
(g) $z_x = \frac{1}{y}$, $z_y = -\frac{x}{y^2} + 6y$ (h) $z_x = 3 - \frac{4y}{x^3}$, $z_y = \frac{2}{x^2}$
(i) $z_x = \frac{4x}{\sqrt{4x^2-y}}$, $z_y = -\frac{1}{2\sqrt{4x^2-y}}$

(22) Using the Product Rule:
$\left[\frac{\partial}{\partial T}\left(\frac{G}{T}\right)\right]_P = \frac{1}{T}\left[\frac{\partial G}{\partial T}\right]_P + G\frac{d(1/T)}{dT} = \frac{1}{T}\left[\frac{\partial G}{\partial T}\right]_P - \frac{G}{T^2}$

If $\frac{1}{T}\left[\frac{\partial G}{\partial T}\right]_P = -S$, then we can substitute it in and get

$-\frac{S}{T} - \frac{G}{T^2} = -\frac{(G+TS)}{T^2} = -\frac{H}{T^2}$ as required

(23) $\frac{\partial V}{\partial T} = \frac{nR}{P}$, $\frac{\partial T}{\partial P} = \frac{V}{nR}$, $\frac{\partial P}{\partial V} = -\frac{nRT}{V^2}$
Product $= \frac{nR}{P} \cdot \frac{V}{nR} \cdot \left(-\frac{nRT}{V^2}\right) = -\frac{nRT}{VP} = -1$

(24) (a) $z_x = 18yx$, $z_y = 6y^2 + 9x^2$, $z_{xx} = 18y$, $z_{yy} = 12y$, $z_{xy} = z_{yx} = 18x$
(b) $z_x = -\frac{x}{\sqrt{y}}$, $z_y = -\frac{x^2}{2\sqrt{y}}$, $z_{xx} = -\frac{x}{\sqrt{y}}$, $z_{yy} = \frac{x^2}{4y^{\frac{3}{2}}}$, $z_{xy} = z_{yx} = -\frac{x}{\sqrt{y}}$
(c) $z_x = 4(2x+3y)$, $z_y = 6(2x+3y)$, $z_{xx} = 8$, $z_{yy} = 18$, $z_{xy} = z_{yx} = 12$
(d) $z_x = \frac{x^2+y^2}{x^2}$, $z_y = -\frac{2y}{x}$, $z_{xx} = -\frac{2y^2}{x^3}$, $z_{yy} = -\frac{2}{x}$, $z_{xy} = z_{yx} = \frac{2y}{x^2}$
(e) $z_x = \frac{1}{y}$, $z_y = -\frac{x}{y^2}$, $z_{xx} = 0$, $z_{yy} = \frac{2x}{y^3}$, $z_{xy} = z_{yx} = -\frac{1}{y^2}$

(f) $z_x = \frac{1}{y^2} + \frac{3y^2}{x^4}$, $z_y = -\frac{2x}{y^3} - \frac{2y}{x^3}$, $z_{xx} = -\frac{12y^2}{x^5}$, $z_{yy} = \frac{6x}{y^4} - \frac{2}{x^3}$,
$z_{xy} = z_{yx} = -\frac{2}{y^3} + \frac{6y}{x^4}$

(25) $\frac{\partial z}{\partial x} = 18x$, $\frac{\partial^2 z}{\partial y^2} = 18y$

(26) $T_x = 2xy - y^2$, $T_y = x^2 - 2xy$, $T_{xx} = 2y$, $T_{yy} = -2x$

(27) (a) $B = \frac{w}{1.7^2}$ (b) $\frac{dB}{dw} = \frac{1}{1.7^2}$ (c) $B = \frac{35}{h^2}$
(d) $\frac{dB}{dh} = -\frac{70}{h^3}$ (e) $\frac{\partial^2 B}{\partial h\,\partial w} = \frac{\partial^2 B}{\partial w\,\partial h} = -\frac{2}{h^3}$

*Abbreviations: Dom = domain; max = maximum; min = minimum; F = inflection point.

Chapter 6

Exp and Log Functions

Preamble

In Chapter 1, Section 1.3.3, we introduced indices and roots, and detailed their manipulation. Here we expand on these concepts and introduce the inverse operation of 'raising to the power': the logarithmic function which is widely used in the Biosciences. We need familiarity with this function as one of the vital building blocks of our mathematical foundations. We will use logarithms base 10 and also so-called 'natural' logarithms. For the latter we learn about the exponential function which is very important in later chapters.

6.1 The Indicial Function

6.1.1 Definition of the indicial function

Before defining logarithmic and exponential functions, we visit the **indicial function** $y = a^x$, as this is the basis of both the functions that we will meet soon.

In the expression a^x, we require that the **base** a is a *positive* real number, *i.e.* $a > 0$.

The **index** x can be any real number.

In order to define a^n, let $a > 0$ be a positive real number and the exponent, $x = n$, be a positive integer.

For the product of a with itself $x = n$ times, we write $a^x = a^n = \underbrace{a \cdot a \cdot \ldots a \cdot a}_{n \text{ factors}}.$

When x is an integer, such as n, then $a^n = a \cdot a \ldots\ldots a \cdot a$, n times.

As shown in Section 1.3.3, when x is the reciprocal of an integer, *i.e.* $\frac{1}{n}$, then $a^{\frac{1}{n}}$ is the nth real root of a: $\sqrt[n]{a}$.

In this way, a^n can be defined whenever x is rational, for instance, if $x = \frac{p}{q}$, then $a^{\frac{p}{q}}$ is the qth root of a multiplied by itself p times:

$$a^{\frac{p}{q}} = \sqrt[q]{\underbrace{a \cdot a \cdot \ldots \cdot a \cdot a}_{p \text{ factors}}}.$$

These definitions allow the computation of a^x whenever x is rational.

When x is irrational, the value of a^x can be approximated using a rational approximation for x.

For example, $\pi = 3.14159\ldots$ is an irrational number, *i.e.* cannot be expressed as a quotient of two integers. The value of a^π can be approximated using $a^{3.1}$, $a^{3.14}$, $a^{3.141}$, ... The accuracy improves with the number of decimal places of π that is used.

We illustrate this in Figure 6.1 where two functions of the type $y = a^x$ are plotted: one with $a > 1$ and the other with $0 < a < 1$. The dotted lines indicate the locations of π and a^π in each graph.

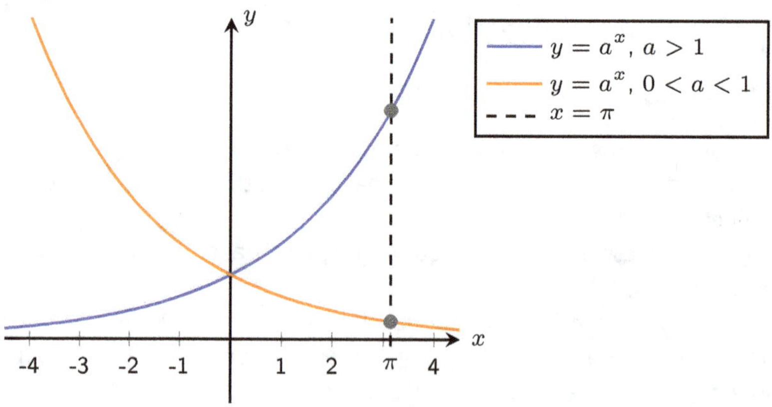

Figure 6.1: Indicial functions

The key information useful when sketching these two cases of indicial functions is as follows:

(i) **Domain:** Every real number ($x \in \mathbb{R}$) in both cases, *i.e.* $(-\infty, +\infty)$.
(ii) **Key points:** $f(0) = 1$ in both cases.
(iii) **Sign:** The indicial function is **always** positive in both cases.
(iv) **Limits and asymptotes:**

(a) If $a > 1$,

- $\lim_{x \to +\infty} f(x) = +\infty$;
- $\lim_{x \to -\infty} f(x) = 0^+$, thus $y = 0$ is a horizontal asymptote of the function as $x \to -\infty$ (*i.e.* for extreme negative values of x).

(b) If $0 < a < 1$,

- $\lim_{x \to +\infty} f(x) = 0^+$, thus $y = 0$ is a horizontal asymptote of the function as $x \to +\infty$ (*i.e.* for extreme positive values of x);
- $\lim_{x \to -\infty} f(x) = +\infty$.

(v) **Continuity:** Indicial functions are continuous on their domain.
(vi) **Stationary and inflection points:** No stationary and inflection points are observed, and concavity is upwards in both cases. Moreover,

(a) if $a > 1$, the function is increasing for all x;
(b) if $0 < a < 1$, the function is decreasing for all x.

(vii) **Co-domain:** The range of y values is $(0, +\infty)$ in both cases.

Graphs of the function $y = a^x$ with $a > 1$ (Figure 6.1) are good models for unconstrained population growth and the spread of certain diseases.

Graphs of the function $y = a^x$ with $0 < a < 1$ (Figure 6.1) are good models for radioactive decay or the concentration of a specific drug in the body after an injection has been given.

Note that rules for manipulating indices were covered in Chapter 1, Section 1.3.3, and practice at handling them was included in the Exercises at the end of Chapter 1.

6.2 The Logarithmic Function

6.2.1 Definition of the logarithmic function

Logarithms were introduced by John Napier in 1614 as a means of simplifying calculations and many fields quickly adopted their use. *Logos* translates to 'reckoning' and *arithmos* to 'number'.

The **logarithm** or **log** is another word for *index* or *power*.

We illustrate this using the graph of $y = 2^x$ (Figure 6.2).

It is clear from this graph that we can represent any positive real number y by raising the number 2 to an appropriate index x.

For example, the number 8 can be represented as $2 \cdot 2 \cdot 2 = 2^3$ and the number $\frac{1}{4}$ can be represented as $\frac{1}{2} \cdot \frac{1}{2} = 2^{-1} \cdot 2^{-1} = 2^{-1-1} = 2^{-2}$.

Put your finger on any value of y along the vertical axis and move it horizontally until you touch the graph (see Figure 6.2). The corresponding x-value of the point will be the power to which you need to raise 2 in order to express y as 2^x.

When you put your finger on $y = 8$, the corresponding x-value is $x = 3$; when you put your finger at $\frac{1}{4}$, the corresponding x-value y is $x = -2$.

This value of x for which $2^x = y$ is called the **logarithm to the base 2** of y and is written as $x = \log_2 y$.

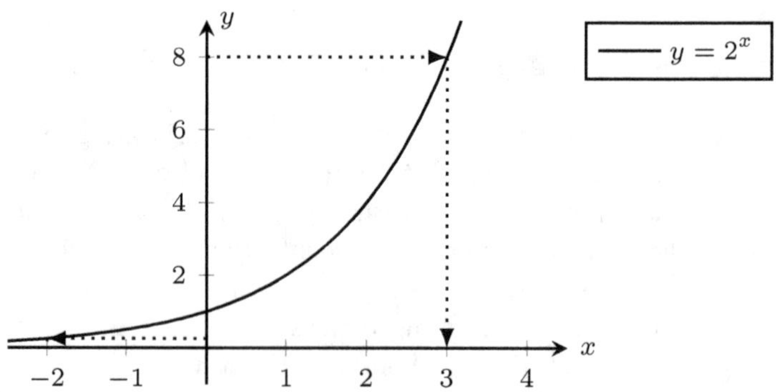

Figure 6.2: Function $y = 2^x$

So we can see that because $2^3 = 8$, we have $\log_2 8 = 3$, and because $2^{-2} = \frac{1}{4}$, we have $\log_2\left(\frac{1}{4}\right) = -2$.

> **Important note:** Effectively taking the logarithm has 'undone' raising the base 2 to the power x.

Example: Using the graph of $y = 2^x$, calculate $\log_2 5$ and $\log_2\left(\frac{1}{3}\right)$.

Solution: Find the points $y = \frac{1}{3}$ and $y = 5$ on the graph of $y = 2^x$ (Figure 6.2). From the graph, we can see that $2^{2.3} \approx 5$ and $2^{-1.6} \approx \frac{1}{3}$.

This means that $\log_2 5 \approx 2.3$ and that $\log_2\left(\frac{1}{3}\right) \approx -1.6$.

We can generalise this notion to include any base $a > 0$ and any positive real number $b > 0$.

The **logarithm to the base *a* of *b*** is the *power* to which the *base* a must be raised in order to obtain b. Thus,

$$b = a^x \iff x = \log_a b \tag{6.1}$$

Therefore, any equation that involves a base raised to an index can be written in terms of logarithms.

Moving comfortably between these two equivalent modes is an important skill to develop.

Here are some examples of the application of equation (6.1):

- $3^2 = 9$ is equivalent to $\log_3 9 = 2$.
- $10^3 = 1000$ is equivalent to $\log_{10} 1000 = 3$.
- $5^{-2} = \frac{1}{25}$ is equivalent to $\log_5\left(\frac{1}{25}\right) = -2$.
- $\left(\frac{1}{5}\right)^{-2} = 25$ is equivalent to $\log_{\frac{1}{5}} 25 = -2$.

Important note: The base of a logarithm may be any real positive number. Commonly, logarithms which have **base 10** can be written simply as log, rather than $\log_{10}$, so for the second example we can write $\log 1000 = 3$. However, **any base other than 10 must** be stated, as shown in the first, third and fourth examples.

6.2.2 Inverse function theorem

These examples illustrate the fact that the operation of raising a to a power and taking the logarithm to the base a are **inverse operations**: each operation undoes the effect of the other.

This means that just like 'taking the square root of x' undoes the effect of 'squaring x' so does 'taking the logarithm to the base a of x' undoes the effect of 'raising a to the power x'.

The inverse function theorem: Let f be a continuous function defined on the interval (a, b) of x and giving y values in the interval (A, B), that is,

$$f : (a, b) \longrightarrow (A, B) \quad \text{and} \quad x \longrightarrow f(x)$$

If the following conditions are satisfied,

(i) $f(x)$ is differentiable on (a, b);
(ii) $f(x)$ is always > 0 (*i.e.* $f(x)$ is always increasing) OR < 0 (*i.e.* $f(x)$ is always decreasing on (a, b),

then the function $y = f(x)$ is **invertible** for all values of $x \in (a, b)$ and $y \in (A, B)$, and the **inverse function** is denoted by $x = f^{-1}(y)$.

Warning! The notation f^{-1} is **not** equivalent to $\frac{1}{f}$!

Three key properties derive from the inverse function theorem:

(1) $f^{-1}(f(x)) = x$ and $f(f^{-1}(y)) = y$;
(2) $x = f^{-1}(y)$ is differentiable on (a, b) allowing us to derive a very important result, which we will need very soon:

$$(f^{-1})'(y) = \frac{1}{f'(x)} = \frac{1}{f'(f^{-1}(y))} \Longleftrightarrow \boxed{\frac{\mathbf{dx}}{\mathbf{dy}} = \frac{1}{\frac{\mathbf{dy}}{\mathbf{dx}}}}$$

(3) The graphs of the functions f and f^{-1} are **reflections of one another** with respect to the straight line $y = x$.

Based on the inverse function theorem, the indicial function $y = a^x$ satisfies the two required conditions (i) and (ii) and is therefore invertible.

The inverse function of $y = a^x$ is then $y = \log_a x$.

Thus, we can summarise the resulting properties:

(1) Taking the logarithm undoes raising to a power:

$$\log_a a^x = x \tag{6.2}$$

Raising to a power undoes taking the logarithm:

$$a^{\log_a x} = x \tag{6.3}$$

(2) The derivative of $y = \log_a x$ can be straightforwardly obtained from the derivative of $y = a^x$, but this topic is further developed in Section 6.5.2 after we have developed a feel for logarithms.

(3) The graphs of the functions $y = a^x$ and $y = \log_a x$ are **reflections of one another** with respect to the straight line $y = x$ (Figure 6.3).

6.2.3 Getting a feel for logarithms

What is x if $10^x = 7$ (*i.e.* the base here is 10)?

It is clear that x must be between 0 and 1, and we can test this on a calculator and find that $x = 0.85$. Thus, $\log 7 \approx 0.85$.

We can see that for $10^x = y$ and $\log y = x$.

- if y is between 1 and 10, $0 \leq x \leq 1$,
- if y is between 10 and 100, $1 \leq x \leq 2$,
- if y is between 0.1 and 1, $-1 \leq x \leq 0$,
- if y is between 0.01 and 0.1, $-2 \leq x \leq -1$,
- *etc.*

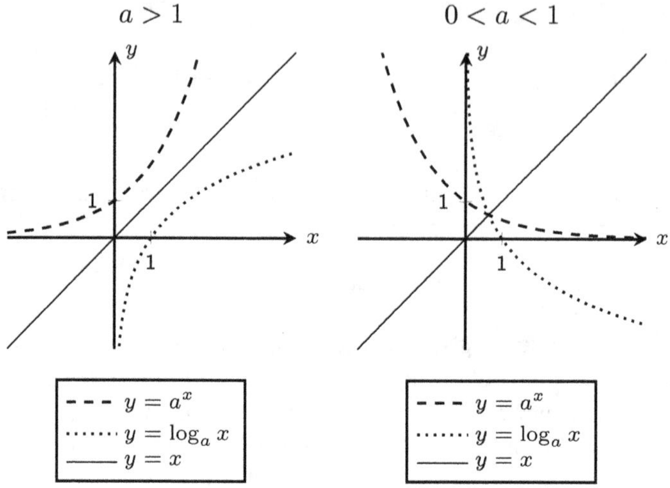

Figure 6.3: Graphs of $y = a^x$ and $y = \log_a x$

Example: Use the graph of $y = \log_{10}(x)$ below to estimate the following: (*i*) $\log 6$, (*ii*) $\log 0.5$.

What is x if (*iii*) $y = 10^x = 4000000000$, or if (*iv*) $y = 10^x = 0.003$?

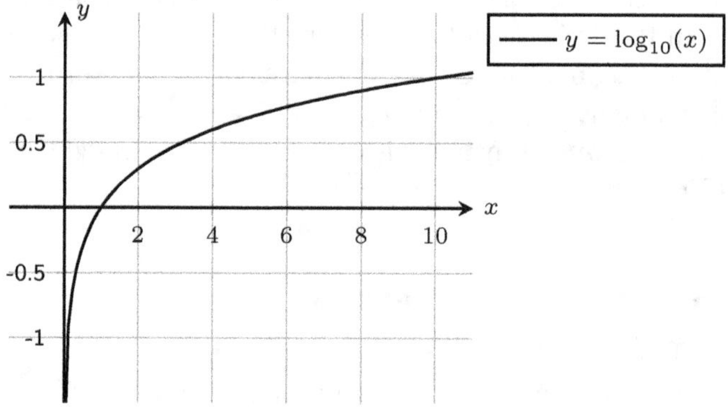

Solution: By putting a finger on the x value (*i*) 6 along the horizontal axis and moving it vertically until the graph is touched, the corresponding y-value of the point will be the logarithm with base 10 of 6, which is 0.8.

(*ii*) Similarly, $\log 0.5 = -0.3$.

(*iii*) From this graph, $4 \approx 10^{0.6}$, so $4 \cdot 10^9 = 10^{0.6} \cdot 10^9 = 10^{9.6}$, so $x = 9.6$.

The '9' gives the 9 'orders of 10' places and the '0.6' gives us approximately the '4'.

(*iv*) From the graph, $3 \approx 10^{0.5}$ so $3 \cdot 10^{-3} = 10^{0.5} \cdot 10^{-3} = 10^{-2.5}$ so $x = -2.5$.

The '-2' gives the two decimal places and the '-0.5' gives us approximately the '3'.

6.2.4 Graphical properties of the logarithmic function

In Figure 6.4, the graphs of $y = f(x) = \log_a x$ for a typical value of $a > 1$ and for $0 < a < 1$ are sketched.

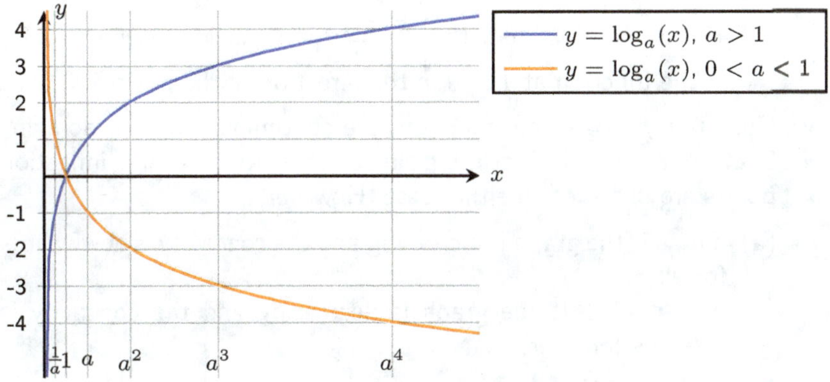

Figure 6.4: Logarithmic functions

The key information useful when sketching the two logarithmic functions above is as follows:

(i) **Domain:** The logarithm is defined *only* for $x > 0$ so the domain is $(0, +\infty)$ in both cases.

(ii) **Key points:** $f(1) = \log_a(1) = 0$ in both cases.

The points $1/a = a^{-1}, 1 = a^0, a, a^2, a^3$ and a^4 along the x-axis respectively correspond to:

(a) their logarithms $-1, 0, 1, 2, 3, 4$ along the y-axis, if $a > 1$;

(b) their logarithms $1, 0, -1, -2, -3, -4$ along the y-axis, if $0 < a < 1$.

(iii) **Sign:** Two distinct cases are possible:

(a) if $a > 1$, the logarithm function is *positive* for $x > 1$ and *negative* for $0 < x < 1$;

(b) if $0 < a < 1$, the logarithm function is *negative* for $x > 1$ and *positive* for $0 < x < 1$.

(iv) **Limits and asymptotes:**

(a) If $a > 1$,

- $\lim_{x \to +\infty} f(x) = +\infty$, the graph has *no* horizontal asymptote;
- $\lim_{x \to 0+} f(x) = -\infty$, thus, $x = 0$ is a vertical asymptote for the function at $x = 0$ in the direction of the negative y-axis.

(b) If $0 < a < 1$,

- $\lim_{x \to +\infty} f(x) = -\infty$, the graph has *no* horizontal asymptote;
- $\lim_{x \to 0+} f(x) = +\infty$, thus, $x = 0$ is a vertical asymptote for the function at $x = 0$ in the direction of the positive y-axis.

(v) **Continuity:** Logarithm functions are continuous on their domain.

(vi) **Stationary and inflection points:** No stationary and inflection points are observed in either case. However,

(a) if $a > 1$, the graph is increasing and the concavity is downwards for all x;

(b) if $0 < a < 1$, the graph is decreasing and the concavity is upwards for all x.

(vii) **Co-domain:** The range of y values is $(-\infty, +\infty)$ in both cases.

In the Biosciences, the function of $y = \log_a(x)$ with $a > 1$ is mainly used to describe physical entities which increase very slowly with increasing x, with x usually representing time.

For example, for $a = 10$, the horizontal axis extends $10^4 = 10000$ units to the right while the vertical axis only extends 4 units upwards!

This feature of the logarithmic function allows us to bring very large numbers back within manageable range.

We thus use logarithmic functions as metrics for a number of physical parameters which have huge **dynamic ranges**. For instance,

- **Volume of sound:** Decibels, with one decibel being one tenth of one bel, named in honour of Alexander Graham Bell, and relating to the power (Watts) in sound waves.

- **Pitch:** One octave is a doubling of frequency, and two octaves is a quadrupling of frequency.
- **Earthquakes:** The Richter scale ranges from 1 for a mini earthquake to >9.0 for total destruction in an earthquake zone, such as the asteroid impact that created the Chicxulub crater in Mexico. This caused the mass extinction which killed the non-avian dinosaurs and triggered an earthquake estimated at magnitude 13, while an earth with a magnitude of 15 might completely destroy the Earth.

  A Richter index, M_L, can be calculated using the Lillie empirical formula: $M_L = \log_{10} A - 2.48 + 2.76 \log_{10} \Delta$, where Δ is the distance from the epicentre in km and A is the amplitude (maximum ground displacement) of the P-wave (first wave to arrive after the quake and travelling directly from the epicentre, not by reflection from the earth's mantle), in micrometers, measured at $0.8\,\mathrm{Hz}$.
- **pH:** It is defined as $-\log[H^+]$, where $[H^+]$ is the ion concentration in $\mathrm{mol.L}^{-1}$, and the usual range is 1–14 in aqueous solutions, *i.e.* $[H^+] = 10^{-1}$–$10^{-14}\ \mathrm{mol\,L}^{-1}$ (or units of Molarity M).
- **Computer programming:** Logarithms are used extensively to improve the speed of computer calculations, as the program can add the logarithms of numbers in order to multiply those numbers together (see Rule **A** below).

6.3 Algebraic Properties of the Logarithm

Each of the properties established for indices in Section 1.3.3 implies that there is an equivalent property for logarithms. We list these below.

Important note: For all the rules listed in this section, the variables p and q **must be positive**.

Rule A. Products: $\boxed{\log_a(p \cdot q) = \log_a p + \log_a q}$

Proof: Let $p = a^c$ and $q = a^d$. Then, $c = \log_a p$ and $d = \log_a q$.
By the multiplication rule for exponents, we have

$$\log_a(p \cdot q) = \log_a(a^c \cdot a^d) = \log_a\left(a^{c+d}\right) = c + d = \log_a p + \log_a q$$

Example: Given that $\log_2 3 \approx 1.6$, calculate $\log_2 48$.

Solution: $\log_2 48 = \log_2(16 \cdot 3) = \log_2(2 \cdot 2 \cdot 2 \cdot 2) + \log_2 3$
$= \log_2 2 + \log_2 2 + \log_2 2 + \log_2 2 + \log_2 3 \approx 4 + 1.6 = 5.6$
(remember that $\log_2 2 = 1$ as $2^1 = 2$).

Rule B. Quotients: $\boxed{\log_a \left(\frac{p}{q}\right) = \log_a p - \log_a q}$

Proof: Let $p = a^c$ and $q = a^d$ so that $c = \log_a p$ and $d = \log_a q$ as above.

By the Quotient Rule for exponents, we have

$$\log_a \left(\frac{p}{q}\right) = \log_a \left(\frac{a^c}{a^d}\right) = \log_a \left(a^{c-d}\right) = c - d = \log_a p - \log_a q$$

Example: Given that $\log_{10} 5 \approx 0.7$, calculate $\log_{10} 2$.

Solution: $\log_{10} 2 = \log_{10} \left(\frac{10}{5}\right) = \log_{10} 10 - \log_{10} 5 \approx 1 - 0.7 = 0.3.$

Rule C. Powers: $\boxed{\log_a (p^r) = r \cdot \log_a p}$

Proof: Let $p = a^c$ so that $c = \log_a p$ as above. By the power rule for exponents, we have

$$\log_a (p^r) = \log_a (a^c)^r = \log_a (a^{c \cdot r}) = c \cdot r = r \cdot \log_a p$$

Example: Given that $\log_{10} 5 \approx 0.7$, calculate $\log_{10} 25$.

Solution: $\log_{10} 25 = \log_{10} 5^2 = \log_{10}(5 \cdot 5) = \log_{10} 5 + \log_{10} 5$
$= 2 \cdot \log_{10} 5 \approx 2 \cdot 0.7 = 1.4.$

Rule D. Inverses: $\boxed{\log_a \left(\frac{1}{p}\right) = -\log_a p}$

Proof: Using the quotient property (Rule **B**), and remembering that, because $a^0 = 1$, $\log_a 1 = 0$,

$$\log_a \left(\frac{1}{p}\right) = \log_a 1 - \log_a p = 0 - \log_a p = -\log_a p$$

or, alternatively, using the power property (Rule **C**),

$$\log_a \left(\frac{1}{p}\right) = \log_a p^{-1} = (-1) \cdot \log_a p = -\log_a p$$

Example: Given that $\log_{10} 5 \approx 0.7$, calculate $\log_{10} \left(\frac{1}{5}\right)$.

Solution: $\log_{10} \left(\frac{1}{5}\right) = -\log_{10} 5 \approx -0.7$.

Rule E. Changing Base: $\boxed{\log_\mathbf{a} \mathbf{p} = \dfrac{\log_\mathbf{b} \mathbf{p}}{\log_\mathbf{b} \mathbf{a}}}$ for any two bases $a > 0$ and $b > 0$

Proof: Let $p = a^c$ so that $c = \log_a p$ as before.
Then, $\log_b p = \log_b a^c = c \cdot \log_b a$.
Dividing both sides of the equation $\log_b p = c \cdot \log_b a$ by $\log_b a$, we get

$$\frac{\log_b p}{\log_b a} = c \cdot \frac{\log_b a}{\log_b a} = c = \log_a p$$

Important note: If $p = b$, Rule **E** becomes

$$\log_a b = \frac{\log_b b}{\log_b a} = \frac{1}{\log_b a}$$

Example: Given that $\log_{10} 5 \approx 0.7$ and $\log_{10} 2 \approx 0.3$, calculate $\log_2 5$.

Solution: $\log_2 5 = \dfrac{\log_{10} 5}{\log_{10} 2} \approx \dfrac{0.7}{0.3} \approx 2.3$.

Warning! We now mention some **non-properties**. These are rules that you may think should be true, but that are actually not correct. These frequently cause difficulties when manipulating logarithms:

- $\log_a(x \pm y) \neq \log_a x \pm \log_a y$. The logarithms of a sum or a difference, $\log_a(x+y)$ and $\log_a(x-y)$, **cannot** be simplified any further and should **always** be left as they are.
- Products and quotients of logarithms, $\log_a x \cdot \log_a y$ and $\dfrac{\log_a x}{\log_a y}$, **cannot** be simplified any further and should also **always** be left as they are.

6.3.1 Manipulating logarithms

The relationships established above can be used to simplify or split up expressions containing logarithms.

Example: Express the equation $\log x = \log p + 2\log q - \log k - 3$ without using logarithms.

Solution: If we want to write the equation $\log x = \log p + 2\log q - \log k - 3$ without using logarithms, we first need to rewrite -3 as $-3\log 10$ (remember that $\log_{10} 10 = 1$) so that then using Rules **A**, **B**, and **C**, we get

$$\log x = \log\left(\frac{pq^2}{k \cdot 10^3}\right), \text{ so } x = \left(\frac{pq^2}{k \cdot 10^3}\right)$$

Example in Biology: Consider the following data for domestic deaths from AIDS in the United States between 1981–1987:

Year since 1980 (t)	1	2	3	4	5	6	7
Total deaths to date (N)	268	1,209	3,826	8,712	17,386	29,277	41,128

These data are plotted on the following graph:

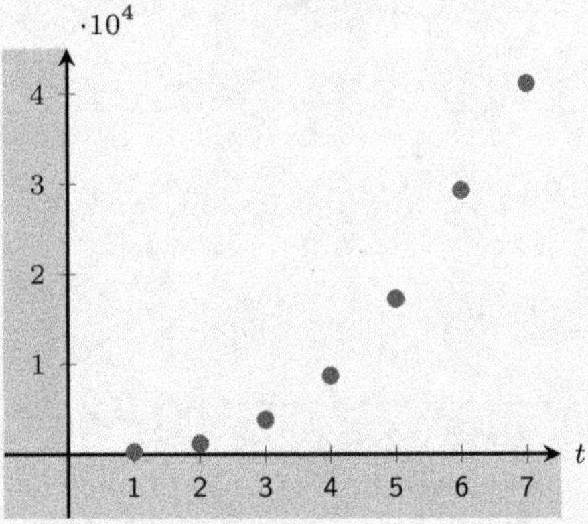

What simple mathematical expression can we use to model these data?

Solution: These data cannot be modelled by using a linear mathematical expression of the form $N = m \cdot t + c$ because the differences between adjacent years are growing from year to year (remember that for the equation of a straight line, $y = mx + c$, if x increases by 1, then y always increases by the constant gradient m).

It cannot be indicial, of the form $N = C \cdot a^t$, because the ratios between adjacent years are decreasing.

It cannot be quadratic of the form $N = a \cdot t^2$ because when we double t we do *not* quadruple N.

If it is of the form $y = a \cdot t^r$ for some power of t other than 2,

- how can we find the correct power of t?
- how can we find the value of a?

One way to address this question is to transform the data and plot $\log N$ versus $\log t$.

$\log t$	0	0.30	0.48	0.60	0.70	0.78	0.85
$\log N$	2.4	3.1	3.6	3.9	4.24	4.5	4.6

When we do this we find that the transformed data points lie on a straight line, as shown in the following graph:

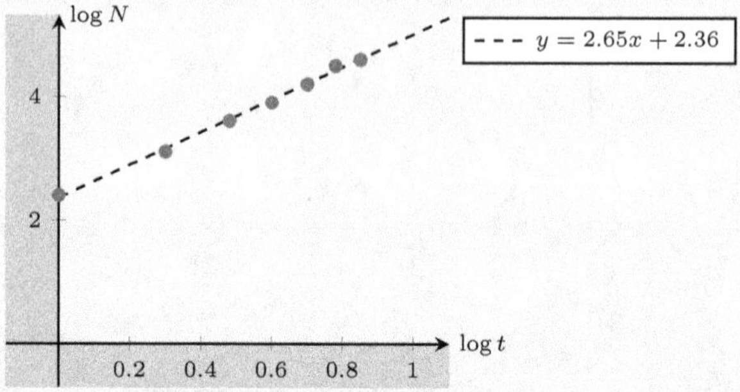

From this analysis, we can see that the line $y = 2.65x + 2.36$ (gradient of 2.65 and intercept with the y-axis of 2.36) is a good fit to the data so that an appropriate model for the transformed data is

$$\log_{10} N = \underbrace{2.65}_{m} \log_{10} t + \underbrace{2.36}_{c} = \log_{10} t^{2.65} + 2.36$$

If we raise 10 to the powers on each side and use the properties of logarithms, we have

$$10^{\log_{10} N} = 10^{(\log_{10} t^{2.65} + 2.36)}$$

$$\iff N = 10^{(\log_{10} t^{2.65})} \cdot 10^{2.36} = t^{2.65} \cdot 10^{2.36} \approx 230 \cdot t^{2.65}$$

Warning! 10 must be raised to the powers on each side simultaneously.

From this we can conclude that $N = 230{\cdot}t^{2.65}$ is a possible model for the original data, *i.e.* deaths *versus* time, and will appropriately fit the points in the initial plot. By extrapolating the line beyond the data points, we could now use this model to, for instance, see how many deaths were predicted to occur in 1988.

Note that in terms of the slope m and the intercept c obtained from the transformed data, this model takes the general form of $N = 10^c \cdot t^m$.

The lesson here is that sometimes we can find a simple pattern by plotting the *logarithm* of one variable against the *logarithm* of another *i.e.* by transforming the data to generate *log–log plots* or if appropriate, *log-linear plots*.

6.4 The Exponential Function

6.4.1 Definition of the exponential function

Let us consider again the graphs of the indicial functions of the form $y = a^x$, in particular $y = 1^x$, $y = 1.5^x$, $y = 2^x$, $y = 3^x$, and $y = 4^x$, as shown in Table 6.1.

All functions pass through the point $A(0, 1)$ on the y-axis, but each has a different gradient at that point (Figure 6.5): *i.e.* the tangent line to each graph $y = a^x$ at the point $A(0, 1)$ have different gradients.

A closer look will reveal these gradients to be increasing upwards from 0, and this suggests the existence of a base a for which the graph of the curve $y = a^x$ crosses the y-axis at $A(0, 1)$ with a gradient **exactly equal to** 1: *i.e.* the gradient of the tangent line to the graph $y = a^x$ at point $A(0, 1)$ is equal to 1 (Table 6.1).

This special base is called e, and the particular indicial function $y = e^x$ is generally known as the **exponential function**.

Table 6.1: Gradient values of exponential functions at $A(0, 1)$

Function	Gradient at $A(0,1)$	Equation of Tangent for Function at $A(0,1)$
$y = 1^x$	0	$y = 1$
$y = 1.5^x$	0.41	$y = 0.41x + 1$
$y = 2^x$	0.69	$y = 0.69x + 1$
$\mathbf{y = e^x}$	**1**	$\mathbf{y = x + 1}$
$y = 3^x$	1.09	$y = 1.09x + 1$
$y = 4^x$	1.39	$y = 1.39x + 1$

Additionally, the real number e, also known as **Euler's number**, was initially found to be the value of this particular limit:

$$e = \lim_{n \to +\infty} \left(1 + \frac{1}{n}\right)^n$$

where n is a real positive integer ($n \in \mathbb{N}$).

This limit was also proved to be true if x is a real number ($x \in \mathbb{R}$) when $x \to +\infty$:

$$e = \lim_{x \to +\infty} \left(1 + \frac{1}{x}\right)^x \tag{6.4}$$

The proof of equation (6.4) can be found in Section 6.7.3.

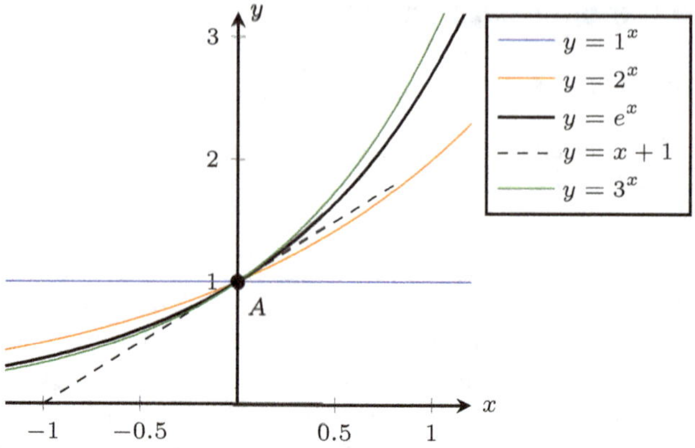

Figure 6.5: Exponential function: $y = e^x$

6.4.2 Definition of the real number e

The real number e, **Euler's number**, is that base for which the graph of the **exponential function** $y = e^x = exp(x)$ passes through the point $(0, 1)$ with gradient exactly equal to 1. Sometimes it is convenient to write $\exp(x)$ for e^x.

Thus, recalling the definitions of gradient and derivative, we have

$$\left.\frac{d}{dy}e^x\right|_{x=0} = \exp'(0) = \lim_{h\to 0}\frac{e^{(0+h)} - e^0}{(0+h) - 0} = \lim_{h\to 0}\frac{e^h - 1}{h} = 1 \qquad (6.5)$$

The number e is irrational, as are $\sqrt{2}$ and π, as they cannot be expressed as the quotient of integers. In fact, e satisfies **no** algebraic equation, but we can see from Figure 6.5 that it lies between 2 and 3.

Its approximate value is $e \approx 2.71828$.

So, for the function $y = e^x$, called the **exponential function**, the gradient at $A(0, 1)$ is 1. Thus, at $A(0, 1)$,

$$y = e^x \Longrightarrow \frac{dy}{dx} = \text{gradient} = e^x$$

This relationship between the function and its derivative turns out to be true at **any point** on the curve. The function $y = e^x$ is the **only** elementary function that is unchanged when differentiated, apart from the function $y = 0$.

Thus, the derivative for the function $y = e^x$ is $y' = e^x$.

Let $f(x) = e^x$. Then, according to the definition of a derivative,

$$f'(x) = \lim_{h\to 0} \frac{f(x+h) - f(x)}{h} = \lim_{h\to 0} \frac{e^{x+h} - e^x}{h} = \lim_{h\to 0} \frac{e^x(e^h - 1)}{h} = e^x$$

since $\lim_{h\to 0} \dfrac{e^h - 1}{h} = 1$ (equation (6.5)).

6.4.3 The Binomial Series and the exponential function

As mentioned above, remembering that e is an irrational number, the function e^x can also be expressed as a limit

$$e^X = \lim_{n\to\infty} \left[1 + \frac{X}{n}\right]^n$$

where n is a real positive integer ($n \in \mathbb{N}$). We can use the Binomial Series (Section 2.1.1) to expand it. The Binomial Expansion states that

$$(1+x)^n = 1 + \frac{nx}{1} + \frac{n(n-1)}{2!}x^2 + \frac{n(n-1)(n-2)}{3!}x^3 + \cdots$$

If we substitute $x = \dfrac{X}{n}$ into this general form, we get

$$\left[1+\frac{X}{n}\right]^n = 1+\frac{nX}{n}+\frac{n(n-1)X^2}{2!\,n^2}+\frac{n(n-1)(n-2)X^3}{3!\,n^3}+\cdots \quad (6.6)$$

where the ! denotes the 'factorial' function: for instance, here $n! = n \cdot (n-1) \cdot (n-2) \cdot (n-3)...3 \cdot 2 \cdot 1$.

In the limit as $n \to \infty$, the coefficients $\dfrac{n(n-1)}{n^2} \longrightarrow 1$, $\dfrac{n(n-1)(n-2)}{n^3} \longrightarrow 1$ and the coefficients of all the other terms behave similarly so that equation (6.6) becomes

$$e^X = 1 + X + \frac{X^2}{2!} + \frac{X^3}{3!} + \frac{X^4}{4!} + \cdots \quad (6.7)$$

Now if $X = 1$, then

$$e = 1 + 1 + \frac{1}{2} + \frac{1}{6} + \frac{1}{24} \cdots = 2.71828\ldots$$

Note that if we differentiate equation (6.7), we get

$$\frac{d}{dx}e^X = 1 + \frac{2X}{2!} + \frac{3X^2}{3!} + \frac{4X^3}{4!} + \cdots = 1 + X + \frac{X^2}{2!} + \frac{X^3}{3!} + \cdots = e^X$$

exactly as expected from the definition of e^X (see Section 6.4.1).

6.5 Moving from the Exponential to Natural Logs

Euler's number, e, has been found to arise in many branches of Mathematics.

It is also used as a **base** for logarithms, $\log_e$, and this is often denoted by $\ln$ instead of $\log_e$.

For example, $y = \log_e 10 = \ln 10$ and $y = \log_e x = \ln x$.

Logs to the base e are called **natural logarithms**.

From the Changing Base Rule (Rule **E**) above (Section 6.3), we have

$$\log_{10} x = \frac{\log_e x}{\log_e 10} = \frac{\ln x}{\ln 10} \quad \Longleftrightarrow \quad \ln x = \ln 10 \cdot \log_{10} x \approx 2.303 \ \log_{10} x$$

This change of base is why the number 2.303 appears frequently in electrochemistry and kinetics.

Thus, the **natural logarithm** of a number is the logarithm of that number *taken using the base e*.

Here are some of its properties which are common to other logarithmic functions:

- $\ln(e) = \log_e(e) = 1$;
- $\ln(10) = \log_e(10) = ?$ means 'Find the power to which I need to raise e in order to get 10': $\ln(10) \approx 2.303$, so $e^{2.303} \approx 10$;
- $\ln(e^b) = \log_e(e^b) = \ln(\exp(b)) = b$;
- $e^{\ln(b)} = e^{\log_e b} = \exp(\ln(b)) = b$.

Note that the last two relationships derive from the properties of the functions '$\exp$' and '$\ln$', as *inverse functions* of one another. Thus, 'taking the natural logarithm of x' undoes the effect of 'raising e to the power x'.

In addition, the graphs of $y = e^x$ and $y = \ln(x)$ are reflections of one another with respect to the straight line $y = x$ (see first graph in Figure 6.3, since $e > 1$).

6.5.1 The derivative of $\ln x$

From the inverse function theorem (see Section 6.2.2),

$$y = \ln x \Longleftrightarrow x = e^y$$

Now, let us consider the function e^y which is unchanged by differentiation with respect to y, so

$$x' = \frac{dx}{dy} = e^y = x$$

which confirms $x' = x$ (we can compare this result with $y = e^x \Longrightarrow \frac{dy}{dx} = e^x = y$).

According to the inverse function theorem (Section 6.2.2: third property),

$$y' = \frac{dy}{dx} = \frac{d(\ln x)}{dx} = \frac{1}{\frac{dx}{dy}} = \frac{1}{e^y} = \frac{1}{e^{\ln x}} = \frac{1}{x}$$

Thus, the derivative of $y = \ln x$ is $y' = \frac{1}{x}$.

This is a **very important result**!

6.5.2 The derivative of $\log_a x$

Recalling the conversion formula $\log_a(x) = \frac{\ln(x)}{\ln(a)}$ (Rule **E**, Section 6.3) and noting that $\ln(a)$ is a constant, we have

$$\begin{aligned}\frac{d}{dx}\log_a(x) &= \frac{d}{dx}\left(\frac{1}{\ln(a)} \cdot \ln(x)\right) \\ &= \left(\frac{1}{\ln(a)}\right) \cdot \frac{d}{dx}\ln(x) = \left(\frac{1}{\ln(a)}\right) \cdot \frac{1}{x} = \frac{1}{\ln(a) \cdot x}\end{aligned}$$

Thus, the derivative of $y = \log_a x$ is $y' = \frac{1}{\ln(a) \cdot x}$.

Example: If $y = \log_2(x)$, find the derivative $\frac{dy}{dx}$.

Solution: If $y = \log_2(x)$, then

$$\frac{dy}{dx} = \frac{d}{dx}\left(\frac{1}{\ln(2)} \cdot \ln(x)\right) = \left(\frac{1}{\ln(2)}\right) \cdot \frac{1}{x} = \frac{1}{0.693 \cdot x}$$

6.5.3 The derivative of a^x

As explained in Section 6.2.1, $y = a^x \iff x = \log_a y$. According to the inverse function theorem (Section 6.2.2: third property),

$$y' = \frac{dy}{dx} = \frac{d(a^x)}{dx} = \frac{1}{\frac{dx}{dy}} = \frac{1}{\frac{1}{\ln(a)\cdot y}} = \ln(a) \cdot y = \ln(a) \cdot a^x$$

Thus, the derivative of $y = a^x$ is $y' = \ln(a) \cdot a^x$.

Note: There are different ways to obtain these derivatives using the Chain and Product Rules.

6.6 Solving Expressions Containing Logs or Indices

Sometimes in Biosciences we need to solve equations or inequalities for x containing logarithms or indices.

Depending on the type of equation or inequality, an appropriate strategy can be used.

Different types of equations and inequalities exist, but we focus only on specific types of expressions mainly used in Biosciences and which we can relate to the types of equations and inequalities already studied in Chapter 2.

6.6.1 Elementary equations

Elementary equations are of the type

(A) exponential $a^{f(x)} = a^{g(x)}$

(B) logarithmic $\log_a f(x) = \log_a g(x)$

where $f(x)$ and $g(x)$ can be linear, polynomial, quadratic, or rational expressions containing the unknown x. In this case, the strategy is as follows:

(i) If the initial expression contains more indices or logarithms, the properties of indices or logarithms can be used to rearrange the initial expression into the corresponding **elementary form (A)** or **(B)**.

(ii) The functions $f(x)$ and $g(x)$ can then be equated: $f(x) = g(x)$.
(iii) The final equation can be solved by using one of the techniques described in Chapter 2.

Warning! For logarithmic equations only, the solutions for x should be checked by putting them back into the initial equation to ensure that they do not give a null or negative argument for any of the logarithms. If any do so, then they are not permitted to be solutions of the logarithmic equation.

Example: Solve the indicial equation $3^{2x+4} \cdot 3 = \left(\frac{1}{3}\right)^{x-2}$ for x.

Solution: Following the strategy described above, we can:

(i) rearrange the equation by using the properties of indices:
$3^{2x+4} \cdot 3 = \left(\frac{1}{3}\right)^{x-2} \iff 3^{2x+5} = 3^{2-x}$;
(ii) equate the exponents: $2x + 5 = 2 - x$;
(iii) solve the linear equation obtained: $x = -1$.

Example: Solve the logarithmic equation

$$1 - \log_{\frac{1}{3}}(x-2) = \log_3(x-5) \text{ for } x$$

Solution: Following the strategy outlined above, we can:

(i) rearrange the equation by using the properties of logarithms:
$1 - \log_{\frac{1}{3}}(x-2) = \log_3(x-5)$
$\iff \log_3(3) - \dfrac{\log_3(x-2)}{\log_3(1/3)} = \log_3(x-5)$
$\iff \log_3(3) + \dfrac{\log_3(x-2)}{\log_3 3} = \log_3(x-5)$
$\iff \log_3(3x-6) = \log_3(x-5)$;
(ii) equate the logarithmic arguments: $3x - 6 = x - 5$;
(iii) solve the linear equation: $x = \dfrac{1}{2}$;

(iv) check the solution by putting it back into the initial equation: $1 - \log_{\frac{1}{3}}\left(\frac{1}{2} - 2\right) = \log_3\left(\frac{1}{2} - 5\right)$ gives negative arguments for both the logarithms, therefore the solution $x = \frac{1}{2}$ is not acceptable. Thus, the equation has no solution for $x \in \mathbb{R}$.

Special cases

If the resulting elementary equations have different bases on each side of the equation:

(A) exponential $a^{f(x)} = b^{g(x)}$

(B) logarithmic $\log_a f(x) = \log_b g(x)$

- for **(A)**, the logarithms of both sides of the equation should be taken and Rule **C** for logarithms (Section 6.3) should be used to simplify the expression
$$\log\left(a^{f(x)}\right) = \log\left(b^{g(x)}\right) \Longleftrightarrow f(x) \cdot \log(a) = g(x) \cdot \log(b)$$
then the resulting equation can be solved using one of the techniques described above;
- for **(B)**, Rule **E** for logarithms (Section 6.3) can be used to change the base of one logarithm to the base of the other logarithm, for instance,
$$\log_a f(x) = \log_b g(x) \Longleftrightarrow \log_a f(x) = \frac{\log_a g(x)}{\log_a b}$$
By substituting $\log_a b = c$, the equation becomes
$$\log_a f(x) = \frac{\log_a g(x)}{c} = \log_a g(x)^{\frac{1}{c}} \Longleftrightarrow f(x) = g(x)^{\frac{1}{c}}$$

Example: Solve the equation $4^{3x-2} = 26^{x+1}$ for x.

Solution: Since the indices we want to equate have different bases, 4 and 26 respectively, we should use the following strategy:

(i) Take logs of both sides: $\log(4^{3x-2}) = \log(26^{x+1})$.
(ii) Rearrange and simplify by using the laws of logarithms: $(3x - 2)\log 4 = (x + 1)\log 26$.

(iii) Multiply out the brackets:
$3x \log 4 - 2\log 4 = x \log 26 + \log 26.$

(iv) Collect terms in x: $3x\log 4 - x\log 26 = \log 26 + 2\log 4$

$$\iff x(3\log 4 - \log 26) = \log(26 \cdot 16)$$

$$\iff x = \frac{\log(26 \cdot 16)}{3\log 4 - \log 26} = \frac{\log 416}{3\log 4 - \log 26} = \frac{2.62}{1.806 - 1.415}$$
$$= 6.7.$$

Example in Chemistry: As mentioned above, the notion of pH uses the slow growth of the logarithmic function to bring very large numbers back within a manageable range. The concentration $[H^+]$ of hydrogen atoms in a solution ranges from very low (10^{-14} M) to very high (1 M), *i.e.* a range of over 14 orders of magnitude.

We need to have a scale which expresses $[H^+]$ over this entire range of concentrations. The solution is to use the logarithmic function. Indeed, if we use $-\log_{10}[H^+]$ we get a number between 0 and 14. This number is defined to be the pH of a solution.

Using this definition,

(1) find the pH of a 0.011 M solution of HCl,
(2) find the H^+ concentration of a solution of HCl with a pH of 3,
(3) and find the pH of 100 mL of a solution containing 9 mg of HCl. The molecular weight of HCl is 34.46.

Solution: Let us answer each question in turn:

(1) Using the definition of pH:

$$\text{pH} = -\log_{10}[H^+] = -\log_{10} 0.011 = -(-1.959) = 1.959$$

(2) Since $[H]^+$ is the unknown: $-\log_{10}[H^+] = 3$ is an elementary logarithmic equation:

(i) Rearrange: $-\log_{10}[H^+] = 3 \iff \log_{10}[H^+] = \log_{10}(10^{-3})$.

(ii) Equate: $[H^+] = 10^{-3} = 0.001$ M.

(3) Since the molarity of HCl $= \dfrac{0.009}{36.46 \cdot 0.1} = 0.0025$ M, then

$$\text{pH} = -\log[H^+] = -\log 0.0025 = -(-2.61) = 2.61$$

6.6.2 Quadratic equations

If the form of the equation contains $a^{2x} = (a^x)^2$ or $\log_a^2(x) = (\log_a x)^2$, a **substitution method** should be used as detailed in the following steps:

(i) Simplify the equation as much as possible by using the properties of indices or logarithms.
(ii) Make the substitution $X = a^x$ or $X = \log_a(x)$ to obtain a quadratic equation.
(iii) Solve this quadratic equation in X (see Section 2.3) to obtain the two possible solutions X_1 and X_2.
(iv) Equate $X_{1 \text{ or } 2} = a^x$ or $X_{1 \text{ or } 2} = \log_a x$ and solve these elementary equations for x, as appropriate.

Warning! $\log_a^2(x) \neq \log_a(x^2)$.

In fact $\log_a^2(x) = (\log_a(x))^2 = \log_a(x) \cdot \log_a(x)$, while $\log_a(x^2) = 2\log_a(x)$.

Example: Solve the quadratic logarithmic equation

$$2(2\sqrt{3}\log x)^2 - 14\log(x^2) + 4 = 0 \text{ for } x$$

Solution: We can adopt the following strategy/steps to solve this equation:

(i) Simplify: $2(2\sqrt{3}\log x)^2 - 14\log(x^2) + 4 = 0$
$\iff 2 \cdot 2^2 \cdot 3 \cdot (\log x)^2 - 28\log x + 4 = 0.$
(ii) Substitute $\log x = X$ to give $24X^2 - 28X + 4 = 0.$
(iii) Solve this quadratic equation by factorisation:
$24X^2 - 28X + 4 = 0 \iff 6X^2 - 7X + 1 = (6X-1)(X-1) = 0,$
so $X_1 = \dfrac{1}{6}$ and $X_2 = 1.$
(iv) Substitute the two roots, X_1 and X_2 for their corresponding logarithms and solve the resulting elementary logarithmic equations:
$\log x_1 = \dfrac{1}{6} \iff x_1 = 10^{\frac{1}{6}} = \sqrt[6]{10} = 1.47$ and
$\log x_2 = 1 \iff x_2 = 10.$
(v) Both solutions $x_1 = 1.47$ and $x_2 = 10$ are acceptable since they do not make the arguments of the logarithms in the initial equation null or negative.

We can check our solutions by substituting them into the original equation in turn:

- $x_1 = 1.47$:
  $(2\sqrt{3}\log 1.47)^2 - 7\log(1.47^2) + 2 = 12 \cdot (0.167^2) - 7 \cdot 0.334 + 2 = 0$
  Works!
- $x_2 = 10$:
  $(2\sqrt{3}\log 10)^2 - 7\log(10^2) + 2 = 12 \cdot 1 - 7 \cdot 2 + 2 = 0$
  Works!

Example in Biology: In a greenhouse culture of fungi, after the initial time $t = 0$, the growth rate of the fungal species follows the function $y = -\log^2(t+1) + 2\log(t+1)$ with t in days. When is the growth rate equal to 0? What is the physical meaning of this time t?

Solution: To find when the growth rate, y, is 0, we put $y = -\log^2(t+1) + 2\log(t+1) = 0$, which is a quadratic logarithmic equation.

We can then follow the strategy explained above:

(1) Since the equation is already simplified, we can make the substitution $\log(t+1) = T$, which gives a quadratic equation in T: $-T^2 + 2T = 0$.
(2) We can solve this equation by factorisation:
$-T^2 + 2T = 0 \Longleftrightarrow T(-T+2) = 0$, which gives $T_1 = 0$ days and $T_2 = 2$ days.
(3) We can now substitute the logarithms back in to find the solutions for t:
- $T_1 = 0 \Longleftrightarrow \log(t_1 + 1) = 0 \Longleftrightarrow t_1 = 0$ which is not later than the initial time, $t = 0$;
- $T_2 = 2 \Longleftrightarrow \log(t_2 + 1) = 2 \Longleftrightarrow \log(t_2 + 1) = \log 10^2$
$\Longleftrightarrow t_2 = 100 - 1 = 99$ days, which is a little over 3 months.

Therefore, after about 3 months, no more new fungi will grow.

6.6.3 Elementary inequalities

- **(A) Exponential**

Elementary inequalities of the type $\boxed{a^{f(x)} > 0}$ are true for all possible real x values of the expression $f(x)$, *i.e.* for all values of the domain of $f(x)$.

Conversely, the inequality $\boxed{a^{f(x)} < 0}$ does **not** have any real values as its solution.

We may also meet other types of elementary inequalities:

$$\textbf{(A}_{>}\textbf{)}\ a^{f(x)} > a^{g(x)} \quad \text{OR} \quad \textbf{(A}_{<}\textbf{)}\ a^{f(x)} < a^{g(x)}$$

In these cases, the inequality ($A_{>}$) or ($A_{<}$) should be solved by following the same strategy described for elementary exponential equations (Section 6.6.1).

However, before comparing the expressions $f(x)$ and $g(x)$ with the symbol $>$ or $<$ in step (ii) in 6.6.1 above, care must be taken with the base a of the indices.

In particular,

- if $a > 1$, since the function $y = a^x$ is increasing over the entire domain of real numbers, we have:
  - ◇ **($A_{>}$)**: $a^{f(x)} > a^{g(x)} \Longleftrightarrow f(x) > g(x)$
  - ◇ **($A_{<}$)**: $a^{f(x)} < a^{g(x)} \Longleftrightarrow f(x) < g(x)$

  ('the inequality symbol is *preserved*');
- if $0 < a < 1$, since the function $y = a^x$ is decreasing over the entire domain of real numbers, we have:
  - ◇ **($A_{>}$)**: $a^{f(x)} > a^{g(x)} \Longleftrightarrow f(x) < g(x)$
  - ◇ **($A_{<}$)**: $a^{f(x)} < a^{g(x)} \Longleftrightarrow f(x) > g(x)$

  ('the inequality symbol is *reversed*').

The final inequality should be solved by using one of the techniques described in Chapter 2 and will give the ranges of solutions for the initial inequality.

Example: Solve the following exponential inequality: $e^{-2} - e^{\frac{x-1}{x}} < 0$.

Solution: As described above, we can solve the inequality by using the following steps:

(1) Rearrange: $e^{-2} < e^{\frac{x-1}{x}}$.

(2) Compare the exponents: $-2 < \dfrac{x-1}{x}$. The sign $<$ is 'preserved' since the base $e > 1$.

(3) Solve the rational inequality obtained by using the technique explained in Chapter 2, giving

$$-2 < \frac{x-1}{x} \Longleftrightarrow \frac{3x-1}{x} > 0 \Longleftrightarrow x < 0 \text{ OR } x > \frac{1}{3}$$

Example in Medicine: A gallium scan is a type of nuclear medicine test which uses a gallium-67 (^{67}Ga) radioactive pharmaceutical to obtain images of a specific type of tissue, or the disease state of tissue.

The amount of radioactive gallium present in a patient's blood after intravenous injection at time t (in hours) is given by the equation $G(t) = G_0 \left(\frac{1}{2}\right)^{\frac{t}{T_{1/2}}}$, where $G(t)$ is the amount of the substance present after time t, G_0 is the amount of ^{67}Ga initially injected, and $T_{1/2}$ is the half-life of ^{67}Ga (*i.e.* $T_{1/2}$ is the length of time it takes for half of the radioactive atoms of a specific radionuclide to decay).

The common injection dose of ^{67}Ga is about 150 megabecquerels (MBq) and the $T_{1/2}$ of ^{67}Ga is $80\,\text{hr}$. How long will it take for at least 90% of such an injected dose of ^{67}Ga to decay?

Solution: In order to estimate the time for at least 90% of an injected dose of ^{67}Ga to decay, we can solve the following inequality:

$$\frac{G(t)}{G_0} < 0.1 \Longleftrightarrow \left(\frac{1}{2}\right)^{\frac{t}{T_{1/2}}} < 0.1 \Longleftrightarrow \left(\frac{1}{2}\right)^{\frac{t}{T_{1/2}}} < \left(\frac{1}{2}\right)^{\log_{\frac{1}{2}}(0.1)}$$

Since this exponential inequality involves a base $a = \frac{1}{2} < 1$, we must reverse the inequality symbol to compare the exponents: $\frac{t}{T_{1/2}} > \log_{\frac{1}{2}}(0.1)$.

Thus, $t > T_{1/2} \cdot \log_{\frac{1}{2}}(0.1) \iff t > 3.32 \cdot 80 = 265.7\,\text{hr}$ which means that it takes just over 11 days for at least 90% of an injected dose of ^{67}Ga to decay.

It is notable that the actual amount of ^{67}Ga that was injected never had to be specified in the calculation.

Note on units: The argument of the logarithm is overall unitless.

- **(B) Logarithmic**

Generally two types of elementary logarithmic inequalities may occur in Biosciences:

(B$_>$) $\log_a f(x) > \log_a g(x)$ OR **(B$_<$)** $\log_a f(x) < \log_a g(x)$

In these cases, the inequality (B$_>$) or (B$_<$) can be solved by using the following strategy:

- if $a > 1$, since the function $y = \log_a x$ is increasing over the whole domain of real numbers, we have:
  - ⋄ **(B$_>$)**: $\log_a f(x) > \log_a g(x) \iff f(x) > g(x)$
  - ⋄ **(B$_<$)**: $\log_a f(x) < \log_a g(x) \iff f(x) < g(x)$

  ('the inequality symbol is *preserved*');
- if $0 < a < 1$, since the function $y = \log_a x$ is decreasing over the whole domain of real numbers, we have:
  - ⋄ **(B$_>$)**: $\log_a f(x) > \log_a g(x) \iff f(x) < g(x)$
  - ⋄ **(B$_<$)**: $\log_a f(x) < \log_a g(x) \iff f(x) > g(x)$

  ('the inequality symbol must be *reversed*').

The final inequality is then combined with the inequalities $f(x) > 0$ and $g(x) > 0$, as follows:

$$f(x) \gtrless^* g(x) \quad \text{AND} \quad f(x) > 0 \quad \text{AND} \quad g(x) > 0$$

(*the symbol $\gtrless$ stands for $>$ or $<$ according to the particular situation).

These compound inequalities can be solved by using the appropriate techniques described in Chapter 2 and give the solution ranges of the initial inequality.

Important note: The inequalities $f(x) > 0$ and $g(x) > 0$ ensure the exclusion of any x values which make the argument of the initial logarithms negative within the solution ranges resulting from the condition $f(x) \gtrless g(x)$.

Warning! Sometimes the right side of the inequality is not an expression in x, like $g(x)$, but a number, such as c. This number can be transformed into a logarithm to the appropriate base a by using equation (6.2) ($c = \log_a a^c$) and this will help in solving the inequality.

Example: Solve the logarithmic inequality.
$\log_{\frac{1}{2}}(x^2 - 1) > -1 + \log_{\frac{1}{2}}(1 - x)$.

Solution: As described above, we can solve the inequality by using the following steps:

(1) Rearrange: $\log_{\frac{1}{2}}(x^2 - 1) > \log_{\frac{1}{2}}(2) + \log_{\frac{1}{2}}(1 - x)$ $\iff \log_{\frac{1}{2}}(x^2 - 1) > \log_{\frac{1}{2}}(2 - 2x)$. Note that the equality $-1 = \log_{\frac{1}{2}}(2)$ used here is true since if $\log_{\frac{1}{2}}(2) = x$, then $(1/2)^x = 2$ so $(2^{-1})^x = 2$ and $x = -1$.

(2) Compare the logarithmic arguments: $x^2 - 1 < 2 - 2x$. The sign $<$ is 'reversed' since the base of the logarithms $0 < \dfrac{1}{2} < 1$.

(3) Solve the quadratic inequality obtained by using the technique explained in Chapter 2, giving

$$x^2 - 1 < 2 - 2x \iff x^2 + 2x - 3 < 0 \iff -3 < x < 1 \quad (i)$$

(4) Since all the logarithmic arguments should be > 0, combine the solution sets obtained from (*i*) with those from the other inequalities in (*ii*) $x^2 - 1 > 0$ and (*iii*) $2 - 2x < 0$ by using the technique explained in Section 2.5.4:
(*i*) $-3 < x < 1$ AND (*ii*) $x < -1$ OR $x > 1$ AND (*iii*) $x < 1$.

All the ranges of the solutions should be plotted on the same number line:

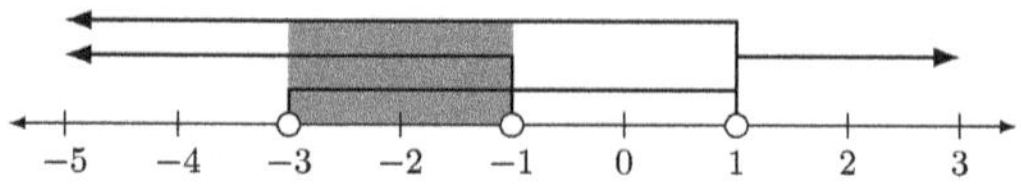

The range of values satisfying the compound inequalities is coloured in grey: $-3 < x < -1$.
All values within this interval satisfy the initial logarithmic inequality.

Example in Biology: In order to correctly prepare a culture of a specific type of algae, the pH of the water tank must be at least 7.8 but must be below 8.2. Determine the corresponding range of H^+ concentration.

Solution: Recalling that pH $= -\log[H^+]$, where $[H^+]$ is the hydrogen ion concentration in moles per litre, the necessary condition is that $7.8 \leq -\log[H^+] \leq 8.2$ for the water tank to be useful for preparing the culture.

We can start solving this logarithmic inequality by multiplying all the parts of it by -1:

$$-7.8 \geq \log[H^+] \geq -8.2 \Longleftrightarrow -8.2 \leq \log[H^+] \leq -7.8$$

We can now compare the arguments of the three logarithms by 'preserving' the inequality symbol since the logarithmic base is

$$10 > 1\text{: } 10^{-8.2} \leq [H^+] \leq 10^{-7.8}$$

Thus, the range of H^+ concentration required for the tank is

$$6.3 \cdot 10^{-9}\,\mathrm{M} \leq [H^+] \leq 1.6 \cdot 10^{-8}\,\mathrm{M}$$

Important note: Sometimes taking the logarithm of both sides of an indicial inequality or raising a number to both sides of a logarithmic inequality can simplify the solution of these indicial or logarithmic inequalities.

In all cases, given the base a of the logarithm or the index, if $a > 1$, the inequality symbol is *preserved*, whereas if $0 < a < 1$, the inequality symbol is *reversed*.

The solutions of the last two problems from Biosciences can be approached in this way, as shown in the following revisited Examples.

Example in Medicine: Solve the inequality $\left(\frac{1}{2}\right)^{\frac{t}{T_{1/2}}} < 0.1$.

Solution: By taking the logarithm of both sides of the inequality, we have:

$$\left(\frac{1}{2}\right)^{\frac{t}{T_{1/2}}} < 0.1 \iff \frac{t}{T_{1/2}} \log\left(\frac{1}{2}\right) <^{*} -1$$

$$\iff \frac{t}{T_{1/2}}(\log 1 - \log 2) < -1 \iff t(-\log 2) < -1 \cdot T_{1/2}$$

$$\iff t > \frac{1 \cdot T_{1/2}}{\log 2} = \frac{80}{0.301} = 265.7 \text{ hr}$$

* Since the base of the logarithm is $10 > 1$, the inequality symbol is preserved.

Example in Chemistry: Solve the inequality $-8.2 \leq \log[\mathsf{H}^{+}] \leq -7.8$.

Solution: If we raise 10 to each side of the compound inequality, we get $10^{-8.2} \leq^{*} 10^{\log[\mathsf{H}^{+}]} \leq^{*} 10^{-7.8}$. So, $10^{-8.2} \leq [\mathsf{H}^{+}] \leq 10^{-7.8}$.

* Since the base of the index is $10 > 1$, the inequality symbol is preserved.

6.7 Plotting Exp and Log Functions

6.7.1 Using translations

The graphs of **exponential and logarithmic functions** should become familiar to you (Figure 6.3) and can be used as 'building blocks' to investigate the form of other exponential or logarithmic functions without plotting them accurately.

The rules given in Section 4.7 for sketching the form of a function can also be applied to exponential or logarithmic functions by using the most appropriate geometric transformation to the graphs in Figure 6.3.

Example: Sketch the function $y = e^{2-x} - 3$ by using the rules of geometric transformations in Section 4.7.

Solution: Initially, we can identify $y = e^x$ as the closest elementary function to $y = e^{2-x} - 3$ and start by plotting it in the Cartesian plane, as shown in the next figure.

Using Rule **D** (Section 4.7), we can sketch $y = e^{-x}$ by reflecting the whole graph of $y = e^x$ in the y-axis.

Then, using Rule **D(i)**, we can sketch $y = e^{2-x}$ by translating the graph of $y = e^{-x}$ by 2 units to the right (positive x direction).

Finally, in order to sketch $y = e^{2-x} - 3$, we can use Rule **A(ii)** by translating the plot of $y = e^{2-x}$ down by 3 units.

The figure below summarises all the intermediate steps involved in sketching $y = e^{2-x} - 3$.

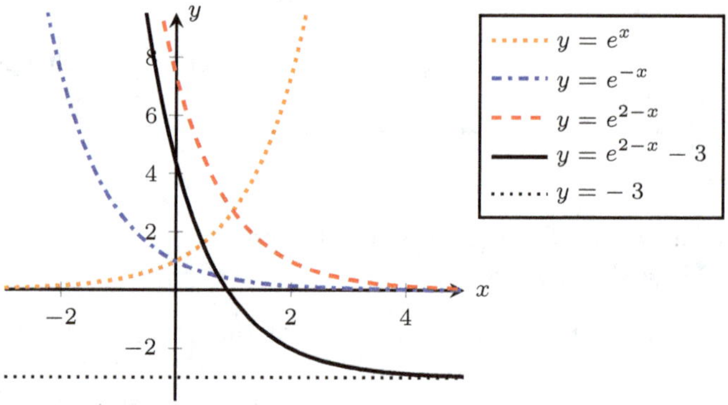

6.7.2 Using derivatives

As described in detail in Chapter 5, derivatives provide useful pieces of information about a function $f(x)$.

The sign of the first derivative $y' = f'(x) = \frac{dy}{dx}$ tells us how $f(x)$ is increasing or decreasing, while the sign of the second derivative $y'' = f''(x) = \frac{d^2y}{dx^2}$ tells us about a function's curvature.

All the rules for differentiating algebraic functions shown in Section 5.3 also apply to exponential or logarithmic functions.

Examples: Differentiate the following functions using the appropriate rules given in Chapter 5 (see Section 5.3):

(1) $y = e^{-ax}$ where $a \in \mathbb{R}$;
(2) $y = \ln(b - x)^2$ where $b \in \mathbb{R}$ and $x \neq b$;
(3) $y = e^{2x} \cdot (3x + 5) + \ln(5x^3 - 2)$.

Solution: The first two functions are composite functions since we have:

(1) $g(x) = -ax$ and $f(g(x)) = e^{g(x)}$;
(2) $v(x) = b - x$, $g(x) = (v(x))^2$ and $f(g(v(x))) = \ln(g(v(x)))$.

Thus, we can use the Chain Rule to get the derivatives:

(1) $\dfrac{dy}{dx} = f'(g(x)) \cdot g'(x) = e^{g(x)} \cdot g'(x) = e^{g(x)} \cdot (-a) = -ae^{-ax}$;

(2) $\dfrac{dy}{dx} = f'(g(v(x)) \cdot g'(v(x)) \cdot v'(x) = \dfrac{1}{g(v(x))} \cdot 2v(x) \cdot (-1)$
$= \dfrac{1}{(b-x)^2} \cdot 2(b - x) \cdot (-1) = -\dfrac{2}{b-x}$.

For the function in example (3), $y = e^{2x} \cdot (3x + 5) + \ln(5x^3 - 2)$, we can differentiate as follows (see Section 5.3.2):

- using Rule **A**, $\dfrac{dy}{dx} = \dfrac{d\,(e^{2x} \cdot (3x + 5))}{dx} + \dfrac{d\,(\ln(5x^3 - 2))}{dx}$;
- using Rules **C**, **D**, and **B**, $\dfrac{d\,(e^{2x} \cdot (3x + 5))}{dx} = e^{2x} \cdot 2 \cdot (3x+5) + e^{2x} \cdot 3$ $= e^{2x} \cdot (6x+13)$, upon simplification by taking out the common factor of e^{2x};
- using Rules **D** and **A**, $\dfrac{d\,(\ln(5x^3 - 2))}{dx} = \dfrac{1}{5x^3 - 2} \cdot 15x^2$.

Thus, the derivative of the whole initial function is

$$\frac{dy}{dx} = e^{2x} \cdot (6x + 13) + \frac{15x^2}{5x^3 - 2}$$

Example in Medicine: The oral route of drug administration represents one of the most popular methods of introducing a drug into the body. When the oral drug absorption follows a model in which the drug is absorbed at a constant rate k_0, the drug concentration in the blood C_P obeys the equation $C_P = \dfrac{k_0}{V_D \cdot k\left(1 - e^{-kt}\right)}$, where t is the time in hr and k is the constant of the drug elimination chemical process (in hr^{-1}). k_0 and V_D are constants.

Differentiate the equation with respect to the time t.

Solution: We can initially rewrite the equation as follows:

$$C_P = \frac{k_0}{V_D \cdot k}\left(1 - e^{-kt}\right)^{-1}$$

We then recognise the product of a constant term, $\dfrac{k_0}{V_D \cdot k}$, and a composite function, $\left(1 - e^{-kt}\right)^{-1}$.

We should initially use Rule **B** (see Chapter 5, section 5.3.2) to take the constant out and then apply the Chain Rule **D** to $\left(1 - e^{-kt}\right)^{-1} = f\left(g\left(v\left(t\right)\right)\right)$, with $v\left(t\right) = -kt$, $g\left(v\left(t\right)\right) = 1 - e^{v(t)}$ giving $f\left(g\left(v\left(t\right)\right)\right) = \left(g\left(v\left(t\right)\right)\right)^{-1}$.

Thus, Rule **D** gives

$$\begin{aligned}\left(f\left(g\left(v\left(t\right)\right)\right)\right)' &= f'\left(g\left(v\left(t\right)\right)\right) \cdot g'\left(v\left(t\right)\right) \cdot v'\left(t\right)\\ &= \left(-1\right) \cdot \left(1 - e^{-kt}\right)^{-1-1} \cdot \underbrace{\left(0 - e^{-kt}\right)}_{\text{rule A}} \cdot \left(-k\right)\end{aligned}$$

The full derivative is thus

$$\begin{aligned}\frac{dC_P}{dt} &= \frac{k_0}{V_D \cdot k} \cdot \left(-1\right) \cdot \left(1 - e^{-kt}\right)^{-2} \cdot \left(-e^{-kt}\right) \cdot \left(-k\right)\\ &= \frac{-k_0 \cdot k e^{-kt}}{V_D \cdot k\left(1 - e^{-kt}\right)^2}\end{aligned}$$

Note on units: The argument of the exponential is overall unitless, which means that k must have units of hr^{-1}.

Example in Biochemistry: The formula for the electrochemical potential across a biological membrane (e.g. a cell) $\widetilde{\mu}$ ($\mathrm{J\,mol^{-1}}$) is

$$\widetilde{\mu} = RT \ln \frac{\left[a_A\right]}{\left[a_B\right]} + ZF\Psi$$

where $[a_A]$ and $[a_B]$ are the concentrations of the migrating species on the inside and outside of the membrane respectively, Z is the charge of the species, F is the Faraday constant ($96{,}485\ \mathrm{C\,mol^{-1}}$), R is the ideal gas constant ($\mathrm{J\,mol^{-1}\,K^{-1}}$), T is the temperature (K), and Ψ is the electrical potential (V).

For protons, $\widetilde{\mu}_{H^+{}_{(A-B)}} = 2.303RT \log \dfrac{[H^+]_A}{[H^+]_B} + ZF\Psi$.

By taking $[H^+]_T = [H^+]_A + [H^+]_B$ where the total H^+ concentration, $[H_T^+]$ is constant, find the rate of change in electromotive force with respect to the H^+ concentration, *i.e.* $\dfrac{d\widetilde{\mu}}{d[H^+]_A}$, for protons exiting the cell (from side A going to side B).

Solution: First we replace the log with ln and substitute $[H^+]_B = [H^+]_T - [H^+]_A$ giving

$$
\begin{aligned}
\widetilde{\mu}_{H^+{}_{(A-B)}} &= RT \ln\left(\frac{[H^+]_A}{[H^+]_T - [H^+]_A}\right) + ZF\Psi \\
&= RT \ln([H^+]_A) - RT \ln([H^+]_T - [H^+]_A) + ZF\Psi
\end{aligned}
$$

To find the rate of change in $\widetilde{\mu}_{H^+{}_{(A-B)}}$ with respect to $[H^+]_A$, we differentiate $\widetilde{\mu}_{H^+{}_{(A-B)}}$ with respect to $[H^+]_A$:

$$
\begin{aligned}
\frac{d\widetilde{\mu}_{H^+{}_{(A-B)}}}{d[H^+]_A} &= \frac{RT}{[H^+]_A} - \frac{RT\cdot(-1)^*}{([H^+]_T - [H^+]_A)} + ZF\frac{d\Psi}{d[H^+]_A} \\
&= \frac{RT}{[H^+]_A} + \frac{RT}{([H^+]_T - [H^+]_A)} + ZF\frac{d\Psi}{d[H^+]_A} \\
&= RT\left(\frac{[H^+]_B + [H^+]_A}{[H^+]_A \cdot [H^+]_B}\right) + ZF\frac{d\Psi}{d[H^+]_A}
\end{aligned}
$$

where $\dfrac{d\Psi}{d[H^+]_A}$ is the change in electrical potential due to the movement of protons across the membrane.

***Note:** (-1) is due to differentiation by the Chain Rule.

6.7.3 L'Hôpital's rule

When we investigate exponential or logarithmic functions, sometimes the calculations of limits at the domain interval extremities and the definition of possible asymptotes require the calculation of particularly difficult indeterminate forms (see Section 4.4.2) of limits.

L'Hôpital's theorem allows the calculation of these indeterminate forms of limits by using derivatives.

L'Hôpital's theorem: Let $f(x)$ and $g(x)$ both be continuous functions on an interval I of real numbers and let the point x_0 belong to I.

If the following conditions are satisfied,

(1) $f(x)$ and $g(x)$ are differentiable in the interval I;
(2) $g'(x_0) \neq 0$;
(3) both functions $f(x)$ and $g(x)$ tend simultaneously to 0 or ∞ as $x \to x_0$, giving the indeterminate form:
$\lim_{x\to x_0} \frac{f(x)}{g(x)} = \frac{0}{0}$ or $\frac{\infty}{\infty}$;
(4) $\lim_{x\to x_0} \frac{f'(x)}{g'(x)} = l$ with $l \in \mathbb{R}$;

then, $\lim_{x\to x_0} \frac{f(x)}{g(x)} = \lim_{x\to x_0} \frac{f'(x)}{g'(x)} = l$.

Important notes:

- If the result of $\lim_{x\to x_0} \frac{f'(x)}{g'(x)}$ is still $\frac{0}{0}$ or $\frac{\infty}{\infty}$, we can keep on differentiating the numerator and the denominator separately till the denominator satisfies the second condition (*i.e.* is non-zero at x_0) and then $\lim_{x\to x_0} \frac{f^n(x)}{g^n(x)}$ equals $l \in \mathbb{R}$.
- x_0 can also tend to $\pm\infty$ if the interval I is unbounded.

Example: Calculate the limit of $\frac{\ln(x+1)}{x}$ as $x \to 0$.

Solution: Upon taking the limit as $x \to 0$ of the numerator and denominator, we obtain $\lim_{x\to 0} \frac{\ln(x+1)}{x} = \frac{\ln(0+1)}{0} = \frac{0}{0}$ which is a limit of indeterminate form.

We can apply L'Hôpital's theorem since the necessary conditions are met:

(1) $f(x)$ and $g(x)$ are differentiable in their domains $\mathbb{R}$: $f'(x) = \dfrac{1}{x+1}$ and $g'(x) = 1$;

(2) $g'(0) = 1 \neq 0$ in $\mathbb{R}$;

(3) $\lim\limits_{x\to 0} \dfrac{f(x)}{g(x)} = \dfrac{0}{0}$;

(4) $\lim\limits_{x\to 0} \dfrac{f'(x)}{g'(x)} = \lim\limits_{x\to 0} \dfrac{\frac{1}{x+1}}{1} = 1.$

We can now conclude that $\lim\limits_{x\to 0} \dfrac{\ln(x+1)}{x} = 1.$

Example: Calculate the following limit: $\lim\limits_{x\to+\infty} \dfrac{e^{3x}}{x^2-2}$.

Solution: Upon taking the limit as $x \longrightarrow \infty$ of numerator and denominator, we obtain $\lim\limits_{x\to+\infty} \dfrac{e^{3\cdot(+\infty)}}{(+\infty)^2-2} = \dfrac{+\infty}{+\infty}$.

We can apply L'Hôpital's theorem since the necessary conditions are met:

(1) $f(x)$ and $g(x)$ are differentiable in their domains $\mathbb{R}$: $f'(x) = 3 \cdot e^{3x}$ and $g'(x) = 2x$;

(2) $\lim\limits_{x\to+\infty} g'(x) = +\infty \neq 0$ in $\mathbb{R}$;

(3) $\lim\limits_{x\to+\infty} \dfrac{f(x)}{g(x)} = \dfrac{+\infty}{+\infty}$;

(4) BUT $\lim\limits_{x\to+\infty} \dfrac{f'(x)}{g'(x)} = \dfrac{+\infty}{+\infty}$.

Thus, we can differentiate the numerator and denominator of the fraction $\dfrac{3\cdot e^{3x}}{2x}$ separately and then calculate $\lim\limits_{x\to+\infty} \dfrac{f''(x)}{g''(x)}$.

$$f''(x) = 9 \cdot e^{3x} \text{ and } g''(x) = 2 \text{ and } \lim_{x\to+\infty} \frac{f''(x)}{g''(x)} = \frac{9\cdot e^{3\cdot(+\infty)}}{2} = +\infty$$

We can now conclude that $\lim\limits_{x\to+\infty} \dfrac{e^{3x}}{x^2-2} = +\infty.$

Example in Chemistry: Let us consider the following series of irreversible reactions, where species A reacts to form an intermediate species, I, which then reacts to form the product, P:

$$\mathrm{A} \xrightarrow{k_1} \mathrm{I} \xrightarrow{k_2} \mathrm{P}$$

where k_1 and k_2 represent the reaction rate constants.

The final product P concentration after time t is given by the following equation: $[P] = [A]_0 \cdot \left[1 + \dfrac{k_2 e^{-k_1 t} - k_1 e^{-k_2 t}}{k_1 - k_2}\right]$, where $[A]_0$ is the initial concentration of A. This equation is valid for the case where $k_1 \neq k_2$.

What happens if $k_2 \to k_1$?

Solution: It can easily be seen that when $k_1 = k_2$, an indeterminate form occurs: $[P] = [A]_0 \cdot \left[1 + \dfrac{0}{0}\right]$.

The corresponding limiting equation of the fraction $\dfrac{k_2 e^{-k_1 t} - k_1 e^{-k_2 t}}{k_1 - k_2}$ can be obtained with L'Hôpital's rule by differentiating with respect to k_2 holding k_1 fixed, since all the necessary conditions are met:

$$\lim_{k_2 \to k_1} \frac{k_2 e^{-k_1 t} - k_1 e^{-k_2 t}}{k_1 - k_2} = \lim_{k_2 \to k_1} \frac{\frac{d}{dk_2}\left(k_2 e^{-k_1 t} - k_1 e^{-k_2 t}\right)}{\frac{d}{dk_2}\left(k_1 - k_2\right)}$$

$$= \lim_{k_2 \to k_1} \frac{e^{-k_1 t} - k_1 e^{-k_2 t} \cdot (-t)}{0 - 1} = -e^{-k_1 t}\left(1 + k_1 t\right)$$

Thus, the product P concentration after time t is given by the limiting equation $[P] = [A]_0 \cdot \left[1 - (1 + k_1 t)\, e^{-k_1 t}\right]$ when $k_2 = k_1$.

Note on units: The argument of the exponential is overall unitless, which means that k_1 must have units of reciprocal time, *e.g.* hr^{-1}.

6.7.4 Euler's number

As stated earlier (equation (6.4)), Euler's number is the result of a particular limit, $e = \lim\limits_{x \to +\infty} \left(1 + \dfrac{1}{x}\right)^x$, and this can now be proved using L'Hôpital's rule.

Proof: Let $L = \lim_{x\to+\infty}\left(1+\frac{1}{x}\right)^x$.

By taking the natural logarithm of both sides of the equation, we have $\ln(L) = \ln\left[\lim_{x\to+\infty}\left(1+\frac{1}{x}\right)^x\right]$ which due to continuity equals $\ln(L) = \lim_{x\to+\infty}\ln\left(1+\frac{1}{x}\right)^x$.

By using Rule **C** for logarithms (Section 6.3), we get $\ln(L) = \lim_{x\to+\infty} x\ln\left(1+\frac{1}{x}\right) = \lim_{x\to+\infty}\frac{\ln\left(1+\frac{1}{x}\right)}{\frac{1}{x}}$ through algebraic rearrangement.

If we now calculate this limit, we obtain $\ln(L) = \lim_{x\to+\infty}\frac{\ln\left(1+\frac{1}{x}\right)}{\frac{1}{x}} = \frac{\ln(1+0)}{0} = \frac{0}{0}$ which is an indeterminate form.

Since this limit meets the conditions for applying L'Hôpital's theorem, we can differentiate the numerator and the denominator to calculate it: $\ln(L) = \lim_{x\to+\infty}\frac{\left(1+\frac{1}{x}\right)^{-1}\cdot\left(-\frac{1}{x^2}\right)}{-\frac{1}{x^2}} = 1$.

Finally, by raising e to both sides of the equation $e^{\ln(L)} = e^1$, we have $L = e$.

6.7.5 Investigating a function

Now we can progress from simply **sketching** to completely **plotting** an exponential or logarithmic function $f(x)$ by using all the various pieces of information detailed in Section 5.4.3.

Note that there are now many software packages available online for characterising and plotting functions, so you can easily check your results.

Example: Investigate the function $f(x) = xe^{-x}$ and plot it in a Cartesian plane.

Solution: Let us follow the list of steps suggested in Section 5.4.3:

(i) **Domain:** $\mathbb{R}$, *i.e.* $(-\infty, +\infty)$, as $f(x)$ is the product of a polynomial function and an exponential function.

(ii) **Key points:** When $f(0) = 0$, $f(1) = \frac{1}{e}$ and $f(-1) = -e$, so we can mark the points $A\ (0, 0)$, $B\ \left(1, \frac{1}{e}\right)$, and $C\ (-1, -e)$ in the Cartesian plane.

$f(x) = 0 \iff xe^{-x} = 0$, which gives $x = 0$ as a unique solution, since the exponential expression is always > 0.
Thus, the intersection point between the function and the x-axis is $A\ (0, 0)$.

(iii) **Possible symmetry:** $f(-x) = (-x)e^{+x}$ which does not equal either $f(x)$ or $-f(x)$. Thus, the function is neither even nor odd.

(iv) **Sign:** $f(x) > 0 \iff xe^{-x} > 0$. Using the method for solving polynomial inequalities (Section 2.5.6), we obtain $f(x) > 0$ when $x > 0$ since the exponential expression is always > 0.

Thus, the function is positive for $x > 0$, so we can shade the area below the x-axis for $x > 0$ and the area above the x-axis for $x < 0$.

The function will not be plotted in the shaded areas in the Cartesian plane.

(v) **Limits and asymptotes:** Now, it is essential to understand the behaviour of the function $f(x)$ at the limiting points of its domain: in this case at $-\infty$ and $+\infty$:

- $\lim_{x \to -\infty} xe^{-x} = -\infty \cdot e^{+\infty} = -\infty$.

Reminder: When we write $f(\pm\infty) = \pm\infty$, we really mean $\lim_{x \to \pm\infty} f(x) = \pm\infty$, but '$\pm$ infinity' ($\pm\infty$) does not lie in the domain of the function.

As $\lim_{x \to -\infty} f(x) = -\infty$, there might possibly be a slanting asymptote of $f(x)$ in the form of a straight line $y = mx + q$ (see Section 3.2.1).

Now, $m = \lim_{x \to -\infty} \frac{f(x)}{x} = \lim_{x \to -\infty} \frac{xe^{-x}}{x} = \lim_{x \to -\infty} e^{-x} = e^{+\infty} = +\infty$.

Thus, no slanting asymptote to the function can be drawn as x approaches $-\infty$.

- $\lim_{x\to+\infty} xe^{-x} = +\infty \cdot e^{-\infty} = +\infty \cdot 0$ which is an indeterminate form. If we rearrange the expression, we have $\lim_{x\to+\infty} \frac{x}{e^x} = \frac{+\infty}{+\infty}$ which is still an indeterminate form, but we can use L'Hôpital's rule (see Section 6.7.3) to calculate the limit by differentiating the numerator and the denominator separately:

$$\lim_{x\to+\infty} \frac{x}{e^x} = \lim_{x\to+\infty} \frac{1}{e^x} = \frac{1}{+\infty} = 0$$

Thus, the straight line $y = 0$ is a horizontal asymptote of $f(x)$ as x tends to $+\infty$.

(vi) **Continuity:** $f(x)$ is continuous as polynomial and exponential functions are continuous in every closed interval $[a, b]$ of their domain, which is $\mathbb{R}$.

(vii) **Stationary and inflection points:** We differentiate $f(x)$ by using the Product Rule **C** and Chain Rule **D** from Chapter 5: $f'(x) = 1 \cdot e^{-x} + xe^{-x} \cdot (-1) = (1 - x)e^{-x}$, and then set $f'(x)$ equal to zero: $(1 - x)e^{-x} = 0$ to give a root at $x = 1$, since the exponential expression is always > 0.

We differentiate again using the same Rules **C** and **D**: $f''(x) = (x - 2)e^{-x}$; at $x = 1$, we have $f''(1) = -\frac{1}{e}$ which is negative, so $f(x)$ has a local maximum at $x = 1$.

We then solve the inequality $f'(x) > 0$ to find in which intervals of the domain $f(x)$ is increasing: using the polynomial inequality method (Section 2.5.6), $(1-x)e^{-x} > 0$ gives the range of solutions $x < 1$, since the exponential expression is always > 0.

Thus, $f(x)$ is increasing for $x < 1$ and conversely is decreasing for $x > 1$.

To find the points of inflection, we set $f''(x) = (x - 2)e^{-x} = 0$ and solve to get $x = 2$, since the exponential expression is always > 0.

By solving the inequality $f''(x) > 0 \iff (x - 2)e^{-x} > 0$, we get the interval of solutions which is $x > 2$.

Thus, $f(x)$ is concave upwards for $x > 2$, while it is concave downwards for $x < 2$.

If we put all these values back into our original function $f(x)$, we can find the y coordinates of the stationary values and inflection points.

Thus, there is a maximum at $max\left(1, \frac{1}{e}\right)$, and one inflection point at $F\left(2, 2e^{(-2)}\right)$.

We can calculate the equation of the tangent line at the inflection point F to show the curvature of the function on either side with respect to the tangent line in F:

$$y - f(2) = f'(2) \cdot (x-2) \iff y - \left(2e^{(-2)}\right) = \left(-e^{(-2)}\right) \cdot (x-2)$$
$$\iff y = -\frac{x}{e^2} + \frac{4}{e^2} = \frac{4-x}{e^2}$$

(viii) **Co-domain:** y values go from $-\infty$ to the maximum $f(1) = \frac{1}{e}$:

$$\left(-\infty, \frac{1}{e}\right]$$

Now we can summarise all the properties found above in a table and use them to plot a graph of the function on a Cartesian plane.

x	$-\infty$	1	2	$+\infty$
$f(x)$	$-\infty$	$\frac{1}{e} \approx 0.37$	$\frac{2}{e^2} \approx 0.27$	0
$f'(x)$	+ve	0	−ve	−ve
$f(x)$'s trend	↗	max	↘	↘
$f''(x)$	−ve	−ve	0	+ve
$f(x)$'s curvature	⌢	⌢	F	⌣

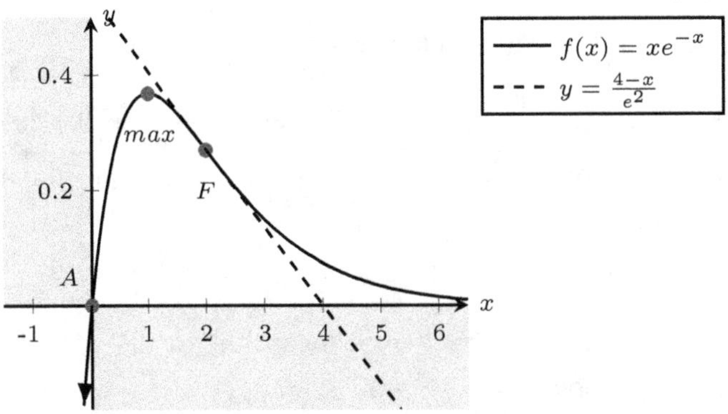

Note: The plot has been drawn to highlight the key points of the function. Obviously, as x tends to $-\infty$, the function goes very rapidly to $-\infty$ as shown by the downward arrow.

Example in Chemistry: An adaption of the Henderson–Hasselbalch equation can be used to show that the pH of $100\,\text{ml}$ of a solution containing $50\,\text{mM}$ acetic acid being titrated against $1\,\text{M}$ sodium hydroxide is given by

$$f(x) = 4.75 + \log_{10}\left(\frac{x}{5-x}\right)$$

where $f(x)$ is the pH and x represents the volume of sodium hydroxide added during titration in ml.

Investigate the overall trend of the function in the domain $[0.1, 4.9]$, and, by finding the inflection point, determine how much of the base solution must be added for the titration curve to reach its shallowest gradient.

Lastly, plot a graph of the function in a Cartesian plane.

Solution:

(i) **Domain:** As $f(x)$ is a logarithmic function, its argument has to be positive, so we must put $\dfrac{x}{5-x} > 0$ which gives $0 < x < 5$ by using the rational inequalities method (Section 2.5.6). Thus, the domain is $(0, 5)$, which includes the domain of investigation $[0.1, 4.9]$.

(ii) **Key points:** Solving $f(x) = 0 \Longleftrightarrow 4.75 + \log_{10}\left(\frac{x}{5-x}\right) = 0$ $\Longleftrightarrow \log_{10}\left(\frac{x}{5-x}\right) = -4.75 \Longleftrightarrow \frac{x}{5-x} = 10^{-4.75}$ gives $x \simeq 0.000089$, which is out of the range of the domain $[0.1, 4.9]$.

(iii) **Sign:** Solving $f(x) > 0 \Longleftrightarrow \log_{10}\left(\frac{x}{5-x}\right) > -4.75$ $\Longleftrightarrow \frac{x}{5-x} > 10^{-4.75}$ (since logarithm to the base $10 > 1$) gives $0.00008 < x < 5$; thus, that $f(x) > 0$ in the domain $[0.1, 4.9]$.

(iv) **Limits:** No limits need to be calculated in the closed interval $[0.1, 4.9]$.

(v) **Continuity:** As $f(x)$ is a logarithmic function, $f(x)$ is continuous in every closed interval $[a, b]$ of the domain $[0.1, 4.9]$.

(vi) **Stationary and inflection points:** Let us differentiate the function twice:

$$f'(x) = 0 + \frac{1}{\ln(10)} \cdot \frac{1}{\frac{x}{(5-x)}} \cdot \left(\frac{1 \cdot (5-x) - x \cdot (-1)}{(5-x)^2}\right)$$

$$= \frac{(5-x)}{x \cdot \ln(10)} \cdot \frac{(5-x+x)}{(5-x)^2} = \frac{5}{2.303} \cdot \frac{1}{x(5-x)}$$

$f'(x) = \frac{5}{2.303} \cdot \frac{1}{x(5-x)}$ which can be shown always to be $\neq 0$ and > 0 in the function domain.

Thus, $f(x)$ has no stationary points and is always increasing in its domain.

By using the Quotient Rule of differentiation,

$$f''(x) = \frac{5}{2.303} \cdot \frac{0 - (5 - 2x)}{x^2(5-x)^2} = \frac{5}{2.303} \cdot \frac{2x - 5}{x^2(5-x)^2}$$

$f''(x) = \frac{5}{2.303} \cdot \frac{2x-5}{x^2(5-x)^2} = 0$ if $x = \frac{5}{2} = 2.5$ and the inflection point is $F(2.5, 4.75)$.

Therefore, $2.5\,\mathrm{mL}$ of base must be added to reach the inflection point.

Lastly, we need to investigate the curvature of the function by solving the following inequality:

$$f''(x) > 0 \Longleftrightarrow \frac{5}{2.303} \cdot \frac{2x-5}{x^2(5-x)^2} > 0$$

By using the method for solving rational inequalities (see Section 2.5.6), we can see that

$$f''(x) > 0 \Longleftrightarrow 2x - 5 > 0 \Longleftrightarrow x > \frac{5}{2}$$

Thus, $f(x)$ is concave upwards for $x > \frac{5}{2}$, while it is concave downwards for $x > \frac{5}{2}$.

(vii) **Co-domain**: $[f(0.1), f(4.9)]$.

We can now make a detailed plot of the pH curve obtained by titrating $100\,\mathrm{ml}$ of a solution containing $50\,\mathrm{mM}$ acetic acid against $1\,\mathrm{M}$ sodium hydroxide.

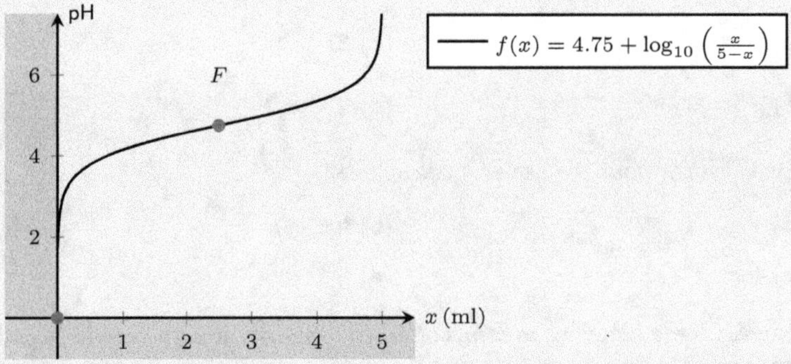

Note that the plot has a gently rising plateau. This shows us that the pH does not change much (i.e. is 'buffered') over a wide range of concentrations of x.

6.8 Exercises

(1) Simplify the following expressions.

(a) $\frac{2^3}{2^{-2}}$ (b) $\frac{(\sqrt{9})^4}{9^2}$ (c) $\frac{27^{\frac{1}{3}}}{16^{-\frac{1}{4}}}$ (d) $125^{\frac{2}{3}}$

(e) $\frac{a^5}{a^{-4}}$ (f) $\frac{(\sqrt{x})^8}{x^3}$ (g) $\frac{y^{\frac{1}{3}}}{y^{-\frac{1}{4}}}$ (h) $(2x^3)^4$

(2) Evaluate the following expressions *(without using your calculator!)*.

(a) 2^{-4} (b) $8^{\frac{2}{3}} + 125^{\frac{1}{3}}$ (c) $\left(\frac{1}{4}\right)^{-2}$

(d) $\left(\frac{27}{64}\right)^{\frac{2}{3}}$ (e) $(3-2)^{10}$ (f) $\sqrt{144+25}$

(g) $(4+6)^5$ (h) $\sqrt[3]{27-19}$

(3) Each of the functions graphed below is of the form $y = C \cdot a^x$. Find the values of C and a.

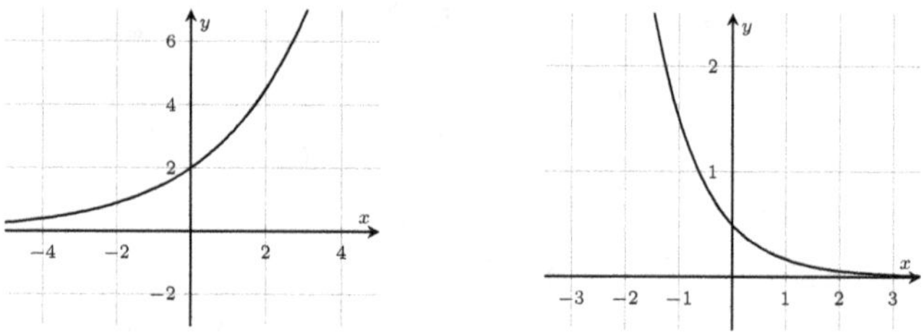

(a) (b)

(4) Convert each of the expressions into its logarithmic equivalent.

(a) $3^4 = 81$ (b) $27^{-\frac{1}{3}} = \frac{1}{3}$

(c) $x^y = 3$ (d) $2^{-5} = \frac{1}{32}$

(5) Evaluate *(without the use of a calculator!)*:

(a) $\log_2 64$ (b) $\log_{\sqrt{2}} 1$

(c) $\log_\pi \pi^3$ (d) $5^{\log_5 \sqrt{2}}$

(6) Express these functions in terms of $\log a$, $\log b$, and $\log c$, where a, b, and c are positive.

(a) $\log\left(\frac{a}{bc}\right)$ (b) $\log(a^3b^2c)$

(c) $\log\sqrt{\frac{b}{a}}$ (d) $\log\left(\frac{\sqrt{ab^4}}{\sqrt[3]{c^2}}\right)$

(7) Simplify:

(a) $\log 6 + \log 5 - \log 3$ (b) $\log(x^2-4) - \log(x+2)$

(c) $2\log_a 3 + \log_a 4 - 2\log_a 6$ (d) $2\log_a \sqrt{a^3} - \frac{1}{2}\log_a(4a)$

(8) Given that $\log_3 4 = 1.26$ and $\log_3 6 = 1.63$, find:

(a) $\log_6 4$ (b) $\log_4 6$ (c) $\log_4 3$ (d) $\log_3 24$

(9) By taking an appropriate logarithm or raising to an appropriate power, make x the subject in each of the following equations.

(a) $\log_3 x = 3$ (b) $4^x = 64$ (c) $\log_5 x = 2t + 1$

(d) $\log_3(x-2) = t - 2$ (e) $\frac{1}{2} = 3 \cdot \left(\frac{1}{4}\right)^x$ (f) $3 = 4 \cdot 5^x$

(10) Consider the following data for terrapin population in different years:

t, year	1	2	3	4	5
P, number of terrapins	200	230	263	301	344

(a) Plot a graph of P versus t; what is the curvature of the trend line? Could it be a straight line?

(b) Plot a graph of $y = \log P$ *versus* $x = t$ and verify that the straight line obtained is of the form $y = \underbrace{0.059}_{m}\, x + \underbrace{2.243}_{c}$.

(c) By raising each side of this latter equation to the power 10, derive the formula $P = 175 \cdot (1.145)^t$ for the terrapin population.

(11) Theory suggests two possible models for the data below. One possibility involves a constant index r with x appearing as the base: $y = Ax^r$. The other possibility involves a constant base a with x appearing as an index: $y = Ca^x$.

x, independent variable	1	2	3	4	5
y, dependent variable	3.5	68.9	394.2	1358.1	3545.2

(a) Explore this theory by examining two plots: $\log y$ *versus* x, and $\log y$ versus $\log x$.

(b) Which of these plots most closely resembles a straight line?

(c) Use your results to decide which model best fits the data and then to determine the values of the constants involved (either A and r if you decide that $y = Ax^r$ is best, or C and a otherwise).

(12) Solve the following indicial or logarithmic elementary equations for x.

(a) $\left(\frac{1}{3}\right)^{(3x-5)} - 3^{(4-x)} = 0$ (b) $\frac{2^4}{2^{x^2}} = \left(\frac{1}{2}\right)^{-3}$

(c) $\left(\frac{1}{2}\right)^{\frac{1}{x}} = 8^{(x-1)}$ (d) $3^{(2x-1)} = 15^{(1-x)}$

(e) $\log_2(7-x) - \log_2(x+1) = 0$ (f) $\log_{\frac{1}{3}}(x^2-1) = \log_{\frac{1}{3}}(x) + 1$

(g) $2\ln(x-1) = \ln(x^2+5)$ (h) $\log_4(5x-6) = \log_2(x)$

(13) Suppose we analyse $300\,\mathrm{mL}$ of a solution containing $5.5\,\mathrm{mg}$ HCl. Assume that the HCl is completely ionised and its molecular mass is $36.46\,\mathrm{Da}$.

(a) What is the pH of this solution?

(b) What would the pH be if the concentration of HCl were doubled?

(14) A terrapin population is described by the equation $P = 175 \cdot (1.145)^t$, where P represents the number of terrapins in the population and t represents time, measured in years.

(a) Find the length of time needed for the terrapin population to double its initial size.

(b) How long does it take the original population of terrapins to triple its initial size?

(15) Solve the following indicial or logarithmic quadratic equations for x.

(a) $4^x + 2 - 3 \cdot 2^x = 0$ (b) $2^{(1-2x)} + 2^{(2-x)} = 6$

(c) $\ln^2(x) - 3\ln(x) + 2 = 0$ (d) $1 - \log_2^2(x) = 0$

(16) A novel radioactive iodine (^{131}I) therapy for hyperthyroidism can be administered through tablets once a month. After intestinal absorption, the amount of radioactive iodine present in a patient's blood at time t (in hours) is given by the equation

$$I(t) = I_0 \left(\frac{1}{2}\right)^{\frac{t}{T_{1/2}}} - I_0 \left(\frac{1}{2}\right)^{\frac{2t}{T_{1/2}}}$$

where $I(t)$ is the amount of the substance in megabecquerels (MBq) present after time t in days, I_0 is the amount of ^{131}I initially swallowed, and $T_{1/2}$ is the half-life of ^{131}I ($T_{1/2} = 8$ days).

(a) The maximum amount of radioactive substance in blood is observed at $t = T_{1/2}$ days. What is the proportion of I_0 present at $t = T_{1/2}$ days?

(b) When, after the first 10 days, is only 5% I_0 observed in the blood?

(17) Solve the following indicial or logarithmic inequalities for x.

(a) $3^{(2-x)} + 3 > 12$ (b) $\left(\frac{1}{2}\right)^{(x^2-x)} - \frac{1}{4} < \frac{3}{4}$

(c) $\log_2(3x-2) < 2\log_2(x)$ (d) $\log_{\frac{1}{2}}(1-2x) > 2$

(18) A bacterial culture contains 2,000 bacteria at time $t = 0$ in hours, and it doubles every half an hour. When will the bacterial population reach at least 100,000 bacteria?

(19) Death is likely if the pH of human blood plasma changes by more than ± 0.4 from its normal value of 7.4. What is the approximate range of molar concentrations of hydrogen ions for which life can be sustained?

(20) Sketch graphs of the following functions by using the transformations detailed in Section 4.7.

(a) $y = -e^{x+3}$ (b) $y = e^{x-4} - 2$ (c) $y = \left|e^{(3-x)} - 4\right|$

(d) $y = \ln(5-x) - 2$ (e) $y = 3 - |\ln(x)|$ (f) $y = |\ln(3+x)|$

(21) Find the derivatives with respect to x of the following functions using the Product, Quotient, Chain Rules as appropriate.

(a) $y = e^x(3x+5)$ (b) $y = e^{5x}$ (c) $y = (x^2+9x)\ln(x)$

(d) $y = e^{3x+\ln(x)}$ (e) $y = \dfrac{\ln(x)}{5x^3}$ (f) $y = \ln(x^2 - x)$

(g) $y = e^{\frac{x^2+4}{3-x}}$ (h) $y = \ln(5x^3 - 2)$ (i) $y = e^{2x^3+x}$

(j) $y = \log_{10}(x+1)$ (k) $y = \ln\left(\dfrac{x}{6x+1}\right)$ (l) $y = 5^x x^8$

(m) $y = \log_3(x^3 - 4x)$ (n) $y = \dfrac{x}{\ln(x^2)}$ (o) $y = 2^{e^x}$

(22) Differentiate the following functions which are used in Biosciences, with respect to the argument of the function, by using the differentiation rules as appropriate. All the other parameters in the formulae should be treated as constants.

(a) $E(C_O) = \dfrac{RT}{nF}\ln\left(\dfrac{C_O}{C_I}\right)$ (b) $pH(x) = pKa - \log\left(\dfrac{x}{K-x}\right)$

(c) $N(p) = \dfrac{\ln(1-P)}{\ln(1-f)}$ (d) $N(g) = \dfrac{\ln(1-P)}{\ln\left(1-\frac{i}{g}\right)}$

(e) $f(\overline{Y}) = \log\left(\dfrac{\overline{Y}}{1-\overline{Y}}\right)$ (f) $C(t) = C_O e^{-rt}$

(g) $L(a) = L_\infty\left(1 - e^{-k(a-a_O)}\right)$ (h) $L(T) = 10^{\frac{T-T_{ref}}{z}}$

(i) $K_E(T) = \dfrac{4\pi a^3}{3000} N_A e^{-\frac{V}{RT}}$ (j) $p(v) = k_1 v^2 e^{k_2 v}$

(k) $P(t) = \dfrac{P_O K}{P_O + (K - P_O)e^{-rt}}$ (l) $p(s) = \dfrac{1-e^{-2s}}{1-e^{-4Ns}}$

(23) Calculate the specified limits of the following functions using L'Hôpital's rule.

(a) $\lim\limits_{x\to 0} \dfrac{e^x - 1}{x}$ (b) $\lim\limits_{x\to 0} \dfrac{\ln x + 1}{x}$ (c) $\lim\limits_{x\to 0} \dfrac{\ln(5x+1)}{e^{2x} - 1}$

(d) $\lim\limits_{x\to +\infty} \dfrac{3x^2 + 5}{6 - 7x^3}$ (e) $\lim\limits_{x\to +\infty} \dfrac{x + e^x}{1 - e^x}$ (f) $\lim\limits_{x\to +\infty} \dfrac{e^{2x} - 3}{x - \ln(x)}$

(g) $\lim\limits_{x\to 3} \dfrac{2^x - 8}{9 - x^2}$ (h) $\lim\limits_{x\to 0} \dfrac{1 - e^{5x}}{\ln(3x+1)}$ (i) $\lim\limits_{x\to -\infty} \dfrac{2 - e^{-x}}{\ln(2 - x)}$

(j) $\lim\limits_{x\to +\infty} \dfrac{x^2 - 4x}{\ln(1+x)}$ (k) $\lim\limits_{x\to 0} \dfrac{2x - e^{3x} + 1}{4x}$ (l) $\lim\limits_{x\to 1} \dfrac{1 - \sqrt{x}}{2x^2 - x - 1}$

(24) In the following irreversible reaction, species A reacts with species B to form the product, P:

$$A + B \xrightarrow{k} P$$

where k is the reaction rate constant.

The concentrations of the product P after time t is given by the following equation:

$$[P](t) = \frac{a \cdot b \cdot \left[e^{(a-b)kt} - 1\right]}{a \cdot e^{(a-b)kt} - b}$$

where a and b are the initial concentration of A and B, respectively. However, this equation is only valid for the case where $b \neq a$.

(a) What happens to the equation for $[P](t)$ if $b \to a$?

(b) At what value of $[P]$ is the reaction complete (*i.e.* as $t \to +\infty$)?

(25) In a spectrophotometric method where Beer's law is obeyed, the uncertainty U in the determination of the concentration of a solution can be expressed by the equation

$$U(T) = -\frac{k}{T \cdot \ln(T)}$$

where T is the transmittance of the solution sample and k is the absolute error in the transmittance.

(a) What happens to the equation $U(T)$ if $T \to 0$? If an indeterminate form occurs, solve it by using L'Hopital's rule on the equivalent expression $U(T) = -\dfrac{k\frac{1}{T}}{\ln(T)}$. What is its physical meaning?

(b) According to the equation for $U(T)$, at what value of U does the uncertainty reach a minimum?

(c) What is the range of physically meaningful values of transmittance, T?

(d) If $T = 10^{-A}$, where A is the absorbance of the solution sample, express A as a function of T, and then the uncertainty, U, as a function of A.

(26) Identify the coordinates of any critical points in the following functions and determine whether they are maxima, minima, or points of inflection.

(a) $y = x + e^{-x}$ (b) $y = e^{3x-x^2}$ (c) $y = 4xe^{-x}$

(d) $y = e^{2x} - 4e^{x}$ (e) $y = \ln(x^2 + 1)$ (f) $y = \ln(x) - \sqrt{x}$

(g) $y = x\ln(x)$ (h) $y = \ln(2 - \ln(x))$

(27) For the following functions, determine whether, at the given point, y is increasing $\nearrow$, decreasing $\searrow$, or flat $\longrightarrow$, and state the curvature (upward $\smile$ or downward $\frown$) of the graph.

(a) $y = 5xe^{-x}$ at $x = 3$ (b) $y = e^{-\frac{3}{x}}$ at $x = 5$

(c) $y = e^{x^2-x}$ at $x = -1$ (d) $y = \ln(2 - 3x)$ at $x = -5$

(e) $y = \ln\left(\dfrac{2-x}{3+x^2}\right)$ at $x = -4$ (f) $y = x^2\ln(x)$ at $x = 3$

(28) Plot graphs of the following functions by following the steps listed in Section 5.4.3.

(a) $y = e^{-\frac{1}{x}}$ (b) $y = e^{\frac{x}{x-1}}$ (c) $y = 4xe^{-2x}$

(d) $y = \ln(x^2 - 4)$ (e) $y = \ln\left(\dfrac{1}{1+x^2}\right)$ (f) $y = \dfrac{1+\ln(x)}{x}$

(29) An adaption of the Henderson–Hasselbalch equation can be used to show that the pH of $200\,\mathrm{ml}$ of a solution containing $40\,\mathrm{mM}$ methylamine being titrated against $2\,\mathrm{M}$ hydrochloric acid is given by $f(x) = 10.6 + \log_{10}\left(\dfrac{8-x}{x}\right)$, where $f(x)$ is the pH and x represents the volume of base in ml.

(a) Study the overall trend of the function in the domain $[0.1, 7.9]$.

(b) By finding the inflection point, determine how much base must be added for the titration curve to reach its shallowest gradient.

(c) Lastly, plot a graph of the function in a Cartesian plane.

(30) At the beginning of the Covid-19 pandemic in Italy, which was the first European country to be seriously hit, increasing numbers of infections were diagnosed at the beginning of March 2020. When the number of cases increased exponentially, a public lockdown was enforced, and everyone wanted to know for how long it would continue. The cumulative number of cases n was modelled against time t (days) using the equation $n(t) = \dfrac{L}{1+e^{-(t-f)}}$, with L and f being positive constants.

(a) For this mathematical model, can the number of cases increase infinitely?

(b) Is the equation appropriate to model the continuous increase of cases?

(c) Show that the first derivative of $n(t)$ can also be written as

$$n'(t) = \frac{L}{\left(e^{\frac{(t-f)}{2}} + e^{-\frac{(t-f)}{2}}\right)^2}$$

(d) When (on which day) is there an inflection point in the function $n(t)$?

(e) Does $n'(t)$ have a 'maximum' if $t = f$? If so, what does this mean physically?

(f) Sketch the graphs of the function $n(t)$ and its derivative, $n'(t)$, in a Cartesian plane.

(31) In the following series of irreversible reactions, species A reacts to form an intermediate species, I, which then reacts to form the product, P:

$$A \xrightarrow{k_1} I \xrightarrow{k_2} P$$

where k_1 and k_2 are the reaction rate constants.
The concentrations of the intermediate product I after time t is given by the following equation which is valid for the case where $k_1 \neq k_2$:

$$[I](t) = \frac{k_1[A]_0}{k_2 - k_1} \cdot \left(e^{-k_1 t} - e^{-k_2 t}\right)$$

where $[A]_0$ is the initial concentration of A.

(a) What happens to the equation for $[I](t)$ if $k_2 \to k_1$?

(b) At what value does the intermediate $[I]$ concentration reach a maximum according to the equation you obtained in (a)?

(c) If $k_1 = 2$ and $[A]_0 = \frac{3}{2}$, what is the graph of the new equation? Investigate and plot the function for $[I](t)$.

Answers

(1) (a) 32 (b) 9 (c) 6 (d) 25
(e) a^9 (f) x (g) $y^{\frac{7}{12}}$ (h) $16x^{12}$

(2) (a) $\frac{1}{16}$ (b) 9 (c) 16 (d) $\frac{9}{16}$
(e) 1 (f) 13 (g) 10^5 (h) 2

(3) (a) $C = 2, a = \frac{3}{2}$ (b) $C = \frac{1}{2}, a = \frac{1}{3}$

(4) (a) $\log_3 81 = 4$ (b) $\log_{27} \frac{1}{3} = -\frac{1}{3}$ (c) $\log_x 3 = y$ (d) $\log_2 \frac{1}{32} = -5$

(5) (a) 6 (b) 0 (c) 3 (d) $\sqrt{2}$

(6) (a) $\log a - \log b - \log c$ (b) $3\log a + 2\log b + \log c$
(c) $\frac{1}{2}(\log b - \log a)$ (d) $2\log b + \frac{1}{2}\log a - \frac{2}{3}\log c$

(7) (a) 1 (b) $\log(x-2)$ (c) 0 (d) $\log_a\left(\frac{\sqrt{a^5}}{2}\right)$

(8) (a) $\frac{1.26}{1.63} = 0.773$ (b) $\frac{1.63}{1.26} = 1.29$
(c) $\frac{1}{1.26} = 0.793$ (d) $\log_3 4 + \log_3 6 = 2.89$

(9) (a) $x = 3^3$ (b) $x = 3\log_4 4$ (c) $x = 5 \cdot 25^t$
(d) $x = \frac{1}{9} \cdot 3^t + 2$ (e) $x = \frac{\log 6}{\log 4}$ (f) $x = \log_5(\frac{3}{4})$

(10) (a) Concave upwards; it is not a straight line
(b) The equation fits the data on the plot
(c) $P = 10^{\log P} = 10^y = 10^{mx+c} = 10^{mx} \cdot 10^c = (10^m)^x \cdot 10^c$
$= \left(10^{.059}\right)^t \cdot 10^{2.243} \approx 1.45^t \cdot 175$

(11) (a) $\log y$ *versus* x is concave down
(b) $\log y$ *versus* $\log x$ resembles a straight line with equation
$\log y = \underbrace{4.3}_{m} \log x + \underbrace{0.54}_{c}$
(c) Raising both sides of the equation for the line in (b) to base 10 we obtain
$y = 10^{\log y} = 10^{m \log x + c} = 10^{\log x^m} \cdot 10^c = x^m \cdot 10^c = x^{4.3} \cdot 3.5$.
Thus, the appropriate model is $y = Ax^r$, where $A = 3.5$ and $r = 4.3$.

(12) (a) $x = \frac{1}{2}$ (b) $x = -1$, $x = 1$ (c) No real solutions
(d) $x = \frac{\ln(45)}{\ln(135)}$ (e) $x = 3$ (f) $x = \frac{1}{+}\sqrt{3}76$
(g) No real solutions (h) $x = 2$, $x = 3$

(13) (a) 3.3 (b) 3.0

(14) (a) $\frac{\log 2}{\log(1.145)} = 5.12$ years (b) $\frac{\log 3}{\log(1.145)} = 8.11$ years

(15) (a) $x = 0$, $x = 1$ (b) $x = 0$ (c) $x = e$, $x = e^2$ (d) $x = \frac{1}{2}$, $x = 2$

(16) (a) $I(1) = \frac{I_0}{4}$ (b) Solve $I(t) = \frac{I_0}{20}$ which gives $t \approx 34\,\text{days}$

(17) (a) $x < 0$ (b) $x < 0$ OR $x > 1$
(c) $\frac{2}{3} < x < 1$ OR $x > 2$ (d) $\frac{3}{8} < x < \frac{1}{2}$

(18) $2000 \cdot 2^t > 100000 \iff t > 5\,\text{hr}\,38\,\text{min}$

(19) $16\,\text{nmol}\,\text{L}^{-1} < [\text{H}^+] < 100\,\text{nmol}\,\text{L}^{-1}$

(20)

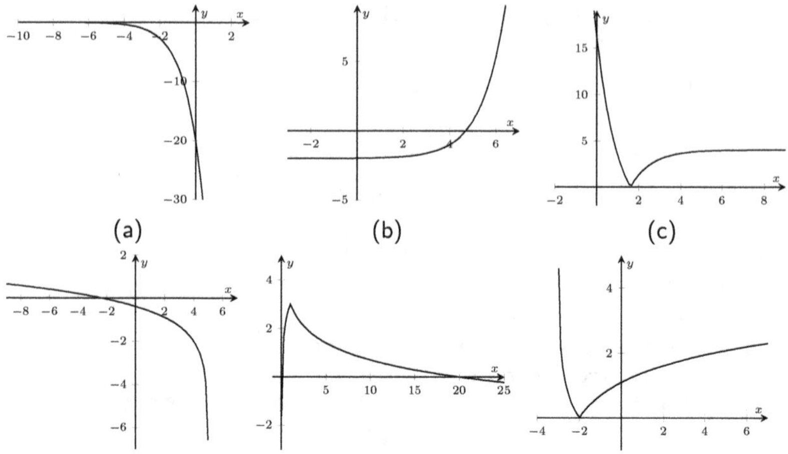

(a) (b) (c) (d) (e) (f)

(21) (a) $y' = 3e^x x + 8e^x$ (b) $y' = 5e^{5x}$
(c) $y' = \ln(x)(2x+9) + x + 9$ (d) $y' = 3e^{3x}x + e^{3x}$
(e) $y' = \frac{1-3\ln(x)}{5x^4}$ (f) $y' = \frac{2x-1}{x^2-x}$
(g) $y' = \frac{e^{\frac{x^2+4}{3-x}}\left(6x-x^2+4\right)}{(3-x)^2}$ (h) $y' = \frac{15x^2}{5x^3-2}$
(i) $y' = e^{2x^3+x}\left(6x^2+1\right)$ (j) $y' = \frac{1}{\ln(10)(x+1)}$
(k) $y' = \frac{1}{x(6x+1)}$ (l) $y' = 5^x \ln(5)x^8 + 8x^7 \cdot 5^x$
(m) $y' = \frac{3x^2-4}{\ln(3)\left(x^3-4x\right)}$ (n) $y' = \frac{\ln(x)-1}{2\ln^2(x)}$
(o) $y' = \ln(2) \cdot 2^{e^x} e^x$

(22) (a) $E'(C_0) = \frac{RT}{nF}\frac{1}{C_0}$ (b) $pH'(x) = -\frac{K}{x(-x+K)}$
(c) $N'(p) = -\frac{1}{\ln(1-f)(1-P)}$ (d) $N'(g) = -\frac{i\ln(1-P)}{x\ln^2\left(\frac{x-i}{x}\right)(x-i)}$
(e) $f'(\overline{Y}) = \frac{1}{\overline{Y}\left(-\overline{Y}+1\right)}$ (f) $C'(t) = -rC_0e^{-rt}$
(g) $L'(a) = L_\infty k e^{-k(a-a_0)}$ (h) $L'(T) = \frac{\ln(10)}{z} \cdot 10^{\frac{T-T_{ref}}{z}}$
(i) $K_E'(T) = \frac{4\pi a^3}{3000}N_A \cdot \frac{Ve^{-\frac{V}{Rx}}}{Rx^2}$ (j) $p'(v) = k_1\left(2ve^{k_2v} + e^{k_2v}k_2v^2\right)$
(k) $P'(t) = \frac{P_0Kre^{-rt}(K-P_0)}{\left(P_0+e^{-rt}(K-P_0)\right)^2}$ (l) $p'(s) = \frac{2e^{-2s}}{1-e^{-4N}}$

(23) (a) 1 (b) 1 (c) $\frac{5}{2}$ (d) 0 (e) -1 (f) $+\infty$
(g) $-\frac{4}{3}\ln(2)$ (h) $-\frac{5}{3}$ (i) $+\infty$ (j) $+\infty$ (k) $-\frac{1}{4}$ (l) $-\frac{1}{6}$

(24) (a) $[P] = \frac{a^2kt}{1+akt}$ (b) $\lim\limits_{t\to+\infty} \frac{a^2kt}{1+akt} = a$

(25) (a) $\lim_{t\to 0^+} U(T) = +\infty$; if the transmittance is very low, the uncertainty in the measurement is very high
(b) $T_{min} = \frac{1}{e}$
(c) $T > 0$ (as argument of ln) AND $\ln(T) < 0$ (as $U > 0$) which give $0 < T < 1$
(d) $A = -\log_{10}(T)$ and $U = \frac{k\cdot 10^A}{2.303\cdot A}$.

(26) (a) $min(0,1)$
(b) $max\left(\frac{3}{2}, e^{\frac{9}{4}}\right)$, $F_1\left(\frac{3-\sqrt{2}}{2}, e^{\frac{7}{4}}\right)$, $F_2\left(\frac{3+\sqrt{2}}{2}, e^{\frac{7}{4}}\right)$
(c) $max\left(1, \frac{4}{e}\right)$, $F\left(2, \frac{8}{e^2}\right)$ (d) $min(\ln(2), -4)$, $F(0,-3)$
(e) $min(0,0)$, $F_1(-1, \ln(2))$, $F_2(1, \ln(2))$
(f) $max(4, 2\ln(2) - 2)$, $F(16, 4\ln(2) - 4)$ (g) $min\left(\frac{1}{e}, -\frac{1}{e}\right)$
(h) $F(e, 0)$

(27) (a) $\searrow, \smile$ (b) $\nearrow, \frown$ (c) $\searrow, \smile$
(d) $\searrow, \frown$ (e) $\nearrow, \smile$ (f) $\nearrow, \smile$

(28) * (a) Dom: $(-\infty,0)\cup(0,+\infty)$; Pos: $(-\infty,0)\cup(0,+\infty)$; Asy: $y = 1$; F: $\left(\frac{1}{2}, \frac{1}{e^2}\right)$
(b) Dom: $(-\infty,1)\cup(1,+\infty)$; Pos: $(-\infty,1)\cup(1,+\infty)$; Asy: $y = e$; F: $\left(\frac{1}{2}, \frac{1}{e}\right)$
(c) Dom: $(-\infty,+\infty)$; Pos: $(0,+\infty)$; Asy: $y = 0$; max: $\left(\frac{1}{2}, \frac{2}{e}\right)$; F: $\left(1, \frac{4}{e^2}\right)$
(d) Dom: $(-\infty,-2)\cup(2,+\infty)$; Pos: $(-\infty,-\sqrt{5})\cup(\sqrt{5},+\infty)$; Asy: $x = -2$, $x = 2$
(e) Dom: $(-\infty,+\infty)$; Pos: None; max: $(0,0)$; F: $(-1,-\ln 2)$, $(1,-\ln 2)$
(f) Dom: $(0,+\infty)$; Pos: $\left(\frac{1}{e},+\infty\right)$; Asy: $y = 0$; max: $(1,1)$; F: $\left(\sqrt{e}, \frac{3}{2\sqrt{e}}\right)$

(29) (a) Dom: $[0.1, 7.9]$; Pos: $[0.1, 7.9]$; F: $(4, 10.6)$
(b) $4\,\mathrm{mL}$ to reach pH $= 10.6$
(c)

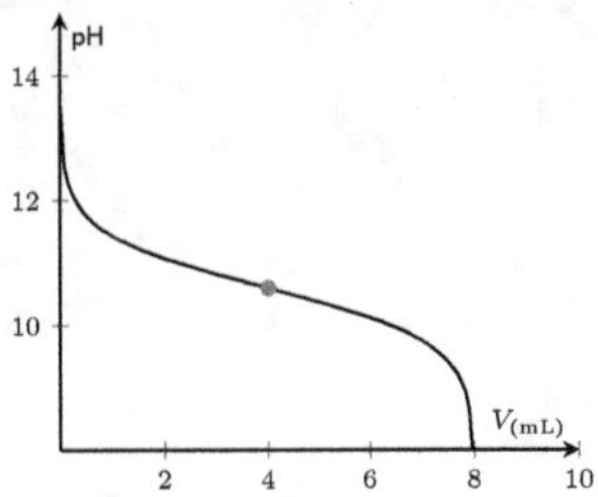

(30) (a) $\lim_{t\to+\infty} n(t) = \frac{L}{1+e^{-(t-f)}} = L$
(b) $n'(t) = \frac{L\cdot e^{-(t-f)}}{\left(1+e^{-(t-f)}\right)^2}$ is always > 0
(c) Hint: $n'(t) = \frac{L}{e^{(t-f)}\cdot\left(1+e^{-(t-f)}\right)^2}$
(d) $\left(f, \frac{L}{2}\right)$
(e) The point $\left(f, \frac{L}{2}\right)$ corresponds to the peak of simultaneous cases

(f) $n(t)$ is plotted as a solid line, whereas $n'(t)$ is a dashed line

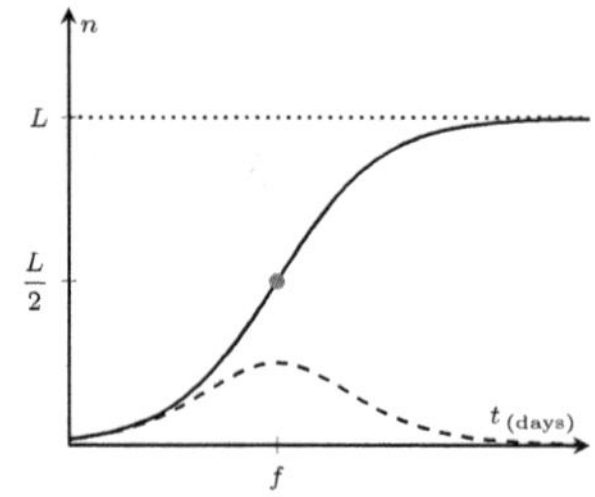

(31) (a) $[I] = k_1[A]_0 \left(te^{-k_1 t}\right)$

(b) $t_{max} = \frac{1}{k_1}$

(c) $I[t] = 2 \cdot \frac{3}{2} \left(te^{-\frac{3}{2}t}\right)$

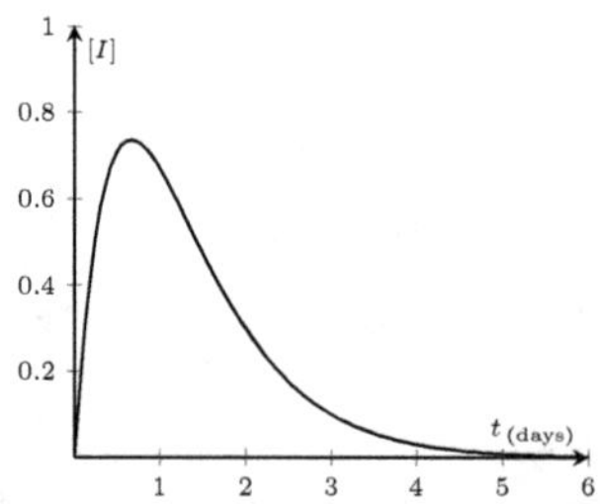

*Abbreviations: Dom = domain; Pos = function positive sign; Asy = asymptotes; max = maximum; min = minimum; F = inflection point.

Chapter 7

Trig Functions and Complex Numbers

Preamble

In this chapter, first we introduce some so-called 'trigonometric functions' that we meet in the Biosciences if we study recurring or periodic trends over time: these are usually called 'sinusoidal' trends. These functions are also vital in describing waves (*e.g.* electromagnetic waves) and oscillations. By adding together sinusoidal functions of different wavelength (repeat distance), phase (starting point), and amplitude (height), we can also describe the electron density in atoms and molecules (so-called Fourier synthesis).

Second, we present a brief glimpse into the world of complex number algebra, based on the mysterious square root of -1, a number called i, which we need in the study of crystallography where the Argand diagram is of great assistance in interpreting diffraction patterns composed of individual 'reflections' from crystals. In addition, we need complex number algebra to understand the link between the intensity of an observed reflection and its 'structure factor' and phase.

The imaginary number i also appears in the Fourier transforms that are a vital part of Nuclear Magnetic Resonance (NMR) spectroscopy, X-ray crystallography, small angle X-ray scattering (SAXS) analyses, and quantum theory in Chemistry.

7.1 Trigonometric Functions

Trigonometry concerns some very particular functions which have angles as independent variables. We derive and study three of them here: sine ($\sin$), cosine ($\cos$), and tangent ($\tan$).

However, first we must think about the units involved in manipulating angles: degrees ($^\circ$) and radians (rad).

7.1.1 Reminder about radians

A circle is divided into 360° and this is also equivalent to 2π radians, so $360^\circ = 2\pi$ radians and $180^\circ = \pi$ radians. Thus, $1^\circ = \dfrac{\pi}{180}$ radians. So,

$$1 \text{ radian} = \frac{180^\circ}{\pi} \approx 57.3^\circ \tag{7.1}$$

This important relationship must not be forgotten!

Example: Convert the right angle ($\theta = 90^\circ$) into radians.

Solution: Since 360° is equivalent to 2π radians, we have $\theta = 90^\circ = \dfrac{360^\circ}{4} = \dfrac{2\pi}{4} = \dfrac{\pi}{2}$ radians.

Example in Biochemistry: A protein sample is placed in an ultracentrifuge at a rotor speed of $59{,}780\,\mathrm{rpm}$ (rpm: revolutions per minute) at $25^\circ\mathrm{C}$. To calculate the mass of the protein from its radial position after ultracentrifugation, the angular velocity ω of the rotor in units of $\mathrm{rad}\cdot\mathrm{s}^{-1}$ is required. Convert the rotor speed to radians per second.

Solution: The rotor speed of $59{,}780\,\mathrm{rpm}$ is in rpm. One revolution corresponds to 2π radians, and we need ω in $\mathrm{rad}\cdot\mathrm{s}^{-1}$ rather than in degrees per minute, so $\omega = \dfrac{59780 \cdot 2\pi}{60} = 6257\,\mathrm{rad}\cdot\mathrm{s}^{-1}$.

This example shows us the importance of the relationship in equation (7.1)!

7.1.2 The sine and cosine functions

A point P on a circle of centre O and unit radius can be projected onto the x-axis or the y-axis (Figure 7.1).

By starting with the radius lined up with the positive x-axis and moving the endpoint P anticlockwise round the circumference of the circle we can see that if we project P onto the x-axis, the length of the projected line will oscillate from one unit at $\theta = 0°$ to zero at $\theta = 90°$, then to unit length again at $\theta = 180°$ and so on for $360°$ round the circle back to the positive x-axis (Figure 7.1).

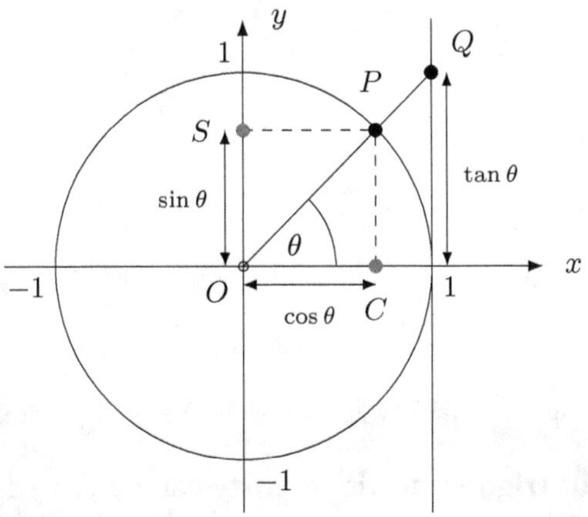

Figure 7.1: $P\,(\cos\theta, \sin\theta)$ and $Q\,(1, \tan\theta)$

Using this idea, the fundamental **trigonometric functions** are defined as follows:

- $\sin\theta$ is equivalent to the projection of OP onto the y-axis going round the circle anticlockwise from the positive x-axis (*i.e.* the y coordinates of P).
- $\cos\theta$ is equivalent to the projection of OP onto the x-axis going round the circle anticlockwise from the positive x-axis (*i.e.* the x coordinates of P).
- $\tan\theta$ corresponds to the y coordinate of point Q in Figure 7.1 as OPQ moves from the x-axis round the circle.
  - $\tan\theta$ is also the gradient of the line OP.
  - Additionally, $\tan\theta$ is defined as $\boxed{\tan\theta = \dfrac{\sin\theta}{\cos\theta}}$.

Additionally, in plane geometry, the values of the trigonometric functions sin, cos, and tan of one acute angle (θ) of a right angle triangle (the right angle is represented by a small square box in the corner of the two sides which are at $90°$ to one another) can be related to the three sides of a triangle which has a hypotenuse (h), an adjacent side (a), and an opposite side (o), as shown in Figure 7.2.

The trigonometric functions are defined as follows:

$$\sin\theta = \frac{o}{h} \quad \text{and} \quad \cos\theta = \frac{a}{h} \quad \Longrightarrow \quad \tan\theta = \frac{\sin\theta}{\cos\theta} = \frac{o}{h}\cdot\frac{h}{a} = \frac{o}{a}$$

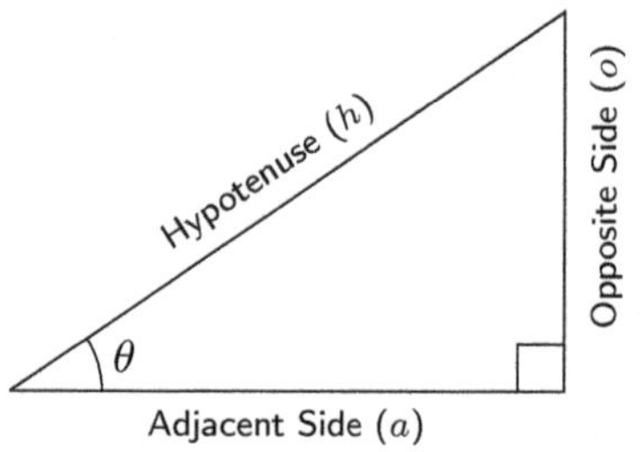

Figure 7.2: Trigonometric functions in a right angle triangle

A very **useful trigonometric identity** can be proved from Pythagoras' theorem which states that, for a right angle triangle with sides a and b and hypotenuse c, $a^2 + b^2 = c^2$.

Putting this together with the information in Figure 7.1, if we take $a = OC$, $b = PC$, and hypotenuse $c = OP$ or with sides $a = PS$ and $b = OS$ and hypotenuse $c = OP$, we can deduce that

$$\sin^2\theta + \cos^2\theta = 1 \tag{7.2}$$

Warning! $\sin^2(\theta) \neq \sin(\theta^2)$.
In fact, $\sin^2(\theta) = (\sin(\theta))^2 = \sin(\theta)\cdot\sin(\theta)$, and $\sin(\theta^2) = \sin(\theta\cdot\theta)$.

The same warning applies to cosine and tangent: $\cos^2(\theta) \neq \cos(\theta^2)$ and $\tan^2(\theta) \neq \tan(\theta^2)$.

Going back to Figure 7.1, by moving the endpoint P anticlockwise round the circumference of circle, we can list the values of the trigonometric functions obtained for different key angles. These are summarised in Table 7.1.

Table 7.1: Trigonometric function values at key angles

θ (degrees)	**0°**	**30°**	**45°**	**60°**	**90°**	**135°**	**180°**	**225°**	**270°**	**315°**	**360°**
θ (radians)	0	$\frac{\pi}{6}$	$\frac{\pi}{4}$	$\frac{\pi}{3}$	$\frac{\pi}{2}$	$\frac{3\pi}{4}$	π	$\frac{5\pi}{4}$	$\frac{3\pi}{2}$	$\frac{7\pi}{4}$	2π
$\sin\theta$	0	$\frac{1}{2}$	$\frac{\sqrt{2}}{2}$	$\frac{\sqrt{3}}{2}$	1	$\frac{\sqrt{2}}{2}$	0	$-\frac{\sqrt{2}}{2}$	-1	$-\frac{\sqrt{2}}{2}$	0
$\cos\theta$	1	$\frac{\sqrt{3}}{2}$	$\frac{\sqrt{2}}{2}$	$\frac{1}{2}$	0	$-\frac{\sqrt{2}}{2}$	-1	$-\frac{\sqrt{2}}{2}$	0	$\frac{\sqrt{2}}{2}$	1
$\tan\theta$	0	$\frac{\sqrt{3}}{3}$	1	$\sqrt{3}$	$\pm\infty$	-1	0	1	$\pm\infty$	1	0

Once we have completed a first circuit of the circumference of the circle in Figure 7.1, we can keep moving the endpoint P anticlockwise, and the values of $\sin\theta$ and $\cos\theta$ will repeat 'periodically' every 2π, while the values of $\tan\theta$ repeat 'periodically' every π.

Figure 7.3 shows a plot of the trigonometric functions $\sin\theta$ and $\cos\theta$ in a Cartesian plane for all possible real values of θ in radians.

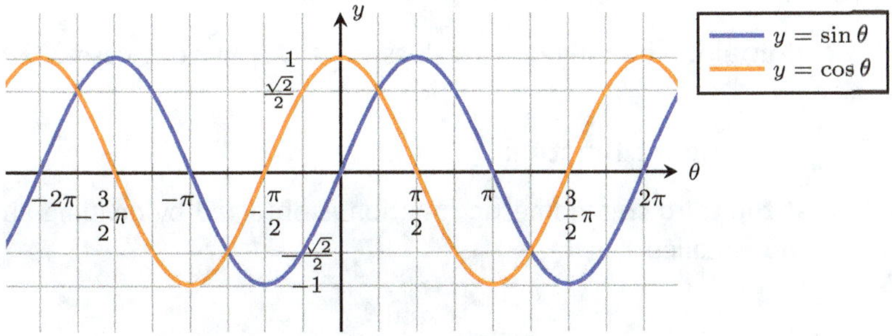

Figure 7.3: Trigonometric functions: $\sin\theta$ and $\cos\theta$

The key information useful for sketching the trigonometric functions $\sin\theta$ and $\cos\theta$ is as follows:

(i) **Domain:** Every real number ($\theta \in \mathbb{R}$) in both cases, *i.e.* $(-\infty, +\infty)$.

Important note: The values of both functions repeat every 2π, so

- $\sin(\theta + 2k\pi) = \sin(\theta)$
- $\cos(\theta + 2k\pi) = \cos(\theta)$

where $k \in \mathbb{Z}$, *i.e.* k can be any positive or negative integer.

Thus, we can highlight the key elements of both the functions in the restricted range of θ values $[0, 2\pi]$.

(ii) **Symmetry:**

(a) $\sin(-\theta) = -\sin(\theta)$, so the function $\sin\theta$ is odd and thus it is symmetrical around the point $O\,(0,0)$, which is the origin of the Cartesian plane;

(b) $\cos(-\theta) = \cos(\theta)$, so the function $\cos\theta$ is even and thus it is symmetrical across the y-axis.

(iii) **Limits and asymptotes:** The limits of both functions as θ tends to $\pm\infty$ do not exist.

(iv) **Continuity:** Both functions are continuous in any closed intervals $[a, b]$ of $\mathbb{R}$.

(v) **Derivative:** By inspection of the gradients of both $\sin\theta$ and $\cos\theta$ as a function of θ, it can be seen that

(a) the derivative of $\sin\theta$ is $y' = \cos\theta$;

(b) the derivative of $\cos\theta$ is $y' = -\sin\theta$.

(vi) **Co-domain:** The range of y values is $[-1, 1]$ in both cases.

7.1.3 The tangent function

Recall that the third trigonometric function is obtained by dividing $\sin\theta$ by $\cos\theta$ and is called $\tan\theta$:

$$\tan\theta = \frac{\sin\theta}{\cos\theta} \tag{7.3}$$

Figure 7.4 shows a plot of $\tan\theta$ in a Cartesian plane for all possible real values of θ in radians.

The key information useful for sketching the trigonometric function $\tan\theta$ is as follows:

(i) **Domain:** Every real number ($\theta \in \mathbb{R}$) **except** $\theta = \frac{\pi}{2} + k\pi$, where $k \in \mathbb{Z}$, *i.e.* k can be any positive or negative integer.

Thus, θ values such as $\ldots, -\frac{3\pi}{2}, -\frac{\pi}{2}, \frac{\pi}{2}, \frac{3\pi}{2}, \ldots$ are **not included** in the domain since these values make the **denominator** of the fraction $\tan\theta = \frac{\sin\theta}{\cos\theta}$ **equal to** 0, which has **no meaning** for the fraction.

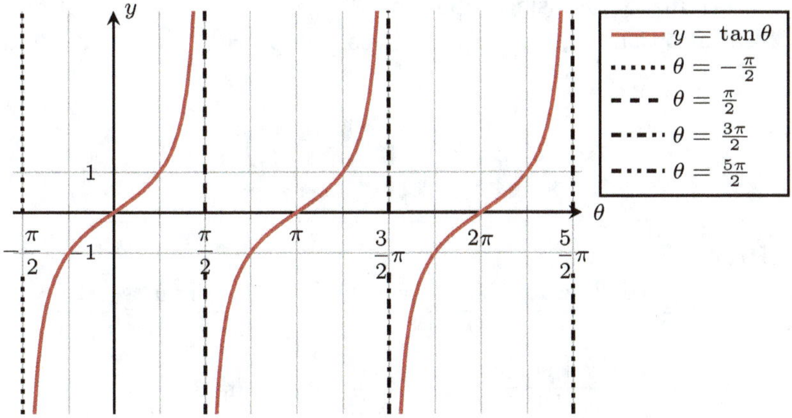

Figure 7.4: Trigonometric function: $\tan\theta$

> **Important note:** The values of $\tan\theta$ functions repeat every π, so
>
> $$\tan(\theta+\pi)=\tan(\theta)$$
>
> Thus, we can highlight the key elements of the function $\tan\theta$ in the restricted range of θ values $[0,\pi]$.

(ii) **Symmetry:** $\tan(-\theta)=-\tan(\theta)$, so the function $\tan\theta$ is odd and thus is symmetrical around the point $O\,(0,0)$.

(iii) **Limits and asymptotes:** The limits of the function as θ tends to $\pm\infty$ do not exist.

However, since the $\lim\limits_{\theta\to\frac{\pi}{2}}\tan\theta=\pm\infty$, the straight line $\theta=\dfrac{\pi}{2}$ is a **vertical asymptote** of $\tan\theta$ around $\theta=\dfrac{\pi}{2}$.

Any straight line of equation $\theta=\dfrac{\pi}{2}+k\pi$, where $k\in\mathbb{Z}$ (*i.e.* k can be any positive or negative integer), is a vertical asymptote of the function $\tan\theta$.

(iv) **Continuity:** The function is continuous in any closed intervals $[a,b]$ of its domain.

(v) **Derivative:** By using equation (7.3), the Quotient Rule of differentiation and equation (7.2), it can be proved that the derivative of $y = \tan\theta = \dfrac{\sin\theta}{\cos\theta}$ is

$$y' = \frac{1}{\cos^2\theta} = 1 + \tan^2\theta$$

Proofs:

- $y' = \dfrac{\cos\theta \cdot \cos\theta - \sin\theta(-\sin\theta)}{\cos^2\theta} = \dfrac{\cos^2\theta + \sin^2\theta}{\cos^2\theta} = \dfrac{1}{\cos^2\theta}$;
- $y' = \dfrac{\cos^2\theta + \sin^2\theta}{\cos^2\theta} = 1 + \dfrac{\sin^2\theta}{\cos^2\theta} = 1 + \tan^2\theta.$

(vi) **Co-domain:** The range of y values is $(-\infty, +\infty)$.

7.1.4 The arctangent function

We now introduce the operation which 'undoes' taking the tangent of a value, the so-called arctangent function (arctan).

We need this function if we want to convert from Cartesian coordinates (x, y) to Polar coordinates (r, θ), and for instance in crystallography when we want to set the distance of an X-ray detector from the crystal or find out the resolution of a diffraction pattern (see example below).

If we restrict the domain of the function $y = \tan\theta$ to the interval $\left(-\dfrac{\pi}{2}, \dfrac{\pi}{2}\right)$, the function satisfies the two conditions necessary to be invertible according to the inverse function theorem (see Section 6.2.2):

(1) $y = \tan\theta$ is differentiable on $\left(-\dfrac{\pi}{2}, \dfrac{\pi}{2}\right)$;

(2) $y' = \dfrac{1}{\cos^2\theta}$ is always > 0 (*i.e.* y is always increasing) on $\left(-\dfrac{\pi}{2}, \dfrac{\pi}{2}\right)$.

The inverse function of $\tan\theta$ is called the **arctangent function** and is denoted

$$y = \tan^{-1}(x) = \arctan(x)$$

where the arctan of x is the angle $\theta \in \left(-\dfrac{\pi}{2}, \dfrac{\pi}{2}\right)$ for which $\tan\theta = x$.

Warning! The notation $\tan^{-1}(x)$ is **not** equivalent to $\dfrac{1}{\tan(x)}$!

Thus,

$$y = \arctan x \iff x = \tan y \quad \textbf{only if} \quad -\frac{\pi}{2} < x < \frac{\pi}{2} \qquad (7.4)$$

We can now highlight some key properties of the function $y = \arctan(x)$ derived from the inverse function theorem:

(1) Taking the arctangent of the tangent of an angle θ gives θ:

$$\arctan(\tan(\theta)) = \theta \quad \text{if } \theta \in \left(-\frac{\pi}{2}, \frac{\pi}{2}\right)$$

Taking the tangent of the arctangent of a real value x gives x:

$$\tan(\arctan(x)) = x \quad \text{if } x \in \mathbb{R}$$

(2) The derivative of $y = \arctan(x)$ is

$$y' = \frac{dy}{dx} = \frac{d(\arctan(x))}{dx} = \frac{1}{\frac{dx}{dy}} = \frac{1}{1+\tan^2(y)} = \frac{1}{1+(\tan(\arctan(x)))^2} = \frac{1}{1+x^2}$$

Thus, the derivative of $y = \arctan x$ is $y' = \dfrac{1}{1+x^2}$.

(3) The graphs of the functions $y = \arctan(x)$ and $y = \tan(\theta)$ are **reflections of one another** across the straight line $y = x$ (Figure 7.5).

The key information needed to sketch the function $\arctan(x)$ is as follows:

(i) **Domain:** Every real number ($x \in \mathbb{R}$).
(ii) **Symmetry:** $\arctan(-x) = -\arctan(x)$, so the function $\arctan(x)$ is odd and thus it is symmetrical around the point $O\,(0, 0)$.
(iii) **Limits and asymptotes:** $\lim\limits_{x\to+\infty} \arctan(x) = +\dfrac{\pi}{2}$, thus, $y = \dfrac{\pi}{2}$ is a horizontal asymptote for the function as x tends to very large positive values; $\lim\limits_{x\to-\infty} \arctan(x) = -\dfrac{\pi}{2}$, thus, $y = -\dfrac{\pi}{2}$ is a horizontal asymptote for the function as x tends to very large negative values.

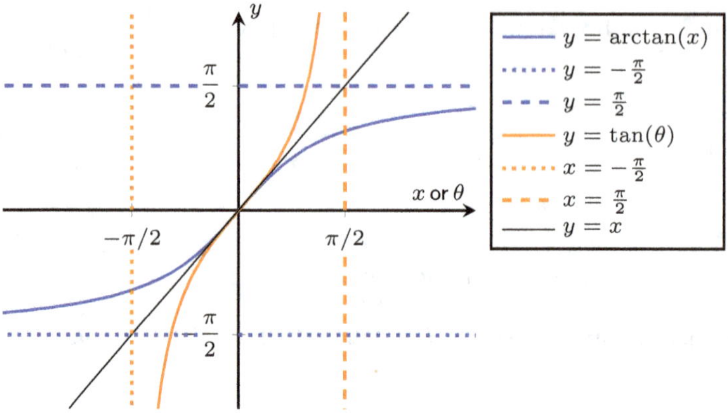

Figure 7.5: Graphs of $\arctan(x)$ and $\tan(\theta)$

(iv) **Continuity:** The function is continuous in any closed intervals $[a, b]$ of its domain, $\mathbb{R}$.

(v) **Co-domain:** The range of y values is $\left(-\frac{\pi}{2}, +\frac{\pi}{2}\right)$.

Example in Biochemistry: In crystallography, the crystal to detector distance, l, is positioned depending on how well the crystal diffracts when it is irradiated by X-rays and then l is set so that the pattern fills the detector, as shown in the following diagram.

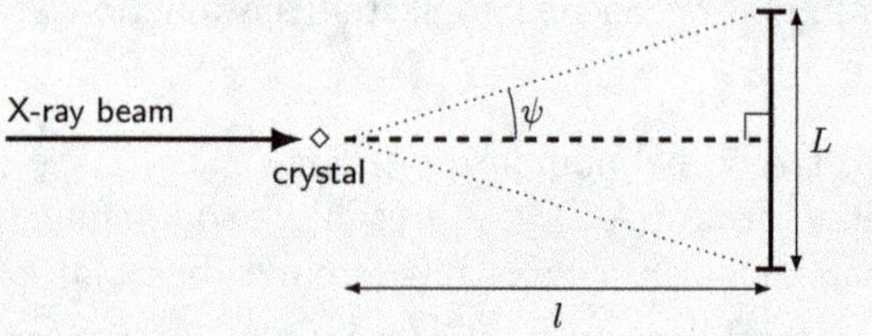

If a circular detector of diameter $L = 345\,\text{mm}$ is centrally placed round the main beam direction at a distance of $l = 184\,\text{mm}$ from the crystal, what is the angle, ψ, subtended by the edge of the detector?

Solution: For this problem, the trigonometric relationships and arctan function are useful.

The detector is 184 mm away and its 345 mm diameter is centrally placed round the beam, so the radius is 172.5 mm.

Thus, we have a right angle triangle and can take the opposite over the adjacent side to find the angle ψ. Since $\tan(\psi) = \frac{\frac{L}{2}}{l}$, thus

$$\psi = \tan^{-1}\left(\frac{\frac{L}{2}}{l}\right) = \tan^{-1}\left(\frac{172.5}{184}\right) = \arctan\left(\frac{172.5}{184}\right) = 48.4^\circ$$

Knowing this angle, we can now calculate the resolution at the edge of the detector, d_{max}, using Bragg's law, which states that

$$n\,\lambda = 2\,d_{max}\sin\left(\frac{\psi}{2}\right)$$

where λ is the incident X-ray wavelength in Å $(1\ \text{Å} = 10^{-10}\,\text{m})$, and n is an integer which we can take as 1.

Taking $\lambda = 1.54\,\text{Å}$, which is the wavelength of copper as for a protein crystallography in-house X-ray generator, we can calculate the resolution d_{max}.

Putting our value of $\psi = 48.4^\circ$ into Bragg's law we can get d_{max}, which tells us the resolution at the edge of the detector:

$$1.54\ \text{Å} = 2\,d_{max}\sin\left(\frac{48.4^\circ}{2}\right) \iff d_{max} = 2.07\ \text{Å}$$

This would be a good level of detail for a protein structure.

7.1.5 Elementary trig equations

Sometimes in Biosciences we need to solve elementary equations for x containing trigonometric functions.

Generally, we usually look for solutions in a specific interval of $x \in [0, 2\pi]$ and the elementary equations are of the type

- **(A)** $\sin(f(x)) = a$,
- **(B)** $\cos(f(x)) = a$,
- **(C)** $\tan(f(x)) = a$,

where $f(x)$ is usually a linear expression containing the unknown θ and a can be any real number.

For all three cases, the solution for θ in $\mathbb{R}$ can be obtained by equating $f(x)$ with the value or the values of angle θ in Table 7.1 for which the trigonometric function gives the value a for common values of θ (see Table 7.1):

- for **(A)** and **(B)** $f(x) = \theta + 2k\pi$,
- for **(C)** $f(x) = \theta + k\pi$,

where $k \in \mathbb{Z}$, *i.e.* k can be any positive or negative integer.

Example: Solve the elementary equation $\sin\left(2x - \frac{\pi}{4}\right) = 0$ in the interval $[0, 2\pi]$.

Solution: If we look at Table 7.1, there are two angles for which the $\sin$ function gives 0 in the interval $[0, 2\pi]$: 0 and π. Therefore,

$$\sin\left(2x - \frac{\pi}{4}\right) = 0 \iff 2x - \frac{\pi}{4} = 0 + 2k\pi \text{ OR } 2x - \frac{\pi}{4} = \pi + 2k\pi$$

$$\iff x = \frac{\pi}{8} + k\pi \text{ OR } x = \frac{5\pi}{8} + k\pi$$

Thus, with $k = 0$ and $k = 1$, we obtain all the possible solutions in the interval $[0, 2\pi]$: $x = \frac{\pi}{8}$, $x = \frac{5\pi}{8}$, $x = \frac{9\pi}{8}$, OR $x = \frac{13\pi}{8}$.

For $k < 0$ or $k > 1$, we can obtain other possible solutions for the initial equation, but these additional solutions would be out of the range $[0, 2\pi]$ required by the exercise.

Example in Biology: In the Darwinian theory of evolution, it is generally assumed that the environments in which life appeared were hydrothermal environments with highly variable conditions in terms of pH, temperature, or redox levels. It has also been supposed that there is a principle of environmental dependence of the evolution processes on hydrothermal parameters.

In particular, the reactivity R of growth catalyst molecules varies following a sinusoidal trend of the type $R = R_{max} \cdot \sin\left(\frac{2\pi}{\alpha} t\right)$ where t is the time in years and $\alpha = 10^6$ years.

When does R reach its maximum value? How often does it reach this value?

Solution: R reaches its maximum value when $\sin\left(\frac{2\pi}{10^6}t\right) = 1$ which is an elementary trigonometric equation. Thus,
$\frac{2\pi}{10^6}t = \frac{\pi}{2} + 2k\pi \Longleftrightarrow t = \left(\frac{\pi}{2} + 2k\pi\right) \cdot \frac{10^6}{2\pi} \Longleftrightarrow \frac{10^6}{4} + k \cdot 10^6$
$= 250{,}000 + k \cdot 10^6$ years, where k can be any positive integer.

Thus, R first reaches its maximum value $250{,}000$ years after the initial count started and then reaches its maximum again every million years after that.

7.2 Investigating Trig Functions

7.2.1 Using translations

Periodical phenomena in Biosciences are generally called 'sinusoidal' and are modelled by using $\sin$ or $\cos$ trigonometric functions of the type

$$y = \alpha \sin\left(\beta \cdot (x + \gamma)\right) + \delta \tag{7.5}$$

where,

- α is the **amplitude**, representing the difference between the maximum y value (y_{max}) and the minimum y value (y_{min}), divided by 2: $\alpha = \frac{y_{max} - y_{min}}{2}$;
- β gives the **period** P of y, which is the difference in x-coordinate between one peak and the next one: $P = \frac{2\pi}{\beta}$;
- γ is the **phase shift**, which is how far the function y is shifted horizontally along the x-axis from the usual position by γ units: if $\gamma > 0$, the function y is translated left, while if $\gamma < 0$, the function y is translated right, and if $\gamma = 0$ it is not shifted at all;

Important note: The phase shift is **always** within the brackets with x.

- δ is the **vertical shift**, which is how far the function y is shifted vertically from the x-axis: $\delta = \dfrac{y_{max} + y_{min}}{2}$.

Sometimes, the $\cos$ function can also be used instead of $\sin$ in equation (7.5).

The graphs of $\sin$ and $\cos$ functions should become familiar to you (Figure 7.3) and can be used as 'building blocks' to investigate the form of other 'sinusoidal' functions without plotting them accurately.

The rules given in Section 4.7 for sketching the form of a function can also be applied to $\sin$ and $\cos$ functions modelling 'sinusoidal' phenomena, and using the most appropriate geometric transformations to the graphs in Figure 7.3.

Example: Sketch the function $y = 3\sin\left(2x + \dfrac{\pi}{2}\right) + 4$ by using the rules of geometric transformations described in Section 4.7.

Then identify the amplitude, period, phase shift, and vertical shift of the function.

Solution: First we can identify $y = \sin x$ as the closest elementary function to $y = 3\sin\left(2x + \dfrac{\pi}{2}\right) + 4$ and start by plotting it in the Cartesian plane, as shown in the next figure.

Using Rule **G(ii)** (Section 4.7), we can sketch $y = \sin(2x)$ by compressing the graph horizontally by a factor of 2.

Then, using Rule **B(i)**, we can sketch $y = \sin\left(2x + \dfrac{\pi}{2}\right) = \sin\left[2\left(x + \dfrac{\pi}{4}\right)\right]$ by translating the graph of $y = \sin(2x)$ left by $\dfrac{\pi}{4}$ of a unit.

Remember that the phase shift is **always** within the brackets with x.

Using Rule **G(ii)**, we can sketch $y = 3\sin\left(2x + \dfrac{\pi}{2}\right)$ by expanding the graph of $y = \sin\left(2x + \dfrac{\pi}{2}\right)$ by a factor of 3 vertically.

Finally, in order to sketch $y = 3\sin\left(2x + \dfrac{\pi}{2}\right) + 4$, we can use Rule **A(i)** by translating the plot of $y = 3\sin\left(2x + \dfrac{\pi}{2}\right)$ up by 4 units.

The figure below summarises all the intermediate steps involved in sketching $y = 3\sin\left(2x + \dfrac{\pi}{2}\right) + 4$.

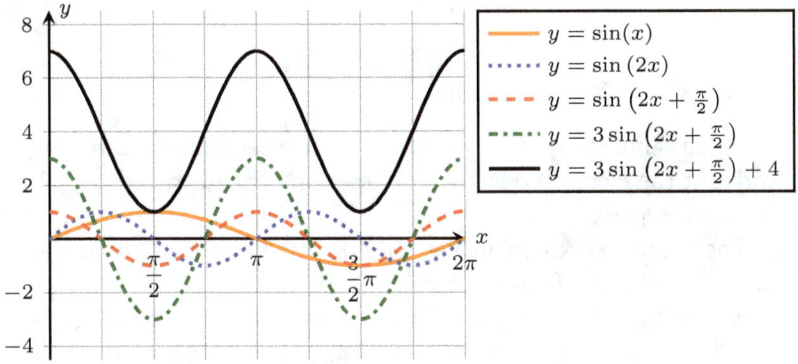

By rearranging the original function as $y = 3\sin\left[2\left(x + \dfrac{\pi}{4}\right)\right] + 4$, we can identify:

- the amplitude $\alpha = 3$;
- the period $P = \dfrac{2\pi}{\beta} = \dfrac{2\pi}{2} = \pi$;
- the phase shift $\gamma = \dfrac{\pi}{4}$;
- the vertical shift $\delta = 4$.

Example in Biochemistry: A simple crystal has a repeat unit (the 'unit cell') containing just two carbon atoms and one oxygen atom in a row: C-C-O. The carbon atoms are 1.5 Å apart, and the oxygen is 2 Å from the second carbon atom.

Show that you can approximately represent the electron density across one dimension in this unit cell in terms of a sum of sine waves of different frequencies (2, 3, and 5), phases, and amplitudes:

$$y = 3\sin\left(2\left(\lambda + \frac{\pi}{2}\right)\right) + 4\sin\left(3(\lambda + \pi)\right) + 6\sin\left(5(\lambda + \pi)\right)$$

where λ is in Å$= 10^{-10}$m.

Solution: If we plot the three different sine waves separately, as shown in the following figure,

- $y = 3\sin\left(2\left(\lambda + \dfrac{\pi}{2}\right)\right)$ (green line),
- $y = 4\sin\left(3(\lambda + \pi)\right)$ (blue line),
- $y = 6\sin\left(5(\lambda + \pi)\right)$ (red line),

and then sum them (black line), we can see that we can almost represent the electron density cross section of the two carbon atoms and one oxygen atom.

There are two extra peaks: on the left-hand sides of the first carbon atom and of the oxygen atom.

By adding higher frequency sine waves of various amplitudes and phases, we could better match the real electron density.

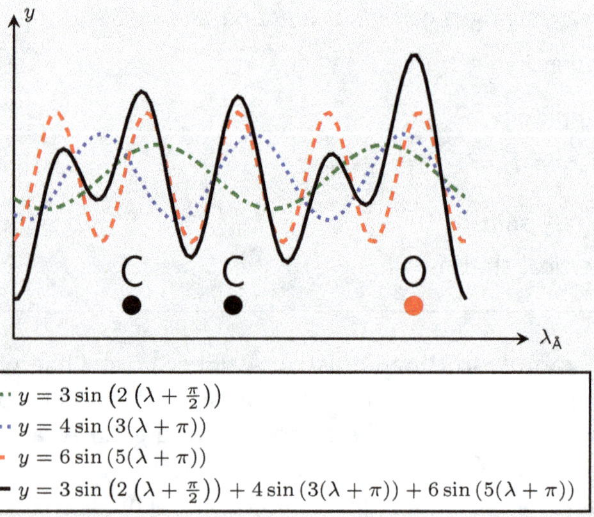

This problem introduces the concepts behind the mathematical tool of the Fourier transform.

By plotting the frequency of the sine waves on the x-axis and their amplitude on the y-axis, we can 'transform' the information contained in the series of sine waves to a different representation.

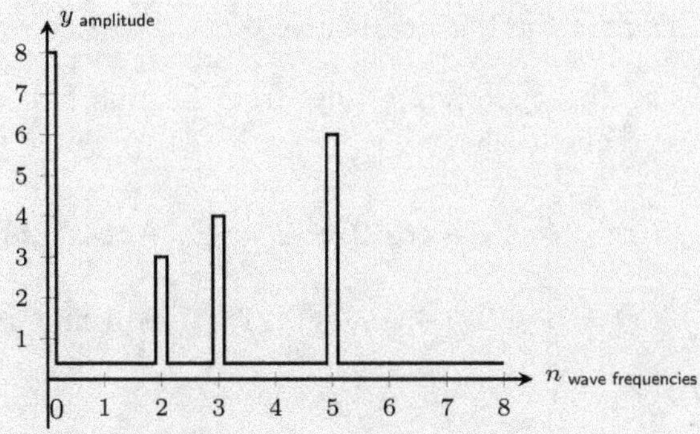

This 'transform' is very useful in interpreting diffraction patterns from protein crystals with many atoms.

Each vertical bar on the plot represents a 'reflection'. The spacing gives information on the size of the repeating unit cell and the peak height (amplitude) tells us something about the electron density and can be used to reconstruct the contents of the unit cell. Fourier transforms are thus extensively used in crystallography.

Note that this example could usefully be solved using a Web-based graphing program.

7.2.2 Using derivatives

As described in detail in Chapter 5, derivatives provide useful information about a function $f(x)$.

All the rules for differentiating algebraic functions shown in Section 5.3 also apply to trigonometric functions.

Example: Differentiate the following function using the appropriate rules: $y = 5\cos(3x) - \sin(x^2 - x)$.

Solution: Using Rule **A** first, then Rule **B**, and finally Rule **D** from Section 5.3.2, we obtain:

$$\begin{aligned}\frac{dy}{dx} &= -5\sin(3x)\cdot 3 - \cos(x^2 - x)\cdot(2x - 1)\\ &= -15\sin(3x) - (2x-1)\cos(x^2 - x)\end{aligned}$$

Example: Calculate the third derivative of $f(x) = \frac{2}{3}x^2 - \sin(2x)$.

Solution: Using the appropriate rules from Section 5.3, we calculate the derivatives sequentially:

$$f'(x) = \frac{2}{3} \cdot 2x - \cos(2x) \cdot 2 = \frac{4}{3}x - 2\cos(2x)$$
$$f''(x) = \frac{4}{3} - 2 \cdot (-\sin(2x)) \cdot 2 = \frac{4}{3} + 4\sin(2x)$$
$$f'''(x) = 4\cos(2x) \cdot 2 = 8\cos(2x)$$

Also for trigonometric functions, the sign of the first derivative $y' = f'(x) = \frac{dy}{dx}$ tells us whether $f(x)$ is increasing or decreasing, while the sign of the second derivative $y'' = f''(x) = \frac{d^2y}{dx^2}$ tells us about its graph's curvature.

For trigonometric functions, L'Hôpital's theorem helps us calculate some indeterminate forms of limits by using derivatives.

Example in Medicine: Following a research study conducted on a panel of ten healthy adults, participants' body temperature was measured as a function of the time of day.

The overall daily variation of body temperature (θ in °C) was modelled using the following cosine function:

$$\theta(t) = M + A\cos(\omega(t + \phi))$$

where M is the constant Mesor (*i.e.* the mean theoretical value around which the cosine model fluctuates), A is the constant amplitude of the fluctuation, ω is the period (*i.e.* usually $\omega = \frac{2\pi}{24}\ \mathrm{rad\,hr}^{-1}$ for a day of 24 hours), t is the time in hr, and ϕ is a constant related to the acrophase (*i.e.* the time at which the maximum theoretical value occurs) in hr. All constants have positive values.

Demonstrate that at time $t = 24 - \phi$ the maximum theoretical value is reached by the cosine curve during a day. What happens at time $t = 12 - \phi$?

Solution: In order to determine whether the maximum theoretical value of daily body temperature is reached at $t = 24 - \phi$, we should differentiate the function $\theta(t)$ with respect to t and set the first derivative equal to 0:

$$\frac{d\theta}{dt} = 0 + A \cdot (-\sin(\omega(t + \phi))) \cdot \omega = 0 \Longleftrightarrow \sin(\omega(t + \phi)) = 0$$

Two solutions are possible:

(i) $\omega(t + \phi) = 0 \Longleftrightarrow t = -\phi \Longleftrightarrow t = 24 - \phi$ as time t can only be positive and $\theta(t)$ has a period of 24 hours;

(ii) $\omega(t + \phi) = \pi \Longleftrightarrow \frac{2\pi}{24}(t + \phi) = \pi \Longleftrightarrow t + \phi = 12 \Longleftrightarrow t = 12 - \phi$.

Then, we differentiate again $\frac{d^2\theta}{dt^2} = -A \cdot (\cos(\omega(t + \phi))) \cdot \omega^2$ and substitute $t = 24 - \phi$ and $t = 12 - \phi$ into the second derivative to see what happens around these values:

(i) $t = 24 - \phi \Rightarrow -A \cdot \left(\cos\left(\frac{2\pi}{24}(24 - \phi + \phi)\right)\right) \cdot \left(\frac{2\pi}{24}\right)^2 = -A \cdot (\cos(2\pi)) \cdot \left(\frac{2\pi}{24}\right)^2 = -A \cdot 1 \cdot \left(\frac{2\pi}{24}\right)^2 < 0$: this is negative and so represents a maximum value. Thus, at $t = 24 - \phi$ the maximum theoretical value of body temperature is reached during a day.

(ii) $t = 12 - \phi \Rightarrow -A \cdot \left(\cos\left(\frac{2\pi}{24}(12 - \phi + \phi)\right)\right) \cdot \left(\frac{2\pi}{24}\right)^2 = -A \cdot (\cos(\pi)) \cdot \left(\frac{2\pi}{24}\right)^2 = -A \cdot (-1) \cdot \left(\frac{2\pi}{24}\right)^2 > 0$: this is positive and so represents a minimum value. Thus, at $t = 12 - \phi$ the minimum theoretical value of body temperature is reached.

7.3 Complex Numbers

We now extend our mathematical tool box by introducing so called 'complex numbers' and the algebraic rules necessary to deal with them.

They are used in many branches of science, including in Biosciences where they assist in analysing crystallographic diffraction patterns and appear in Argand diagrams and Fourier transforms. They are also frequently used in the field of electrical engineering in the treatment of circuits and alternating currents.

7.3.1 Arithmetic: the set of complex numbers

- **The imaginary number i**

  The polynomial equation $x^2 - 1 = 0$ has two real roots: 1 and -1, but the polynomial equation $x^2 + 1 = 0$ has *no* real roots.

  To obtain a solution to this equation, we introduce an **imaginary root** i so that the polynomial equation x^2+1 has two imaginary roots: i and $-i$.

  That is, i has the property that $i^2 = -1$.

- **The complex plane – Cartesian coordinates**

  The complex numbers comprise the set of all expressions of the form $a + bi$, where $i^2 = -1$ and a and b are real numbers:

$$\mathbb{C} = \{a + bi \mid a, b \text{ real numbers}\}$$

  For the number $z = a+bi \in \mathbb{C}$, we define the **real** ($\Re$) and **imaginary** ($\Im$) parts of z to be

$$\Re(z) = a \quad \text{and} \quad \Im(z) = b$$

The imaginary number i has no place on the real number line. Instead, we locate i in the **complex plane** at the point $(0, 1)$.

Just as we think of any real number as being on the *real line*, we can represent any complex number $z = a + bi$ as the point (a, b) in the **complex plane**.

The coordinates (a, b) are usually called the *Cartesian* coordinates for z.

In the complex plane shown in Figure 7.6, the **real numbers** lie on the horizontal axis, and we refer to this horizontal axis as the **real axis**, while the **imaginary numbers** of the type $z = ib$ lie on the vertical axis, which is referred to as the **imaginary axis** of the complex plane.

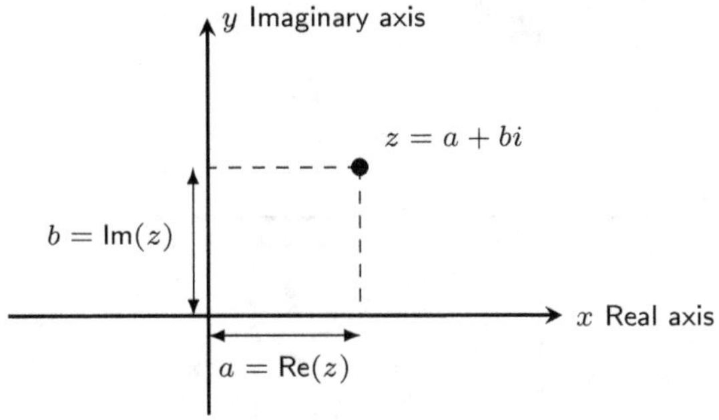

Figure 7.6: Complex plane

Just as the real number line contains all the square roots of the positive integers, the complex plane $\mathbf{C}$ contains *all* of the roots of *every* polynomial.

This is because of the following fundamental theorems.

> The **fundamental theorem of algebra** states that every non-constant single variable polynomial with complex coefficients has at least one complex root.
>
> The number of these roots cannot exceed the degree of the polynomial.
>
> The **complex conjugate root theorem** states that if a **polynomial** in one variable with **real coefficients** has a root of the type $a + bi$ where a and b are real numbers, then its complex conjugate $a - bi$ is also a root of the polynomial.

Example: Locate the roots of $x^2 - 8x + 25$ in the complex plane.

Solution: Applying the quadratic formula to the equation (see Section 2.3.2) $x^2 - 8x + 25 = 0$, we get

$$\begin{aligned} x^2 - 8x + 25 &= ax^2 + bx + c = 0 \\ \iff x &= \frac{-b \pm \sqrt{b^2 - 4ac}}{2a} = \frac{8 \pm \sqrt{64 - 100}}{2} = \frac{8 \pm \sqrt{-36}}{2} \\ &= \frac{8 \pm 6i}{2} = 4 \pm 3i \end{aligned}$$

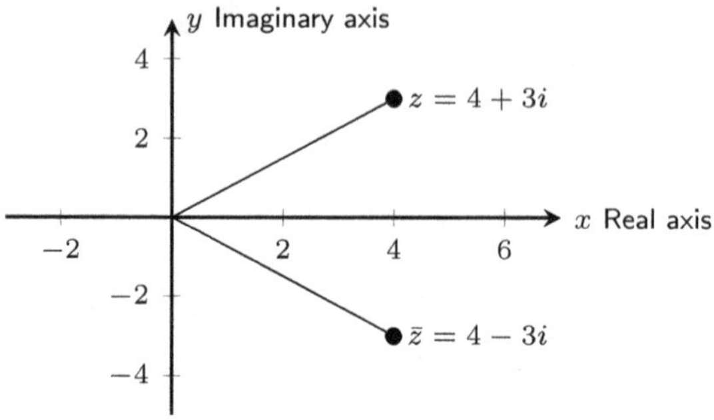

Note that these two roots are reflections of one another in the real axis.

We call numbers related in this way the *conjugate* of one another.

Note also that the distance from these two points, z and $\bar{z}$, to $O(0,0)$ is $\sqrt{4^2+(\pm 3)^2} = \sqrt{16+9} = \sqrt{25} = 5$ units.

We call this distance the *modulus* of the complex numbers $z = 4+3i$ and $\bar{z} = 4-3i$, and denote them as $|z|$ and $|\bar{z}|$, respectively.

In general, let $z = a + bi$.

The **complex conjugate** of z is the number $\bar{z} = a - bi$.

The **modulus** of z is the real number $|z| = \sqrt{a^2+b^2}$, as demonstrated earlier in Section 3.1.1.

The **modulus** is connected to its **complex conjugate** by means of the formula

$$z \cdot \bar{z} = |z|^2 \tag{7.6}$$

Proof: Let $z = a + bi$,

$$\begin{aligned} z \cdot \bar{z} &= (a+bi)(a-bi) = a^2 - (bi)^2 = a^2 - b^2 \cdot i^2 = a^2 - b^2(-1) \\ &= a^2 + b^2 = |z|^2 \end{aligned}$$

Example: If $z = 5+12i$, calculate its complex conjugate $\bar{z}$ and modulus $|z|$.

Solution: If $z = 5 + 12i$, then $\bar{z} = 5 - 12i$.

So, $z \cdot \bar{z} = (5+12i)(5-12i) = 5^2 - 12^2 \cdot i^2 = 5^2 + 12^2 = |13|^2$.

Thus, $|z| = |\bar{z}| = 13$.

Important note: We can compute the square of the modulus of a complex number by multiplying it by its complex conjugate.

Example: Find $|z|^2$ if $z = 6 - 7i$.

Solution: We need to multiply z by its complex conjugate $\bar{z} = 6 + 7i$ so that $|z|^2 = (6-7i)(6+7i) = 36 - 42i + 42i - 49i^2 = 36 + 49 = 85$.

7.3.2 Arithmetic operations with complex numbers

- **Addition and subtraction**

The operations of addition and subtraction of complex numbers (*e.g.* $z = a + bi$ and $w = c + di$) are conveniently expressed by using their Cartesian coordinates and adding/subtracting the real and imaginary parts separately:

$$(a + bi) \pm (c + di) = (a \pm c) + (b \pm d)i$$

These operations behave well with respect to Cartesian coordinates:

$$\Re(z \pm w) = \Re(z) \pm \Re(w) \qquad \text{and} \qquad \Im(z \pm w) = \Im(z) \pm \Im(w)$$

Also, note that $-(a + bi) = -a - bi$.

Example: Calculate $(3 + 2i) + (1 - i)$.

Solution: $(3 + 2i) + (1 - i) = (3 + 1) + (2 - 1)i = 4 + i$.

- **Multiplication**

Multiplication is not quite as straightforward in Cartesian coordinates and the property that $i^2 = -1$ should be remembered:

$$\begin{aligned}(a + bi)(c + di) &= ac + adi + bci + bdi^2 = ac + adi + bci - bd \\ &= (ac - bd) + (ad + bc)i\end{aligned}$$

Example: Calculate $(1 + 2i)(4 + 4i)$.

Solution: $(1 + 2i)(4 + 4i) = 1 \cdot 4 + 4i + 8i + 8i^2 = 4 + 12i - 8 = -4 + 12i$.

- **Division**

Division is even more awkward in Cartesian coordinates since we need to eliminate the imaginary part of the denominator so that it is all real. To achieve this we have to multiply the numerator and the denominator by the complex conjugate of the denominator. Again we need the important property that $i^2 = -1$.

$$\frac{a+bi}{c+di} = \frac{(a+bi)(c-di)}{(c+di)(c-di)} = \frac{(ac+bd)+(bc-ad)i}{c^2-d^2i^2}$$
$$= \left(\frac{ac+bd}{c^2+d^2}\right) + \left(\frac{bc-ad}{c^2+d^2}\right) i$$

Example: Calculate $\dfrac{7+2i}{8+4i}$.

Solution: $\dfrac{7+2i}{8+4i} = \dfrac{(7+2i)(8-4i)}{(8+4i)(8-4i)} = \dfrac{56+16i-28i-8i^2}{64+16}$
$$= \frac{64-12i}{80} = \frac{16-3i}{20}.$$

Note that there is another way to carry out division of two complex numbers, $w = a+bi$ and $z = c+di$.

We know that

$$z \cdot \bar{z} = |z|^2 \iff \frac{1}{z} = \frac{\bar{z}}{|z|^2} \implies \frac{w}{z} = \frac{w \cdot 1}{z} = \frac{w \cdot \bar{z}}{|z|^2}$$

which for $w = 7+2i$ and $z = 8+4i$ gives
$\dfrac{7+2i}{8+4i} = \dfrac{56+16i-28i-8i^2}{64+16} = \dfrac{64-12i}{80} = \dfrac{16-3i}{20}$ as before.

A way of overcoming the inconvenience of Cartesian coordinates in complex number multiplication and division is to express the complex numbers in terms of their polar coordinates instead.

7.3.3 Geometry: the complex plane and polar coordinates

It is often convenient to represent the complex number $z = a+bi$ in terms of its polar coordinates, as shown in Figure 7.7, which is called an Argand diagram after the mathematician Jean-Robert Argand (1768–1822).

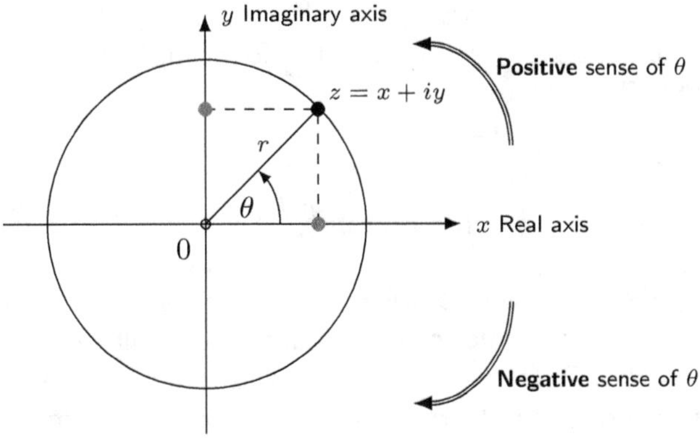

Figure 7.7: The Argand diagram

The angle θ is called the **argument** of z and is sometimes denoted $\arg(z)$.

The principal values of θ are usually taken to lie between $-\pi$ $(-180°)$ and π $(180°)$ in order to span all possible values of θ in the Cartesian plane relative to the positive x-axis.

The real number $r = |z|$ is the **modulus** and is sometimes denoted $\mathrm{mod}(z)$.

Based on the relationship between the trigonometric functions sin, cos, and tan for one acute angle (θ) and the three sides of a right angle triangle (see Section 7.1.2), a well-defined connection between the Cartesian and the polar coordinates for $z = x + iy$ can be made.

If we are given the polar coordinates of $z\,\langle r, \theta\rangle$, but want to express it in Cartesian coordinates $z\,(x, y)$, we use the following equations:

- $x = r\cos\theta$,
- $y = r\sin\theta$,

which give $z = r\cos\theta + ri\sin\theta \iff z = r(\cos\theta + i\sin\theta)$.

The positive x-axis is defined as having $\theta = 0$ and positive θ goes in an anticlockwise direction around $(0, 0)$ in the xy plane (see Figure 7.7).

Negative θ goes in a clockwise direction from the positive x-axis around $(0, 0)$ in the xy plane (see Figure 7.7).

If we are given the Cartesian coordinates $z\,(x, y)$ and want to find the polar coordinates $z\,\langle r, \theta\rangle$, we use the following equations and those given in Table 7.2:

- $r = \text{mod}(z) = |z| = \sqrt{x^2 + y^2}$;
- $\theta = \arg(z) = \tan^{-1}\left(\dfrac{y}{x}\right) = \arctan\left(\dfrac{y}{x}\right)$.

In applying these Cartesian to polar coordinate conversion formulae, the signs of x and y determine in which quadrant θ will sit. The rules are detailed in Table 7.2. Beware of the sign and the value of the argument $\arg(z)$ as they critically determine the correct quadrant for z.

Table 7.2 summarises the way to calculate $\arg(z) = \theta$ for a complex number $z = x + iy$.

Table 7.2: Rules for calculating $\arg(z)$

x	y	$\arg(z) = \theta$	Position of z in the Cartesian plane
>0	>0	$\arctan\left(\frac{y}{x}\right)$	Quadrant I (top right)
>0	<0	$\arctan\left(\frac{y}{x}\right)$	Quadrant IV (bottom right)
$=0$	>0	$\frac{\pi}{2}$	positive y-axis
$=0$	<0	$-\frac{\pi}{2}$	negative y-axis
<0	>0	$\arctan\left(\frac{y}{x}\right) + \pi$	Quadrant II (top left)
<0	<0	$\arctan\left(\frac{y}{x}\right) - \pi$	Quadrant III (bottom left)
>0	$=0$	0	positive x-axis
<0	$=0$	$+\pi$ OR $-\pi$	negative x-axis

Example: Find the Cartesian coordinates for the complex number z which has polar coordinates $r = 2$ and $\theta = \dfrac{\pi}{6}$.

Solution: Let us start by calculating $\Re(z)$ and $\Im(z)$:

$$\Re(z) = x = r\cos\theta = 2\cos\left(\frac{\pi}{6}\right) = 2\left(\frac{\sqrt{3}}{2}\right) = \sqrt{3}$$

$$\Im(z) = y = r\sin\theta = 2\sin\left(\frac{\pi}{6}\right) = 2\left(\frac{1}{2}\right) = 1$$

Thus, $z = \sqrt{3} + i$.

Example: Find the polar coordinates for the complex number $z = -3 + 4i$.

Solution: Let us start by calculating $\mathrm{mod}(z)$ and $\arg(z)$:

$$\mathrm{mod}(z) = r = \sqrt{(-3)^2 + 4^2} = \sqrt{25} = 5$$
$$\arg(z) = \theta = \arctan\left(\frac{4}{-3}\right) = -0.93 + \pi \text{ rad} \approx 127^\circ$$

Thus, $z = 5\left[\cos(127^\circ) + i\sin(127^\circ)\right]$ which confirms its position in the top left quadrant (II) of the Cartesian plane since $x < 0$ and $y > 0$.

Example: Find the polar coordinates for the complex number $z = -2i$.

Solution: Let us start by calculating $\mathrm{mod}(z)$ and $\arg(z)$:

$$\mathrm{mod}(z) = r = |z| = 2 \text{ and } \arg(z) = \theta = -\frac{\pi}{2}$$

We can check this result: $z = 2\left[\cos\left(-\frac{\pi}{2}\right) + i\sin\left(-\frac{\pi}{2}\right)\right]$ $= 2\cos\left(-\frac{\pi}{2}\right) + 2i\sin\left(-\frac{\pi}{2}\right) = -2i$, since $\cos\left(-\frac{\pi}{2}\right) = 0$ and $\sin\left(-\frac{\pi}{2}\right) = -1$.

- **Multiplication in polar coordinates**

First, we need to highlight three useful and important trigonometric identities. We have already met the first one (equation (7.2)), but the second two are new:

- $\cos^2\theta + \sin^2\theta = 1$;
- $\sin(\theta_1 + \theta_2) = \sin\theta_1\cos\theta_2 + \sin\theta_2\cos\theta_1$;
- $\cos(\theta_1 + \theta_2) = \cos\theta_1\cos\theta_2 - \sin\theta_1\sin\theta_2$.

Now let us compute the product of two complex numbers, z_1 and z_2.

Let $z_1 = r_1\cos\theta_1 + ir_1\sin\theta_1$ and $z_2 = r_2\cos\theta_2 + ir_2\sin\theta_2$. We first find the real part of the product $z_1 \cdot z_2$:

$$\begin{aligned}\Re(z_1 \cdot z_2) &= r_1\cos\theta_1 \cdot r_2\cos\theta_2 - r_1\sin\theta_1 \cdot r_2\sin\theta_2 \\ &= r_1r_2(\cos\theta_1\cos\theta_2 - \sin\theta_1\sin\theta_2) \\ &= r_1r_2\cos(\theta_1 + \theta_2) \qquad (7.7)\end{aligned}$$

Note that for the real part, the moduli have been **multiplied** but the arguments have been **added**.

Now we compute the imaginary part of $z_1 \cdot z_2$:

$$\begin{aligned}\Im(z_1 \cdot z_2) &= r_1 \sin\theta_1 \cdot r_2 \cos\theta_2 + r_2 \sin\theta_2 \cdot r_1 \cos\theta_1 \\ &= r_1 r_2(\sin\theta_1 \cos\theta_2 - \sin\theta_2 \cos\theta_1) \\ &= r_1 r_2 \sin(\theta_1 + \theta_2) \qquad (7.8)\end{aligned}$$

For the imaginary part too, the moduli are **multiplied** together, while the arguments are **added**.

The result of equations (7.7) and (7.8) gives a relatively compact and highly geometric result for the product of $z_1 \cdot z_2$:

$$z_1 \cdot z_2 = r_1 r_2[\cos(\theta_1 + \theta_2) + i \sin(\theta_1 + \theta_2)] \qquad (7.9)$$

The multiplication of complex numbers in polar coodinates is **multiplicative** in the modulus and **additive** in the argument:

- $\mathrm{mod}(z_1 \cdot z_2) = \mathrm{mod}(z_1) \cdot \mathrm{mod}(z_2)$;
- $\arg(z_1 \cdot z_2) = \arg(z_1) + \arg(z_2)$.

This means that when we multiply by z, we are **rotating** through the angle $\arg(z)$ and **radially stretching** by a factor of $\mathrm{mod}(z)$.

Example: Calculate the following product by using polar coordinates:

$$5(\cos\pi + i\sin\pi) \cdot \left(\cos\frac{\pi}{2} + i\sin\frac{\pi}{2}\right)$$

Solution: We can use equation (7.9) to calculate the product:

$$\begin{aligned}&5(\cos\pi + i\sin\pi) \cdot \left(\cos\frac{\pi}{2} + i\sin\frac{\pi}{2}\right) \\ &= 5 \cdot 1 \cdot \left[\cos\left(\pi + \frac{\pi}{2}\right) + i\sin\left(\pi + \frac{\pi}{2}\right)\right] \\ &= 5\left[\cos\left(\frac{3\pi}{2}\right) + i\sin\left(\frac{3\pi}{2}\right)\right] = 5\,[0 + i \cdot (-1)] = -5i\end{aligned}$$

7.3.4 A remarkable connection of z with e^x

Multiplication can be most effectively carried out if we make full use of a remarkable extension of the additive law of exponents for complex numbers.

It can be proved by using differential equations (see Section 9.2.1) that a **general complex number** can now be written as

$$z = x + iy = r\,(\cos\theta + i\sin\theta) = r\,e^{i\theta}$$

This formula has several important consequences:

(i) Any complex number can be written in the polar form $z = r\,e^{i\theta}$, where $r = \text{mod}(z)$ and $\theta = \arg(z)$.

(ii) A unit circle in $\mathbb{C}$ consists of those complex numbers that have a modulus equal exactly to 1, *i.e.* the numbers $z = e^{i\theta}$.

(iii) The exponential notion of the conjugate complex can be written as $\boxed{\bar{z} = r\,e^{-i\theta}}$; since $\cos(-\theta) = \cos\theta$ (even function) and $\sin(-\theta) = -\sin\theta$ (odd function) (see Section 7.1.2), thus

$$\bar{z} = r\,(\cos\theta - i\sin\theta) = r\,(\cos(-\theta) + i\sin(-\theta)) = r\,e^{-i\theta}$$

(iv) Adding z and $\bar{z}$ in their polar coordinates gives $\cos\theta = \dfrac{e^{i\theta} + e^{-i\theta}}{2}$, while subtracting z and $\bar{z}$ gives $\sin\theta = \dfrac{e^{i\theta} - e^{-i\theta}}{2i}$.

(v) Multiplication on the unit circle ($r = 1$) can be carried out by adding the angles:

$$z_1 \cdot z_2 = e^{i\theta_1} \cdot e^{i\theta_2} = e^{i(\theta_1 + \theta_2)}$$

(vi) Exponentiation on the unit circle ($r = 1$) can be done by multiplying the angle by the index:

$$z^n = \left(e^{i\theta}\right)^n = e^{i\theta n} = e^{i(n\theta)}$$

(vii) This result is known as **De Moivre's theorem**, and it is usually stated as

$$z^n = (\cos\theta + i\sin\theta)^n = \cos(n\theta) + i\sin(n\theta)$$

We close this brief look at complex number algebra with a formula attributed to Leonhard Euler (1707–1793) that puts all the important numbers of mathematics together into one expression.

It is said that when he discovered it, Euler was so struck by its profound beauty that he claimed it as a proof for the existence of God:

$$e^{\pi i} + 1 = 0$$

Example: Find values for r and θ in the exponential notation $z = re^{i\theta}$ corresponding to $z = 1 - i$.

Solution: The complex number $z = 1 - i$ can be expressed by using polar coordinates as $z = \sqrt{2}\left[\cos\left(-\frac{\pi}{4}\right) + \sin\left(-\frac{\pi}{4}\right)\right]$.

Thus, the exponential notion of $z = 1 - i$ is $z = \sqrt{2}e^{-i\frac{\pi}{4}}$.

Example in Biochemistry: In crystallography for protein structure determination, when a protein crystal is irradiated with X-rays to produce diffraction, the intensity of a reflection, I, is proportional to the magnitude of the square of the 'structure factor' which can be expressed as $Fe^{i\phi}$, where F is the amplitude of the diffracted wave and ϕ is its phase. Find a general expression for I if it is defined as the square of the modulus of the structure factor.

Find F and ϕ if in Cartesian coordinates the reflection can be written as $z = 3 + 4i$.

Solution: In order to find I, we must calculate the square of the structure factor by multiplying it by its complex conjugate:

$$I = Fe^{i\phi} \cdot Fe^{-i\phi} = |F|^2$$

For a reflection which can be written as $z = 3 + 4i$, the amplitude $F = \sqrt{(3^2 + 4^2)} = 5$, while the phase is $\arctan\left(\frac{4}{3}\right) = 58.9°$.

So the reflection can be written both in polar coordinates as $z = 5(\cos(58.9°) + i\sin(58.9°))$ or in exponential form as $z = 5\,e^{i58.9}$.

Note that this example shows how the phase, ϕ, has 'disappeared' when we square $Fe^{i\phi}$ to get I.

7.4 Exercises

(1) Convert the following angles from radians to degrees.

(a) $\frac{\pi}{6}$ (b) $\frac{\pi}{3}$ (c) $\frac{5\pi}{4}$ (d) $\frac{4\pi}{3}$ (e) $\frac{11\pi}{6}$

(2) Convert the following angles from degrees to radians.

(a) $75°$ (b) $120°$ (c) $210°$ (d) $250°$ (e) $315°$

(3) An ultracentrifuge is used in Biosciences to measure the sedimentation velocity and equilibrium of a sample (*e.g.* a protein) in order to calculate its shape and mass.

(a) If the rotor speed is $100{,}000\,\text{rpm}$ ($\text{rpm} =$ revolutions per minute), what is the angular velocity of the rotor, ω, in $\text{rad} \cdot \text{s}^{-1}$?

(b) If ω in a different experiment is $8000\,\text{rad} \cdot \text{s}^{-1}$, what is the rpm of the rotor?

(4) Solve the following elementary trigonometric equations for x values in radians in the interval $[2, \pi]$.

(a) $\sin\left(2x - \frac{\pi}{4}\right) = -1$ (b) $\cos\left(\frac{\pi}{3} - \frac{x}{2}\right) = 0$

(c) $\sin(x + \pi) = -\frac{\sqrt{2}}{2}$ (d) $\cos(2x - \pi) = 2$

(e) $\tan\left(\frac{\pi}{4} - 2x\right) = -1$ (f) $\tan\left(\frac{x}{2} + \frac{\pi}{3}\right) = 0$

(5) Circadian rhythms are internally controlled biological changes with an approximately 24-hour time period. They can be modelled by using trigonometric functions. A particular species of dog exhibits daily regular fluctuations in body temperature T (°C) that can be approximated by the equation

$$T(t) = A + B \cdot \sin\left(\frac{2\pi}{24}(t + 11)\right)$$

with t in hours from 0 to 24, and the constants $A = 38.75$ and $B = 0.45$ both have units of °C.

(a) What is the body temperature at $t = 1$ p.m.?

(b) What time does the body temperature reach its maximum value?

(c) What time does the body temperature reach its minimum value?

(6) Sketch the following functions by using the geometric transformations described in Section 4.7. Then, identify the amplitude, period, phase shift, and vertical shift for each function.

(a) $\cos(2x) - 3$ (b) $3 - 2\sin(\pi - x)$ (c) $3\cos\left(\frac{x}{2}\right) + 1$

(7) Circadian rhythms help organisms adapt to their cyclic environment and are important for health in humans. These rhythms are driven by a master clock located in the suprachiasmatic nucleus (SCN).

A recent high temporal resolution analysis of the transcript level of two core clock genes, *genA* and *genB*, and a control gene, *genC*, in the SCN has been performed in rats.

The relative mRNA levels of expression of all three genes, denoted (y_A), (y_B), and (y_C), respectively, follow a diurnal rhythm in the SCN and have a sinusoidal trend, as shown in the following figure, but the phase and amplitude of the rhythms of each gene vary.

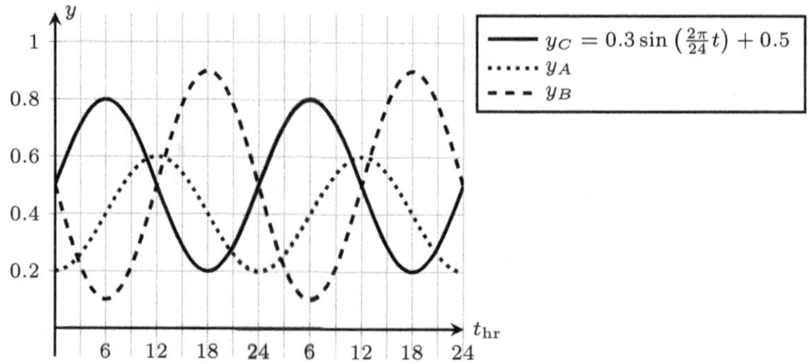

In particular, the temporal organisation of the expression of the control gene *genC* follows the equation $y_C = 0.3\sin\left(\frac{2\pi}{24}t\right) + 0.5$ with time t in hours.

(a) What is the period of each function shown in the graph above?

(b) Which gene shows maximum variability (amplitude) in mRNA expression levels during the day?

(c) What is the phase shift of *genA* and *genB* expressions compared to that of *genC*?

(d) What is the vertical shift of the *genA* and *genB* expressions?

(e) Form the sinusoidal equations $y_A = f(t)$ and $y_A = f(t)$ as described in Section 7.2.1.

(8) Find the derivatives of the following functions using the Product/Quotient/Chain Rules as appropriate.

(a) $y = 5\cos x$ (b) $y = \sin(2x)$ (c) $y = (\sin x)^2$

(d) $y = \sin x \cos x$ (e) $y = \sqrt{\cos x}$ (f) $y = \sin^2 x + \cos^2 x$

(g) $y = \dfrac{1}{\tan x}$ (h) $y = -(\cos x)^2$ (i) $y = -\sin\sqrt{x}$

(j) $y = \tan(x^2 + 3)$ (k) $y = \sin(x^2 - x)$ (l) $y = \arctan(3x^2)$

(9) By using the appropriate differentiation rules, find the derivatives of the following functions used in Biosciences with respect to their argument, All the other parameters in the formulae should be treated as constants.

(a) $T(t) = 72 - 18\cos\left(\frac{\pi}{12}(4-t)\right)$ (b) $d(\theta) = \dfrac{n\lambda}{2\sin(\theta)}$

(c) $I_0(\theta) = I_e\left(\dfrac{q^4}{m^2c^4}\right)\dfrac{1+\cos^2(2\theta)}{2}$

(10) Calculate the specified limits of the following functions using L'Hôpital's rule.

(a) $\lim\limits_{x\to 0}\dfrac{\sin(-4x)}{3x}$ (b) $\lim\limits_{x\to 0}\dfrac{\cos(2x)-1}{5x}$ (c) $\lim\limits_{x\to 0}\dfrac{1-\cos x}{x^2}$

(d) $\lim\limits_{x\to 0}\dfrac{\tan(-3x)}{7x}$ (e) $\lim\limits_{x\to 0}\dfrac{\arctan(4x)}{5x}$ (f) $\lim\limits_{x\to 0}\dfrac{\cos(2x)-1}{\sin^2(3x)}$

(g) $\lim\limits_{x\to 0}\dfrac{\sin(3x)+6x}{e^x-1}$ (h) $\lim\limits_{x\to 0}\dfrac{1-\cos(3x)}{5x^2}$ (i) $\lim\limits_{x\to 0}\dfrac{\ln(1-5x)}{\tan(2x)}$

(11) The population N_L of lynxes in a certain area of Canada varies sinusoidally with time and can be approximately described by the equation

$$N_L = 40{,}000 + 35{,}000\sin\left(\frac{2\pi t}{T}\right)$$

where $T = 11$ years and t is the elapsed time in years.
Measurements started in January 1820, so t is zero at this date.

(a) How many lynxes were there (N_L) in January 1820?

(b) What were the maximum and minimum values that the population ever reached?

(c) In which year (after 1820) did the population first have its maximum value?

(d) In which year did it have its minimum value?

(12) During a research study conducted on a panel of 35 full-term eight-week-old healthy infants, their 24-hr daily rhythm of urinary excretion of the melatonin metabolite 6-sulfatoxymelatonin (6SMT) was measured as a function of the time of day in order to investigate the overall daily variation of melatonin production. The urinary 6SMT excretion rate (U in $\mathrm{ng\,hr^{-1}}$) was modelled using the following cosine function:

$$U(t) = M + A\cos\left(\omega(t-\phi)\right)$$

where $M = 230\,\mathrm{ng\,hr^{-1}}$, $A = 50\,\mathrm{ng\,hr^{-1}}$, $\omega = \dfrac{2\pi}{24}$, t is the time in hr since midnight, and $\phi = 4\,\mathrm{hr}$.

(a) Identify the Mesor (*i.e.*, the mean theoretical value around which the cosine model fluctuates), amplitude, and period of the function $U(t)$.

(b) What is the value of U at midnight?

(c) When (what time of day) is the maximum theoretical value of urinary 6SMT excretion reached?

(d) When is the minimum value reached?

(13) Solve the following quadratic equations.

(a) $x^2 - 6x + 10 = 0$ (b) $2x^2 + 2x + 5 = 0$

(c) $x^2 + 2x + 2 = 0$ (d) $17x^2 + 2x + 1 = 0$

(14) Find $(u+v)$, uv, $|u|$, and $|v|^2$.

(a) $u = 2 + 3i, \quad v = 1 - i$ (b) $u = 3 + 5i, \quad v = 1 + 2i$

(c) $u = 4 - i, \quad v = 2 + 6i$ (d) $u = 16 + i, \quad v = 12 - 3i$

(e) $u = 5i - 2, \quad v = 3 + 4i$ (f) $u = \frac{1}{2} + i, \quad v = 6 + \frac{i}{2}$

(15) Calculate:

(a) $\frac{2+3i}{4+3i}$ (b) $\frac{1+2i}{3-i}$ (c) $\frac{5-6i}{1+2i}$

(16) Find the polar coordinates of the following complex numbers.

(a) $z = 1 + i$ (b) $z = -1 - i$ (c) $z = -i$

(17) Express the following complex numbers as polar coordinates.

(a) $z = 2 + 3i$ (b) $z = i - 1$ (c) $z = 4 + i$

(d) $z = 5 - 2i$ (e) $z = \frac{1}{2} + 6i$ (f) $z = i - 12$

(g) $z = 1 + 7i$ (h) $z = 8 - 15i$ (i) $z = 2i - 1$

(18) Express the following complex numbers $z\langle r, \theta\rangle$ as Cartesian coordinates.

(a) $r = 3, \quad \theta = \pi$ (b) $r = 1, \quad \theta = \frac{\pi}{6}$ (c) $r = 2, \quad \theta = \frac{\pi}{4}$

(d) $r = 6, \quad \theta = -\pi$ (e) $r = 5, \quad \theta = \frac{\pi}{3}$ (f) $r = 8, \quad \theta = \frac{\pi}{2}$

(g) $r = 4, \quad \theta = \pi$ (h) $r = 7, \quad \theta = 0$ (i) $r = 2, \quad \theta = -\frac{3\pi}{2}$

(19) Simplify the following complex numbers by finding the products of the brackets.

(a) $z = 2\left(\cos\frac{\pi}{3} + i\sin\frac{\pi}{3}\right) \cdot 3\left(\cos\frac{\pi}{6} + i\sin\frac{\pi}{6}\right)$

(b) $z = 4\left(\cos\frac{\pi}{2} + i\sin\frac{\pi}{2}\right) \cdot 2(\cos\pi + i\sin\pi)$

(c) $z = 7(\cos 45^\circ + i\sin 45^\circ) \cdot (\cos 67^\circ + i\sin 67^\circ)$

(d) $z = (\cos 10^\circ + i\sin 10^\circ) \cdot 4(\cos 90^\circ + i\sin 90^\circ)$

(e) $z = 10(\cos 0^\circ + i\sin 0^\circ) \cdot (\cos 35^\circ + i\sin 35^\circ)$

(20) Find values for r and θ in the exponential notion $z = re^{i\theta}$ corresponding to the following complex numbers.

(a) $z = 2i + 1$ (b) $z = 3 - 2i$ (c) $z = 4 + 6i$

(d) $z = 9 + 5i$ (e) $z = 8 - 3i$ (f) $z = 5 + 4i$

(g) $z = 0.5 + i$ (h) $z = -2 - i$ (i) $z = -3i$

(21) Derive the Euler equation $e^{\pi i} + 1 = 0$ starting from $e^{i\theta} = \cos\theta + i\sin\theta$.

(22) Show the following X-ray diffraction reflections, $Fe^{i\phi}$, on an Argand diagram where ϕ is given in radians.

(a) $5e^{i1.3}$ (b) $9e^{i5.6}$

(c) $2e^{i3.8}$ (d) $7e^{i2.1}$

(23) Complex numbers are often used to represent wavefunctions. A *wavefunction* $\Psi(x,t)$ is a function describing the properties of a wave in relation to space (x) and time (t). Max Born (1882–1970) postulated that the probability of finding a quantum particle is related to the square of the magnitude of its wavefunction, which is equal to the product of the wavefunction and its complex conjugate:

$$|\Psi(x,t)|^2 = \Psi(x,t) \cdot \bar{\Psi}(x,t)$$

Consider the following complex wavefunction for a particle:
$\Psi(x,t) = \psi_0\, e^{-a\left(\frac{m}{\hbar}x + it\right)}$, where m is the particle mass, $\hbar$ is the Plank's constant, and ψ_0 and a are positive constants.

(a) Write down the real part $\Re(\Psi)$ of the complex wavefunction.

(b) Calculate $|\Psi(x,t)|^2$. What is its physical meaning?

(24) A simplified representation of a three-dimensional wave function for electrons is given by the equation

$$\psi_{(n,l,m)} = R_{(n,l)}(r) \cdot Y_{(l,m)}(\theta, \phi)$$

where $R_{(n,l}(r)$ is the radial contribution, $Y_{(l,m)}(\theta,\phi)$ is the angular contribution, and n, l, m are the quantum numbers.

Consider the time-dependent radial contribution of the wavefunction ψ of a particle confined to a region between 0 and L: $R(r,t) = Ae^{-i\omega t}\sin\left(\frac{\pi r}{L}\right)$ where ω is the angular frequency and A is the positive constant representing the amplitude.

(a) Write down the real part $\Re(R(r,t))$ of the complex radial contribution to the wavefunction ψ.

(b) Calculate $|R(r,t)|^2$.

Answers

(1) (a) $30°$ (b) $60°$ (c) $225°$ (d) $240°$ (e) $330°$

(2) (a) $\frac{5\pi}{12}$ (b) $\frac{2\pi}{3}$ (c) $\frac{7\pi}{6}$ (d) $\frac{25\pi}{18}$ (e) $\frac{7\pi}{4}$

(3) (a) $\omega = 10,466\,\mathrm{rad}\cdot\mathrm{s}^{-1}$ (b) $76,432\,\mathrm{rpm}$

(4) (a) $x = \frac{7\pi}{8}$, $x = \frac{15\pi}{8}$
(b) $x = \frac{5\pi}{3}$
(c) $x = \frac{\pi}{4}$, $x = \frac{3\pi}{4}$
(d) no real solutions, since cos function gives only values in the interval $[-1,1]$
(e) $x = \frac{\pi}{4}$, $x = \frac{3\pi}{4}$, $x = \frac{5\pi}{4}$, $x = \frac{7\pi}{4}$
(f) $x = \frac{4\pi}{3}$

(5) (a) $38.75°$ (b) 7 p.m. (c) 7 a.m.

(6)

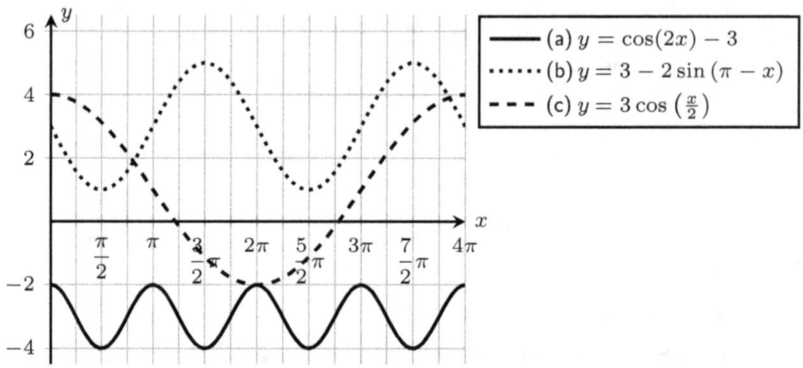

(7) (a) $\beta = 24\,\mathrm{hr}$
(b) *genB*, since $\alpha_A = 0.2$, $\alpha_B = 0.4$ and $\alpha_C = 0.3$
(c) $\gamma_A = 6\,\mathrm{hr}$ and $\gamma_A = 12\,\mathrm{hr}$
(d) $\delta_A = 0.4$ and $\delta_B = 0.5$
(e) $y_A = 0.2\sin\left(\frac{2\pi}{24}(t-6)\right) + 0.4$ and $y_B = 0.4\sin\left(\frac{2\pi}{24}(t-12)\right) + 0.5$

(8) (a) $y' = -5\sin(x)$ (b) $y' = \cos(2x)\cdot 2$ (c) $y' = 2\sin(x)\cos(x)$
(d) $y' = \cos^2(x) - \sin^2(x)$ (e) $y' = -\frac{\sin(x)}{2\sqrt{\cos(x)}}$ (f) $y' = 0$

(g) $y' = -\frac{1}{\sin^2(x)}$ (h) $y' = 2\cos(x)\sin(x)$ (i) $y' = -\frac{\cos(\sqrt{x})}{2\sqrt{x}}$
(j) $y' = \frac{1}{\cos^2(x^2+3)} \cdot 2x$ (k) $y' = \cos(x^2 - x)(2x - 1)$ (l) $y' = \frac{6x}{1+9x^4}$

(9) (a) $T'(t) = -\frac{3\pi}{2}\sin\left(\frac{\pi}{12}(4-t)\right)$ (b) $d'(\theta) = -\frac{\lambda n \cos(\theta)}{2\sin^2(\theta)}$
(c) $I_0(\theta) = -I_e\left(\frac{q^4}{m^2c^4}\right)\cos(2\theta)\sin(2\theta)$

(10) (a) $-\frac{4}{3}$ (b) 0 (c) $\frac{1}{2}$ (d) $-\frac{3}{7}$ (e) $\frac{4}{5}$
(f) $-\frac{2}{9}$ (g) 9 (h) $\frac{9}{10}$ (i) $-\frac{5}{2}$

(11) (a) $N_L = 40{,}000$
(b) maximum value: $N_L = 75{,}000$, minimum value $N_L = 5000$
(c) towards the end of 1828
(d) at the beginning of 1823

(12) (a) Mesor $M = 230\,\mathrm{ng\,hr}^{-1}$, amplitude $A = 50$, period $P = 24\,\mathrm{hr}$
(b) $U = 255\,\mathrm{ng\,hr}^{-1}$ (c) $t = 4\,\mathrm{hr}$ (d) $t = 16\,\mathrm{hr}$

(13) (a) $3 \pm i$ (b) $-0.5 \pm 1.5i$ (c) $-1 \pm i$ (d) $-\frac{1}{17} \pm \frac{4}{17}i$

(14) (a) $u + v = 3 + 2i,\ \ uv = 5 + i,\ \ |u| = \sqrt{13},\ \ |v|^2 = 2$
(b) $u + v = 4 + 7i,\ \ uv = 11i - 7,\ \ |u| = \sqrt{34},\ \ |v|^2 = 5$
(c) $u + v = 6 + 5i,\ \ uv = 14 + 22i,\ \ |u| = \sqrt{17},\ \ |v|^2 = 40$
(d) $u + v = 28 - 2i,\ \ uv = 195 - 36i,\ \ |u| = \sqrt{257},\ \ |v|^2 = 153$
(e) $u + v = 1 + 9i,\ \ uv = 7i - 26,\ \ |u| = \sqrt{29},\ \ |v|^2 = 25$
(f) $u + v = \frac{13}{2} + \frac{3i}{2},\ \ uv = \frac{5}{2} + \frac{25}{4}i,\ \ |u| = \sqrt{\frac{5}{4}},\ \ |v|^2 = \frac{145}{4}$

(15) (a) $\frac{17}{25} + \frac{6}{25}i$ (b) $\frac{1}{10} + \frac{7}{10}i$ (c) $-\frac{7}{5} - \frac{16}{5}i$

(16) (a) $z = \sqrt{2}(\cos 45^\circ + i\sin 45^\circ)$ (b) $z = \sqrt{2}(\cos 225^\circ + i\sin 225^\circ)$
(c) $z = 1(\cos 270^\circ + i\sin 270^\circ)$

(17) (a) $r = \sqrt{13},\ \ \theta = 56.3^\circ$ (b) $r = \sqrt{2},\ \ \theta = 135^\circ$
(c) $r = \sqrt{17},\ \ \theta = 14^\circ$ (d) $r = \sqrt{29},\ \ \theta = -21.8^\circ$
(e) $r = \sqrt{36.25},\ \ \theta = 85.2^\circ$ (f) $r = \sqrt{145},\ \ \theta = 175.2^\circ$
(g) $r = \sqrt{50},\ \ \theta = 81.9^\circ$ (h) $r = 17,\ \ \theta = -61.9^\circ$
(i) $r = \sqrt{5},\ \ \theta = 116.5^\circ$

(18) (a) $z = -3$ (b) $z = \frac{\sqrt{3}}{2} + \frac{i}{2}$ (c) $z = \sqrt{2} + i\sqrt{2}$ (d) $z = -6$
(e) $z = \frac{5}{2} + i\frac{5\sqrt{3}}{2}$ (f) $z = 8i$ (g) $z = -4$ (h) $z = 7$
(i) $z = -2i$

(19) (a) $6i$ (b) $-8i$ (c) $-2.59 + 6.51i$
(d) $-0.68 + 3.92i$ (e) $8.2 + 5.7i$

(20) (a) $\sqrt{5}e^{63.4^\circ i}$ (b) $\sqrt{13}e^{-33.7^\circ i}$ (c) $\sqrt{52}e^{56.3^\circ i}$ (d) $\sqrt{106}e^{29.1^\circ i}$
(e) $\sqrt{73}e^{-20.6^\circ i}$ (f) $\sqrt{41}e^{38.7^\circ i}$ (g) $1.25e^{63.4^\circ i}$ (h) $\sqrt{5}e^{26.6^\circ i}$
(i) $3e^{-90^\circ i}$

(21) $z = \cos\theta + i\sin\theta$; $z = e^{i\theta}$ therefore $e^{i\theta} = \cos\theta + i\sin\theta$. If we take $\theta = \pi$, then $\cos\pi = -1$ and $\sin\pi = 0$. Thus $e^{i\pi} = -1$ and $e^{i\pi} + 1 = 0$.

(22) (a) Angle with x-axis $= 1.3$ rad $= 34.4°$. So the reflection can be represented in the right-hand top quadrant of the Argand diagram by a line length 5 making an angle of $34.4°$ with the x-axis
(b) Similarly, length 9, angle $-59.0°$, bottom right quadrant
(c) Length 2, angle $217.6°$, bottom left quadrant
(d) Length 7, angle $120.4°$, top left quadrant

(23) (a) $\Re(\Psi) = \psi_0\, e^{-a\left(\frac{m}{\hbar}x\right)} \cos(-at)$
(b) $|\Psi(x,t)|^2 = \psi_0^2\, e^{-2a\left(\frac{m}{\hbar}x\right)}$, which indicates that the probability of finding this quantum particle Ψ depends only on its position in the space (x) but not on the time t

(24) (a) $\Re(R(r,t)) = A\cos(-\omega t)\sin\left(\frac{\pi r}{L}\right)$ (b) $|R(r,t)|^2 = A^2\sin^2\left(\frac{\pi r}{L}\right)$

Chapter 8

Integration

Preamble

In Biosciences, we often need to sum up a number of small amounts of a particular variable to find out the total quantity. In each case, we can do this by using the rate at which each of these variables changes during a small increment of time. For instance, we may want the final size of a population which grows or declines over a defined time, the amount of oxygen generated by photosynthesis in a given time period, or the availability of a drug in blood after a certain period, *etc.*

As for many other mathematical operations which are matched by an inverse process (*e.g.* addition and subtraction, multiplication and division, raising to a power and taking a root, and putting a number into an exponent and taking the logarithm), differentiation can be reversed or 'undone'. If taking the derivative allows us to determine the instantaneous rate at which a quantity changes, then 'integration' allows the estimation of a physical quantity by knowing its rate of change, since we can sum up the instantaneous contributions.

Graphically, differentiation gives us the gradient of a curve at a particular point, whereas integration gives the area between the x-axis and the particular curve of a function.

8.1 Area under the Curve and Integration

In order to introduce integration as a useful tool for calculating the area between the curve of a function $f(x)$ and the x-axis, we have to assume that $f(x) \geq 0$ on its domain (*i.e.* the curve of $y = f(x)$ is above the x-axis in a Cartesian plane) and that it is continuous on its domain.

If we want to find the **area under the curve** (**AUC**), we can divide this area into strips, calculate the area of each strip, and sum these areas together (see Figure 8.1).

Let us take the function $y = x^2$ and calculate the **AUC** between $x = 2$ and $x = 4$.

The area of each strip is ΔA and $\Delta A \cong y \cdot \Delta x$ (Figure 8.1).

However, each strip can have a height which is either slightly greater or slightly lower than the y value of the function, so the value of the AUC can be either an **overestimate** or an **underestimate** (Figure 8.1). Generally, the smaller the width of the strips, the lower the error in the estimation of the area.

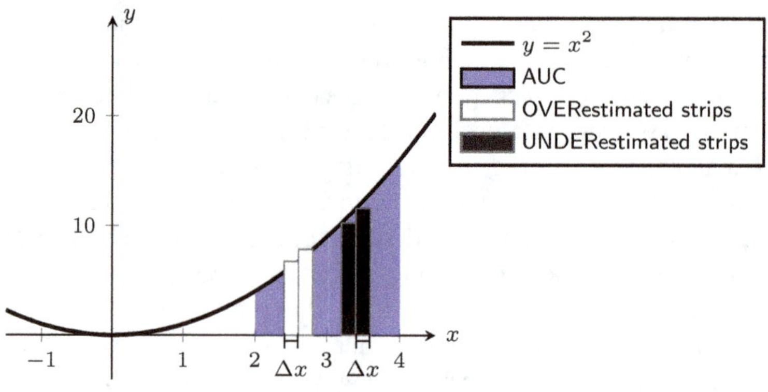

Figure 8.1: Area under the curve

We can sum up the areas of all the small strips from $x = 2$ to $x = 4$ to give the total required AUC:

$$\text{AUC} \cong \sum_{x=2}^{x=4} \Delta A \cong \sum_{x=2}^{x=4} y \, \Delta x \tag{8.1}$$

Let us take $A(x)$ as being the area between 2 and x with $2 < x < 4$ (Figure 8.2).

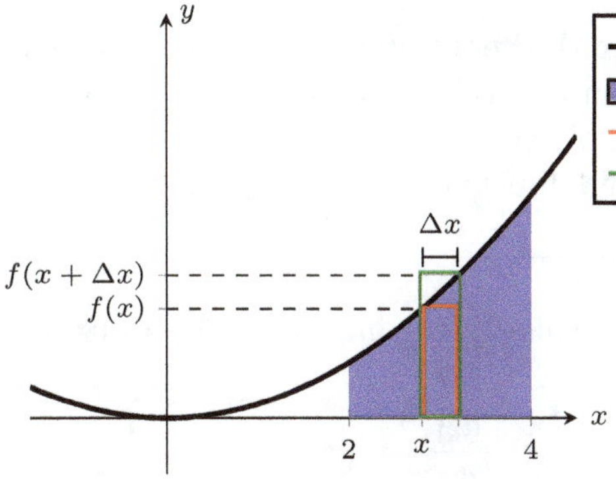

Figure 8.2: Area under the curve

To differentiate the function $A(x)$ with respect to x, we need to look at the gradient of $A(x)$ at every x value of the interval $[2, 4]$. Let us put:

$$\text{gradient} = \frac{A(x + \Delta x) - A(x)}{\Delta x}$$

The **numerator** of this fraction is the area under the curve between x and $x + \Delta x$.

If we denote the area of the smaller rectangle in Figure 8.2 (red-outlined rectangle) by $Min_{\Delta x} = f(x) \cdot \Delta x$ and that of the larger rectangle (green-outlined rectangle) by $Max_{\Delta x} = f(x + \Delta x) \cdot \Delta x$, we have the inequality

$$Min_{\Delta x} \leq A(x + \Delta x) - A(x) \leq Max_{\Delta x}$$
$$f(x) \cdot \Delta x \leq A(x + \Delta x) - A(x) \leq f(x + \Delta x) \cdot \Delta x$$

Dividing by Δx, we see that

$$f(x) \leq \frac{A(x + \Delta x) - A(x)}{\Delta x} \leq f(x + \Delta x)$$

As $\Delta x \to 0$, then $\dfrac{A(x + \Delta x) - A(x)}{\Delta x} \to A'(x)$, and $f(x + \Delta x) \to f(x)$ (as $f(x)$ is continuous for all x within $[2, 4]$). Using the squeeze (or sandwich) theorem, we can say that $A'(x) \to f(x)$.

Squeeze (or sandwich) theorem: Let $g(x)$, $f(x)$, and $v(x)$ be functions defined for an interval I, such as $g(x) \leq f(x) \leq v(x)$ for all x in I, except possibly at point x_0 within I.
If $\lim_{x \to x_0} g(x) = L$ and $\lim_{x \to x_0} v(x) = L$, then $\lim_{x \to x_0} f(x) = L$.

Therefore, for every x value in the interval $[2, 4]$, we have

$$A'(x) = \frac{dA}{dx} = f(x) = y$$

In order to calculate the AUC, we need to look for a new function $A(x)$ whose derivative $A'(x)$ equals $f(x)$.

If we could use strips of an **infinitely small width**, the estimation of the area under the curve would have **no error**: this mathematical summation is effectively what integration does.

Going back to the AUC of $f(x) = x^2$ between 2 and x with $2 < x < 4$ (Figure 8.2), we can now put

$$\text{AUC} = \lim_{\Delta x \to 0} \sum_{x=2}^{x=4} (y \cdot \Delta x) = \int_{x=2}^{x=4} y \, dx \tag{8.2}$$

where $\int$ is the old English letter 'S', which stands for the phrase 'Sum over'.

It tells us that we need to integrate, while dx tells us to integrate with respect to the variable x.

Note that it is **completely meaningless** to have an $\int$ sign with no matching dx.

The process of 'summing over' is is called **integration** and the $\int ...dx$ is called an **integral**.

An integral with defined upper and lower bounds or **'limits'** for the variable x as shown at the top and bottom of the $\int$ sign respectively (e.g. $x = 2$ and $x = 4$ in equation (8.2)) is called a **definite integral**.

Now, using equation (8.2), the area under the curve of $y = f(x)$ and the x-axis between $x = 2$ and $x = 4$ can be calculated by using the following expression:

$$\text{AUC} = \int_{x=2}^{x=4} y\,dx = \int_{x=2}^{x=4} \frac{dA}{dx}\,dx = \left[A(x)\right]_{x=2}^{x=4} = A(4) - A(2) \quad (8.3)$$

where $A(x)$ has to be a differentiable function whose derivative, $A'(x)$, is equal to the original function $f(x)$. This new type of function $A(x)$ is called the **antiderivative** of $f(x)$.

In equation (8.3), we can see that the function $A(x)$ has been differentiated and then integrated with respect to x to give us the AUC and this shows that **INTEGRATION reverses the process of DIFFERENTIATION**.

Finally, in order to calculate the area under the curve of $f(x) = x^2$ between $x = 2$ and $x = 4$ (Figure 8.2), we have the antiderivative $A(x) = \frac{1}{3}x^3$, which has derivative $A'(x) = x^2 = f(x)$, and so we can calculate

$$\int_2^4 x^2\,dx = \left[\frac{1}{3}x^3\right]_2^4 = \left[\frac{1}{3}4^3 - \frac{1}{3}2^3\right] = \frac{64}{3} - \frac{8}{3} = \frac{56}{3}$$

giving us the area between the x^2 curve and the x-axis from $x = 2$ to $x = 4$.

In general, since from Chapter 5 we know that

$$\frac{d(a \cdot x^{n+1})}{dx} = (n+1) \cdot a \cdot x^n$$

so if we 'reverse' the effects of this expression, we get

$$\int_{x_1}^{x_2} a \cdot x^n\,dx = \left[\frac{a}{(n+1)} x^{(n+1)}\right]_{x_1}^{x_2} \quad (8.4)$$

Important note: This rule works for all integer values of n **apart from** $n = -1$.

In fact, if we used equation (8.3) with $n = -1$, we would obtain an integrated function equal to $\frac{a \cdot x^0}{0} = \frac{1}{0}$, which has **no meaning**! (see Section 1.2).

8.1.1 Fundamental theorem of calculus

More generally, if we want to find the area between the curve and the x-axis for positive (≥ 0) functions between the two limits $x = x_1$ and $x = x_2$, the following notation is used:

$$\int_{x_1}^{x_2} f(x)\,dx = [F(x)]_{x_1}^{x_2} = F(x_2) - F(x_1)$$

provided that $f(x)$, called the **integrand**, is a real-valued **continuous** function on $[x_1, x_2]$ and $F(x)$ is an **antiderivative** of $f(x)$ in $[x_1, x_2]$.

An important note on signs:

- If $f(x) \leq 0$ in the interval $[x_1, x_2]$ (*i.e.* the curve of $f(x)$ is below the x-axis), the integration gives the area bounded by the curve and the x-axis, BUT the result of the integration is **negative**.

  Thus, the **modulus** of the integrated result should be taken to calculate the area.

  For instance, if we want to calculate the area between $f(x) = \cos(x)$ and the x-axis in the interval $\left[\frac{\pi}{2}, \frac{3\pi}{2}\right]$, we should note that $\cos(x) \leq 0$ in the interval $\left[\frac{\pi}{2}, \frac{3\pi}{2}\right]$.

  Therefore,

$$\int_{\frac{\pi}{2}}^{\frac{3\pi}{2}} \cos(x)\,dx = \left|\int_{\frac{\pi}{2}}^{\frac{3\pi}{2}} \cos(x)\,dx\right|$$

- When $f(x)$ assumes both positive and negative values on an interval of integration I, we should split the integral into the sum of integrals of $f(x)$ where in each part the sign of the function is constant (either ≥ 0 or ≤ 0) and take the modulus of each piece.

  For instance, if we want to calculate the area between $f(x) = \sin(x)$ and the x-axis in the interval $I = [0, 2\pi]$, we should note that $\sin(x) \geq 0$ in the interval $[0, \pi]$ and $\sin(x) \leq 0$ in the interval $[\pi, 2\pi]$.

  Since $\sin(x)$ can have positive and negative values in I, the integral should be split into two integrals to be subsequently summed together after taking their modulus:

$$\int_{0}^{2\pi} \sin(x)\,dx = \left|\int_{0}^{\pi} \sin(x)\,dx\right| + \left|\int_{\pi}^{2\pi} \sin(x)\,dx\right|$$

- If $\mathbf{A_1} = \int_a^b f(x)\,dx = [F(x)]_a^b = F(b) - F(a)$ and $\mathbf{A_2} = \int_b^a f(x)\,dx = [F(x)]_b^a = F(a) - F(b)$, then $\mathbf{A_1} = -\,\mathbf{A_2}$.

Example: Find the antiderivatives of the following functions and then check the answers by differentiating the resulting antiderivatives. Then, use the antiderivative to evaluate the definite integrals by inserting the limits into them: $\int_{-1}^{3} x^4\,dx$ and $\int_0^2 x^{\frac{3}{2}}\,dx$.

Solution: By using equation (8.3)

$$\int_{-1}^{3} x^4\,dx = \left[\frac{1}{4+1}x^{4+1}\right]_{-1}^{3} = \left[\frac{1}{5}x^5\right]_{-1}^{3}$$

and by differentiating this antiderivative

$$\frac{d}{dx}\left(\frac{1}{5}x^5\right) = \frac{5}{5}x^4 = x^4$$

we can see that this agrees with our original function. We can now evaluate $\left[\frac{1}{5}x^5\right]_{-1}^{3} = \frac{3^5}{5} - \frac{(-1)^5}{5} = \frac{243+1}{5} = \frac{244}{5}$.

Likewise, $\int_0^2 x^{\frac{3}{2}}\,dx = \left[\frac{1}{\frac{5}{2}}x^{\frac{5}{2}}\right]_0^2 = \left[\frac{2}{5}x^{\frac{5}{2}}\right]_0^2$; this can be checked by differentiation: $\frac{d}{dx}\left(\frac{2}{5}x^{\frac{5}{2}}\right) = \frac{2}{5}\cdot\frac{5}{2}x^{\frac{3}{2}} = x^{\frac{3}{2}}$. Again this agrees, so we can evaluate $\left[\frac{2}{5}x^{\frac{5}{2}}\right]_0^2 = \frac{2}{5}2^{\frac{5}{2}} - \frac{2}{5}0^{\frac{5}{2}} = \frac{2}{5}\sqrt{2^5} - 0 = \frac{2\cdot 2^2\sqrt{2}}{5} = \frac{8\sqrt{2}}{5}$.

8.1.2 Indefinite integrals

If we now consider the functions $y_1 = x^2 + 7$ and $y_2 = x^2 - 100$ and we differentiate them both, we obtain $\frac{dy_1}{dx} = 2x$ and $\frac{dy_2}{dx} = 2x$, *i.e.* we obtain the same differentiation result.

This implies that for an integral without limits, such as $\int 2x\,dx$, the antiderivative must include an **unknown constant** usually called κ:

$$\int 2x\,dx = x^2 + \kappa$$

Such an integral without limits is called an **indefinite integral**. We see that although the definite integral is a single real number, *e.g.* an area, the indefinite integral is a family of functions.

When limits are imposed, the constant κ is **always** eliminated because for all real numbers, such as a and b:

$$\int_a^b 2x\,dx = \left[x^2 + \kappa\right]_a^b = b^2 + \kappa - [a^2 + \kappa] = [b^2 - a^2] + \kappa - \kappa = b^2 - a^2$$

since $\kappa - \kappa$ will always be zero.

Generally, if we want to integrate a function $f(x)$ to obtain only the corresponding family of **antiderivatives** $F(x)$, the following mathematical notation is used:

$$\int f(x)\,dx = F(x) + \kappa$$

where κ can be any real number.

Example: Calculate the following indefinite integrals $\int x^6\,dx$ and $\int x^{-\frac{1}{2}}\,dx$, and check the answers by differentiating the antiderivatives obtained.

Solution: By using equation (8.3) $\int x^6\,dx = \dfrac{1}{6+1}x^{6+1} + \kappa = \dfrac{1}{7}x^7 + \kappa$, and by differentiating the answer $\dfrac{d}{dx}\left(\dfrac{1}{7}x^7 + \kappa\right) = \dfrac{7}{7}x^6 = x^6$, we can see that this agrees.

Likewise, $\int x^{-\frac{1}{2}}\,dx\kappa = \dfrac{1}{\frac{1}{2}}x^{\frac{1}{2}} + \kappa = 2x^{\frac{1}{2}} + \kappa = 2\sqrt{x} + \kappa$, which can again be checked by differentiation $\dfrac{d}{dx}(2x^{\frac{1}{2}} + \kappa) = 2 \cdot \dfrac{1}{2} \cdot x^{-\frac{1}{2}} = x^{-\frac{1}{2}}$, and this too agrees.

8.1.3 Properties of integrals

It helps to know a few properties of integrals to aid in their calculation.

When we integrate a function with a scalar multiple (a), the constant can be taken out of the integral.

Rule A. Scalar Multiple: $\boxed{\int \mathbf{a} \cdot \mathbf{f(x)}\,\mathbf{dx} = \mathbf{a} \cdot \int \mathbf{f(x)}\,\mathbf{dx}}$

Example: Calculate $\int \frac{5}{\sqrt[3]{x^2}}\,dx$.

Solution: Using Rule **A** and our knowledge of indices, we can rewrite the integral: $\int 5 \cdot x^{-\frac{2}{3}}\,dx$.

We can now use Rule **A**: $5 \cdot \int x^{-\frac{2}{3}}\,dx$ and then use equation (8.3) to obtain the antiderivative: $5 \cdot \frac{x^{\frac{1}{3}}}{\frac{1}{3}} + \kappa = 15 \cdot \sqrt[3]{x} + \kappa$.

Example in Biochemistry: To find the *net total change in a quantity* (Q), we can *always* integrate its rate of change ($q(t)$):

$$\text{net total change in } Q = Q(b) - Q(a) = \int_a^b q(t)\,dt$$

A biochemical process produces NaCl at the rate of $3\sqrt{t}$ grams per minute.

What is the rate of production four minutes into the process?

How much NaCl has been produced by that time? (Assume that no NaCl is present at $t = 0$.)

Solution: Let $Q(t)$ denote the grams of NaCl produced after t minutes.

The rate of production is $q(t) = 3\sqrt{t}$, so four minutes into the process, this is $q(4) = 3\sqrt{4} = 6\,\text{g}\,\text{min}^{-1}$.

The quantity of NaCl produced over the first four minutes can now be found by integration and is

$$Q(4) - Q(0) = \int_0^4 q(t)\,dt = \int_0^4 3\sqrt{t}\,dt = 3 \cdot \int_0^4 t^{\frac{1}{2}}\,dt = \left[\left(3 \cdot \frac{2}{3} t^{\frac{3}{2}}\right)\right]_0^4$$

$$= \left[2t^{\frac{3}{2}}\right]_0^4 = 2 \cdot 4^{\frac{3}{2}} - 2 \cdot 0 = 16\,\text{g}$$

When we calculate the integral of a sum (or difference) of functions, we can split the integral into a sum (or difference) of separate integrals of the components:

Rule B. Sum or Difference:

$$\int [\mathbf{f(x)} \pm \mathbf{g(x)}]\, \mathbf{dx} = \int \mathbf{f(x)\, dx} \pm \int \mathbf{g(x)\, dx}$$

Example: Calculate $\int \left(4x^3 + 2\sqrt{x} - \frac{3}{x^2}\right) dx$.

Solution: We can start by writing the integral as:

$$\int (4 \cdot x^3 + 2 \cdot x^{\frac{1}{2}} - 3 \cdot x^{-2})\, dx$$

We can then split it up by using Rule **B**:

$$\int 4 \cdot x^3\, dx + \int 2 \cdot x^{\frac{1}{2}}\, dx - \int 3 \cdot x^{-2}\, dx$$

Finally using Rule **A** and equation (8.3), we can calculate the antiderivative: $4 \cdot \frac{1}{4} x^4 + 2 \cdot \frac{2}{3} x^{\frac{3}{2}} - 3 \cdot \frac{1}{-1} \cdot x^{-1} + \kappa = x^4 + \frac{4}{3}\sqrt{x^3} + \frac{3}{x} + \kappa$.

Example in Biology: The rate of change of the population of a colony of insects in a laboratory experiment can be described by the function

$$r(t) = 24t^2 - 4t^3$$

with t in months. When is the maximum net total change observed? What is the function, $P(t)$, which describes the net total change in population, if at $t = 0$ months there are 729 insects?

Solution: Let the population rate of change be $r(t) = 24t^2 - 4t^3$.

The maximum net total change is observed when $r(t) = 0$: $24t^2 - 4t^3 = 0 \Longleftrightarrow 4t^2 \cdot (6 - t) = 0$, which gives $t = 6$.

Thus, the maximum net total change in population is observed in the sixth month.

The function describing the net total change in population is

$$P(t) = \int (24t^2 - 4t^3)\, dt = \int 24t^2\, dt - \int 4t^3\, dt = 24\frac{t^3}{3} - 4\frac{t^4}{4} + \kappa$$
$$= 8t^3 - t^4 + \kappa$$

If at $t = 0$ there are 729 insects, $P(0) = 0 - 0 + \kappa = 729 \iff \kappa = 729$.
Thus, $P(t) = 8t^3 - t^4 + 729$.

8.1.4 Antiderivatives of common fundamental functions

• Logarithmic function

So far, we have seen that integration reverses the process of differentiation and equation (8.3) has **no** meaning if $n = -1$, since we obtain the meaningless antiderivative $\frac{1}{0}x^0 + \kappa$. However, we can solve the integral $\int x^{-1}\, dx$ by remembering that $\frac{d(\ln x)}{dx} = \frac{1}{x}$ (see Section 6.5.1). Therefore,

$$\int \frac{1}{x}\, dx = \ln|x| + \kappa \tag{8.5}$$

Important note: The modulus $|x|$ is required, since x can assume any real value, positive or negative, BUT the argument of a logarithm needs to be > 0 (see Chapter 6.2.4).

The constant κ can also be written as $\kappa = \ln(B)$, where B is a positive constant. Thus, $\int \frac{1}{x}\, dx = \ln|x| + \kappa = \ln|x| + \ln(B) = \ln|B \cdot x|$.

• Exponential function

Remembering that the exponential function, e^x, has the unique property that $\frac{d}{dx}(e^x) = e^x$, we can solve the integral $\int e^x\, dx$ as follows:

$$\int e^x\, dx = e^x + \kappa \tag{8.6}$$

Example: Calculate $\int_1^e \frac{1}{x}\, dx$ and highlight the calculated area in an appropriate plot.

Solution: $\int_1^e \frac{1}{x}\, dx = [\ln(x)]_1^e = \ln e - \ln 1 = 1 - 0 = 1$ and this area is shown in the following figure.

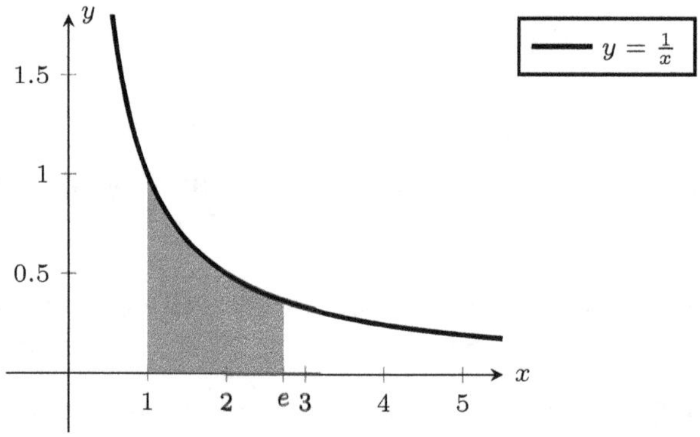

• **Trigonometric functions**

Remembering that (see Sections 7.1.2 and 7.1.3)

$$\frac{d(\sin(x))}{dx} = \cos(x), \quad \frac{d(\cos(x))}{dx} = -\sin(x), \quad \text{and} \quad \frac{d(\tan(x))}{dx} = \frac{1}{\cos^2(x)}$$

we can calculate the following antiderivatives of trigonometric functions:

$$\int \sin(x)\, dx = -\cos(x) + \kappa \tag{8.7}$$

$$\int \cos(x)\, dx = \sin(x) + \kappa \tag{8.8}$$

$$\int \frac{1}{\cos^2(x)}\, dx = \tan(x) + \kappa \tag{8.9}$$

Remembering that $\dfrac{d(\arctan(x))}{dx} = \dfrac{1}{1+x^2}$ (see Section 7.1.4), we can also calculate the antiderivative of this particular rational function $\dfrac{1}{1+x^2}$:

$$\int \frac{1}{1+x^2}\,dx = \arctan(x) + \kappa \tag{8.10}$$

Example: Calculate $\int_0^{\frac{\pi}{2}} \cos(x)\,dx$ and highlight the calculated area in an appropriate plot.

Solution: $\int_0^{\frac{\pi}{2}} \cos(x)\,dx = [\sin(x)]_0^{\frac{\pi}{2}} = 1 - 0 = 1$ and this area is shown in the following figure.

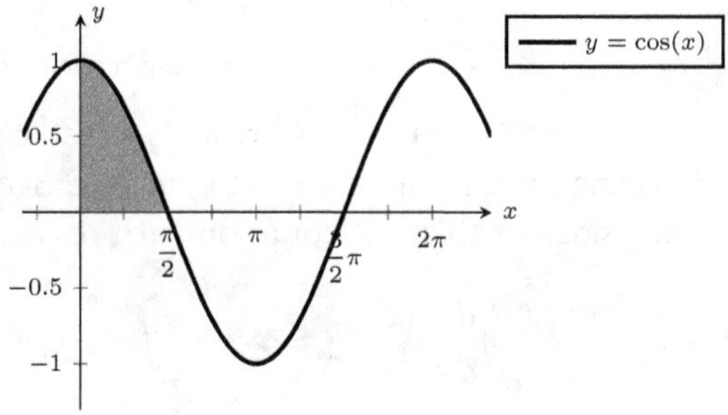

8.2 Techniques of Integration

In some cases, the function to be integrated is more complicated than the polynomial form of equation (8.3). For instance, it might be the function of a function and/or the product of two functions. We thus need some additional methods to deal with these compound integrals.

8.2.1 Integration by substitution

The **substitution method for a function of a function** can be thought of as an **integral version of the Chain Rule** (see Section 5.3.2) and can be used to integrate functions such as $\int_1^2 (3-4x)^{-5}\,dx$ which contain an $f(g(x))$ with $g(x) = 3 - 4x$ and $f(g(x)) = (g(x))^{-5}$.

In general, suppose we wish to integrate

$$I = \int_a^b f(g(x))\,dx$$

Since integration reverses the process of differentiation, we can differentiate both sides of the equation, obtaining $\frac{dI}{dx} = f(g(x))$.

As a shorthand, we substitute $g(x) \equiv g$ so that $\frac{dI}{dx} = f(g)$, which indicates that $\frac{dI}{dx}$ is a composite function f of $g(x)$, where $g(x) \equiv g$.

By applying the Chain Rule (see Section 5.3.2), we have $\frac{dI}{dx} = \frac{dI}{dg}\frac{dg}{dx}$, which can be rearranged to be $\frac{dI}{dg} = \frac{\frac{dI}{dx}}{\frac{dg}{dx}} = \frac{dI}{dx}\frac{dx}{dg}$.

Note that $\frac{dx}{dg}$ means 'the derivative of x with respect to g.' Thus, x is a function of g, and for this to be true, g must be an invertible function. This uses the inverse function theorem (see Section 6.2.2).

Integrating both sides of this last equation with respect to g gives us

$$\int_{g(a)}^{g(b)} \frac{dI}{dg}\,dg = \int_{g(a)}^{g(b)} \frac{dI}{dx}\frac{dx}{dg}\,dg \Longrightarrow I = \int_{g(a)}^{g(b)} f(g)\frac{dx}{dg}\,dg$$

Thus, we can now integrate I by using the substitution method.

Rule C. Substitution Method:

$$\mathbf{I = \int_a^b f(g(x))\,dx = \int_{g(a)}^{g(b)} f(g)\,\frac{dx}{dg}\,dg = [F(g)]_{g(a)}^{g(b)} = [F(g(x))]_a^b}$$

Important note: The expression to be substituted by g should be **appropriately chosen** so that the integral $\int f(g)\frac{dx}{dg}\,dg$ allows straightforward calculation of the antiderivative where one of the components is simpler and might include one of the common fundamental functions shown in previous sections.

Thus, we should first look at the integral I and see if the **integrand** contains a multiplier of $f(g(x))$ that could be the derivative of the bracketed function: *i.e.* I can be rearranged to the form

$$I = \int_a^b f(g(x)) \cdot g'(x)\, dx$$

If so, then substitution is an appropriate strategy and we can use the following **protocol for integration by substitution** for the general integral

$$I = \int_a^b f(g(x))\, dx$$

(i) Decide on the function g for substitution, find dg/dx and thus its reciprocal, $dx/dg = 1/(dg/dx)$.
(ii) Substitute g into the integral, multiply the integral by dx/dg and simplify by cancelling appropriate terms.
(iii) If it is a definite integral, two options are possible:
   - replace the upper and lower limits by $g(b)$ and $g(a)$, respectively, and then perform the integration with respect to g;
   - substitute g into the result so that x is again the variable, and then put the specified limits for x into your result.
(iv) If it is a definite integral, put in the appropriate limits and simplify the result.
(v) If it is an indefinite integral, it is good Mathematical etiquette to transform the answer back to the original variable (*e.g.* x). Also, do NOT forget the constant of integration!

Example: Calculate $I_i = \int (2x+3)^4\, dx$ and $I_{ii} = \int_1^2 (3-4x)^{-5}\, dx$.

Solution: In order to calculate I_i, we could multiply out $(2x+3)^4$ and then integrate it using equation (8.3), but this would be very tedious. Instead we can try a substitution.

We first note that the derivative of $g(x) = 2x+3$ is $g'(x) = 2$, which could be a constant multiplier of $f(g(x)) = (2x+3)^4$ following an appropriate rearrangement. Thus, the substitution method is an appropriate strategy to find the integral I_i.

Let $g = 2x + 3 \Longrightarrow \dfrac{dg}{dx} = 2 \Longleftrightarrow \dfrac{dx}{dg} = \dfrac{1}{2}$.

Now, from the substitution method, Rule **C**, we know that if $f(g(x)) = f(g)$, we can write $\int f(g(x))\,dx = \int f(g)\,\dfrac{dx}{dg}\,dg$.

Using this, substituting into I_i above and using Rule **A**, we have

$$\int g^4 \cdot \frac{1}{2}\,dg = \frac{1}{2}\int g^4\,dg = \frac{1}{2}\left[\frac{1}{5}g^5\right] = \frac{1}{2}\left[\frac{1}{5}(2x+3)^5\right] + \kappa$$
$$= \frac{1}{10}(2x+3)^5 + \kappa$$

The answer can be checked by differentiation using the Chain Rule.

For similar reasons as noted for the previous example, to calculate

$$I_{ii},\ \text{let } g = 3 - 4x \Longrightarrow \frac{dg}{dx} = -4 \Longleftrightarrow \frac{dx}{dg} = -\frac{1}{4}$$

Substitute into I_{ii} above,

$$\int_{g(1)}^{g(2)} g^{-5} \cdot \left(-\frac{1}{4}\right)\,dg$$
$$= -\frac{1}{4}\int_{g(1)}^{g(2)} g^{-5}\,dg = -\frac{1}{4}\left[-\frac{1}{4}g^{-4}\right]_{g(1)}^{g(2)} = \frac{1}{4}\left[\frac{1}{4}(3-4x)^{-4}\right]_1^2$$
$$= \frac{1}{4}\cdot\frac{1}{4}\left[(3-4\cdot 2)^{-4} - (3-4\cdot 1)^{-4}\right] = \frac{1}{16}\left[(-5)^{-4} - (-1)^{-4}\right]$$
$$= \frac{1}{16}\left[\frac{1}{(-5)^4} - \frac{1}{(-1)^4}\right] = \frac{1}{16}\left[\frac{1}{5^4} - 1\right] = -\frac{39}{625}$$

Note that we cannot solve this integral by multiplying out the function in I_{ii}, and so we must use the substitution method.

Note also that the value of the definite integral is < 0, since the integrand function $(3-4x)^{-5}$ is negative in the interval $[1, 2]$. Therefore, the modulus of the integral should be taken to calculate the area between the integrand function and the x-axis: AUC $= \left|-\dfrac{39}{625}\right| = \dfrac{39}{625}$.

Example: Calculate $\int_0^2 x \cdot e^{(1-x^2)}\, dx$.

Solution: To calculate this definite integral, we can use the substitution method since the integrand contains an x as a multiplier of $f(g(x)) = e^{(1-x^2)}$. The derivative of $g(x) = 1 - x^2$ is $-2x$, so substitution is an appropriate integration method here.

So, let $g = 1 - x^2 \Longrightarrow \dfrac{dg}{dx} = -2x \Longleftrightarrow \dfrac{dx}{dg} = \dfrac{1}{-2x}$.

As the value of x goes from 0 to 2, the value of g goes from $g(0) = 1$ to $g(2) = -3$ and we must change the limits to match our substituted function.

Using the substitution method, Rule **C**, we know that if $f(g(x)) = f(g)$, we can write:

$$\int_0^2 x \cdot e^{(1-x^2)}\, dx = \int_1^{-3} x \cdot e^g \frac{1}{-2x}\, dg = -\frac{1}{2}\int_1^{-3} e^g\, dg$$

$$= -\frac{1}{2}\left[e^g\right]_1^{-3} = -\frac{1}{2}\left[e^{-3} - e\right] = \frac{1}{2}\left[e - e^{-3}\right]$$

Important note: Although some integrals, such as $\int \cos(e^x)\, dx$ or $\int \sin(x^2)\, dx$, contain a compound integrand function of the type $f(g(x))$, they cannot be solved using this substitution method, since they do not include a multiplier of $f(g(x))$ which is the derivative of $g(x)$, *i.e.* $g'(x) = e^x$ or $g'(x) = 2x$, respectively, and these multipliers do not appear upon rearrangement of the integrand. Other more complex techniques must be used to solve them, but these are far beyond the scope of this textbook.

Two useful formulae can be derived from Rule **C** in order to simplify the calculation of some common integrals:

(C.i) If $F'(x) = f(x)$, then $\int f(ax+b)\, dx = \dfrac{1}{a}F(ax+b) + \kappa$, where f is usually a power, exponential, sine, or cosine function.

(C.ii) $\int \dfrac{f'(x)}{f(x)}\, dx = \ln|f(x)| + \kappa$.

These can both be checked by differentiating the results with respect to x.

Example: Calculate $I_i = \int e^{-3x}\,dx$ and $I_{ii} = \int \tan(x)\,dx$.

Solution: Using Rule **C(i)**, $I_i = \int e^{-3x}\,dx = -\frac{1}{3}e^{-3x} + \kappa$.

For I_{ii}, we can re-express the integral as

$$I_{ii} = \int \tan(x)\,dx = \int \frac{\sin(x)}{\cos(x)}\,dx$$

We can then multiply by $(-1)\cdot(-1)$: $I_{ii} = (-1)\cdot \int \frac{(-1)\cdot\sin(x)}{\cos(x)}\,dx$.

Now by using Rule **C(ii)**, we can see that $I_{ii} = -\ln|\cos(x)| + \kappa$.

Example in Biology: A new coal-fired power plant releases particulate matter into the atmosphere at the rate of $\frac{t}{t^2+1}$ tonnes per month. How many tonnes are released during the first month of operation?

Solution: To find the amount of particulate matter, we can integrate the rate of change. Using the substitution $g = t^2 + 1$, we have $\frac{dg}{dt} = 2t \Longleftrightarrow \frac{dt}{dg} = \frac{1}{2t}$. As the value of t goes from 0 to 1, the value of g goes from $g(0) = 1$ to $g(1) = 2$.

Thus, using Rule **C** above we have

$$\begin{aligned}\int_0^1 \frac{t}{t^2+1}\,dt &= \int_1^2 \frac{t}{g}\frac{1}{2t}\,dg = \frac{1}{2}\int_1^2 \frac{1}{g}\,dg \\ &= \left[\frac{1}{2}\ln|g|\right]_1^2 = \frac{1}{2}\ln|2| - \frac{1}{2}\ln|1| \\ &= \frac{1}{2}\ln|2| - 0 = \frac{1}{2}\ln|2| \approx \frac{1}{2}\times 0.7 = 0.35 \text{ tonnes}\end{aligned}$$

Example in Biochemistry (Part 1): The probability, $f(v)$, that a molecule of mass m in a gas at temperature, T, has speed, v, is given by the Maxwell–Boltzmann distribution:

$$f(v) = 4\pi\left(\frac{m}{2\pi kT}\right)^{\frac{3}{2}} v^2 e^{-mv^2/2kT}$$

where k is Boltzmann's constant.

The mean speed (the 'expectation value': *i.e.* the expected value of a continuous random variable) is the weighted average of all possible values where the weight is the probability of landing on value v.

Thus, each possible value that can be taken by the random variable is multiplied by its probability of occurring, and the resulting products are summed (integrated) to produce the expected (or 'expectation') value:

$$\overline{v} = \int_0^{+\infty} v\, f(v)\, dv$$

Determine this integral.

Important note: The infinity symbol ∞ can also be used as a limit of an integral, thus making the integral 'improper'. This topic is covered later in Section 8.3.

Solution: In order to determine the integral, we first need to use the method of substitution with $z = g(v)$:

$$\int_{v_1}^{v_2} f(g(v))\, dv = \int_{z_1}^{z_2} f(z)\,\frac{dv}{dz}\, dz$$

Let $z = -\dfrac{mv^2}{2kT} \Longrightarrow \dfrac{dz}{dv} = -\dfrac{m2v}{2kT} = -\dfrac{mv}{kT} \Longleftrightarrow \dfrac{dv}{dz} = -\dfrac{kT}{mv}$.

Because we have changed the variable from v to z, we also have to change the limits from $v_1 = 0$ and $v_2 = +\infty$ to $z_1 = 0$ and $z_2 = -\infty$, respectively.

To simplify the algebra, we call $4\pi\left(\dfrac{m}{2\pi kT}\right)^{\frac{3}{2}} = Q$.

So $f(v) = 4\pi\left(\dfrac{m}{2\pi kT}\right)^{\frac{3}{2}} v^2 e^{-mv^2/2kT} = Q\,v^2 e^{-mv^2/2kT} = Q\,v^2 e^z$,
and we obtain:

$$\overline{v} = \int_0^{+\infty} v\,f(v)\,dv = \int_0^{-\infty} v\,f(z)\,\frac{dv}{dz}dz = \int_0^{-\infty} v\,(Qv^2e^z)\left(\frac{-kT}{mv}\right)dz$$

which gives

$$\overline{v} = Q\left(\frac{-kT}{m}\right)\int_0^{-\infty}\frac{v^3}{v}\,e^z\,dz = Q\left(\frac{-kT}{m}\right)\int_0^{-\infty} v^2\,e^z\,dz$$

Rearranging $z = \dfrac{-mv^2}{2kT}$ so that v^2 is the subject, we can make the substitution for $v^2 = \dfrac{-2kTz}{m}$:

$$\overline{v} = Q\left(\frac{-kT}{m}\right)\int_0^{-\infty} v^2\,e^z\,dz = Q\left(\frac{-kT}{m}\right)\int_0^{-\infty}\frac{-2kT}{m}ze^z dz$$

which gives $\overline{v} = 2Q\left(\dfrac{kT}{m}\right)^2\displaystyle\int_0^{-\infty} z\,e^z dz$.

So far in this chapter, we have not covered any technique that will enable us to integrate this.

Since our integral now looks like a product of two functions of the same variable z, namely $f(z) = z$ and $g(z) = e^z$, we use the 'integration by parts' method explained in the following section.

8.2.2 Integration by parts

The **integration by parts** method is the **integral form of the Product Rule for differentiation** and requires that we represent the integrand (*i.e.* what we want to integrate) as one function multiplied by the **derivative** of another function:

$$\int_a^b f(x)\,g'(x)\,dx$$

Recalling that

$$\frac{d\,[f(x)\,g(x)]}{dx} = f(x)\,g'(x) + g(x)\,f'(x)$$

we can integrate both sides of the equation on the closed interval $[a, b]$:

$$\int_a^b \frac{d[f(x)\,g(x)]}{dx}\,dx = \int_a^b f(x)\,g'(x)\,dx + \int_a^b g(x)\,f(x)'\,dx$$

$$\Longrightarrow \qquad [f(x)\,g(x)]_a^b = \int_a^b g(x)\,f'(x)\,dx + \int_a^b f(x)\,g'(x)\,dx$$

$$\Longleftrightarrow \qquad [f(x)\,g(x)]_a^b - \int_a^b g(x)\,f'(x)\,dx = \int_a^b f(x)\,g'(x)\,dx$$

This is known as the formula for **'integrating by parts'** and is written as follows:

Rule D. By Parts Method:

$$\mathbf{I} = \int_{\mathbf{a}}^{\mathbf{b}} \mathbf{f(x)\,g'(x)\,dx} = [\mathbf{f(x)\,g(x)}]_{\mathbf{a}}^{\mathbf{b}} - \int_{\mathbf{a}}^{\mathbf{b}} \mathbf{g(x)\,f'(x)\,dx}$$

which can also be written for simplicity as

$$\int_a^b f\,g'\,dx = [fg]_a^b - \int_a^b g\,f'\,dx$$

with $f(x) \equiv f$ and $g(x) \equiv g$.

Note that when we integrate $g'(x)$ to get $g(x)$, we do not need to include an unknown constant of integration, such as κ, because this always cancels out during the integration (this important result is shown in the second example below).

Additionally, **care** must be exercised in the choice of $f(x)$ to ensure that $g(x)\,f'(x)$ is simpler to integrate than $f(x)\,g'(x)$.

In using this method, we aim to make the right-hand side a little easier to handle than the left.

A few **guidelines** are useful when choosing f and g':

(i) Whatever you assign to g' be, you need to be able to find g.
(ii) It helps if f' is simpler than f (or at least no more complicated than f).
(iii) It helps if g is simpler than g' (or at least no more complicated than g').

(iv) Choose f in the priority order:

(a) **L**-logarithmic: $\log(x)$, $\ln(x)$, ...

(b) **A**-algebraic: x^n, $\dfrac{x^n}{x^m}$, ...

(c) **T**-trigonometric: $\sin(x)$, $\cos(x)$, ...

(d) **E**-exponential: $\exp(x) = e^x$, 10^x, ...

which spells **LATE**.

(v) Choose g' in the reverse order: *i.e.* **ETAL**.

Important note: Practice definitely pays off!

Example: Calculate $\int x\sqrt{x+1}\,dx$.

Solution: We first see that the integrand consists of the product of two functions: $x \cdot \sqrt{x+1}$. Thus, we can try to use the integration by parts method.

Recalling that $\sqrt{x+1} = (x+1)^{\frac{1}{2}}$, we start by identifying the most appropriate functions for f and g'.

If we initially choose $f = x$ and $g' = \sqrt{x+1}$, so that $f' = 1$ and $g = \frac{2}{3}(x+1)^{\frac{3}{2}}$, we have:

$$\begin{aligned}\int x\sqrt{x+1}\,dx &= \left[x \cdot \frac{2}{3}(x+1)^{\frac{3}{2}}\right] - \int 1 \cdot \frac{2}{3}(x+1)^{\frac{3}{2}}\,dx \\ &= \left[\frac{2}{3}x\,(x+1)^{\frac{3}{2}}\right] - \left[\frac{2}{3} \cdot \frac{2}{5}(x+1)^{\frac{5}{2}}\right] + \kappa\end{aligned}$$

On the other hand, if we had chosen the other option for f and g': $f = \sqrt{x+1}$ and $g' = x$, so that $f' = \frac{1}{2}(x+1)^{-\frac{1}{2}}$ and $g = \frac{1}{2}x^2$, we would have had:

$$\int x\sqrt{x+1}\,dx = \left[(x+1)^{\frac{1}{2}} \cdot \frac{x^2}{2}\right] - \int \frac{1}{2}\,\frac{1}{\sqrt{x+1}} \cdot \frac{x^2}{2}\,dx$$

The second term is WORSE than the integral we started with and so does not help us at all! This shows the vital importance of making the correct choice for f and g' from the outset.

Important note: This integral can also be solved by using the substitution method, but more algebra is involved to obtain the same answer.

Important question: In the integration by parts method when g' is integrated to get g and then the second term is integrated too, why is only one arbitrary constant of integration κ required for an indefinite integral?

Example: Repeat $\int x\sqrt{x+1}\,dx$ without omitting the constants of integration at each step.

Solution: As above we use $f = x$ and $g' = \sqrt{x+1}$, so that $f' = 1$ and $g = \frac{2}{3}(x+1)^{\frac{3}{2}} + \kappa_1$ and so putting $I = \int x\sqrt{x+1}\,dx$ we get:

$$
\begin{aligned}
I &= \left[x \cdot \left(\frac{2}{3}(x+1)^{\frac{3}{2}} + \kappa_1\right)\right] - \int 1 \cdot \left(\frac{2}{3}(x+1)^{\frac{3}{2}} + \kappa_1\right)\,dx \\
&= \left[\frac{2}{3}\,x\,(x+1)^{\frac{3}{2}}\right] + \kappa_1 x - \left[\frac{2}{3}\cdot\frac{2}{5}(x+1)^{\frac{5}{2}} + \kappa_2\right] - (\kappa_1 x + \kappa_3) \\
&= \left[\frac{2}{3}\,x\,(x+1)^{\frac{3}{2}}\right] - \left[\frac{2}{3}\cdot\frac{2}{5}(x+1)^{\frac{5}{2}}\right] + \kappa_4
\end{aligned}
$$

where $\kappa_4 = -\kappa_2 - \kappa_3$.

Thus, we see that all the intermediate 'working' constants with multipliers ($\kappa_1 x$ here) cancel out, and that for an indefinite integral we are left with just one arbitrary constant of integration as with other methods of integration.

Example in Biochemistry (Part 2): In order to estimate the average speed $\overline{v}$ of a molecule m in a gas at temperature, T, as given by the Maxwell–Boltzmann equation, now integrate

$$\overline{v} = 2Q\left(\frac{kT}{m}\right)^2 \int_0^{-\infty} z\,e^z\,dz$$

Solution: We can integrate this by parts: $\int_0^{-\infty} z\,e^z dz$ by using Rule **D**.

We choose $f = z$ and $g' = e^z$ so that $f' = 1$ and $g = e^z$:

$$\int_0^{-\infty} z\,e^z\,dz = [z\,e^z]_0^{-\infty} - \int_0^{-\infty} e^z\,dz = [z\,e^z]_0^{-\infty} - [e^z]_0^{-\infty}$$

Thus, $\overline{v} = \int_0^{+\infty} v\,f(v)\,dv = 2Q\left(\frac{kT}{m}\right)^2 \int_0^{-\infty} z\,e^z dz$

$$= 2Q\left(\frac{kT}{m}\right)^2 \left\{[z\,e^z]_0^{-\infty} - [e^z]_0^{-\infty}\right\}.$$

We now need to understand whether using this improper integral to calculate $\overline{v}$ gives a finite value. We cover this question below in Part 3 of this problem.

Special Example A: Calculate $\int \ln(x)\,dx$.

Solution: There appears to be only one function here, but we can make it suitable for solution using integration by parts if we consider the integral as a product: $\int 1{\cdot}\ln(x)\,dx$, so we can let $f = \ln(x) \Longrightarrow f' = \frac{1}{x}$ and $g' = 1 = x^0 \Longrightarrow g(x) = x$.

Putting these into our expression, we now have:

$$\int \ln(x)\,dx = x\ln(x) - \int x\cdot\frac{1}{x}\,dx = x\ln(x) - \int 1\,dx = x\ln(x) - x + \kappa$$

Special Example B: Calculate $I = \int_a^b \cos(x)\cdot e^{-x}\,dx$.

Solution: Sometimes, when integrating by parts, we end up where we began, as in this example.

In order to calculate $I = \int \cos(x)\cdot e^{-x}\,dx$ by parts, let

$f = \cos(x) \Longrightarrow f' = -\sin(x)$ and $g' = e^{-x} \Longrightarrow g(x) = -e^{-x}$

Thus,

$$I = [-\cos(x)\cdot e^{-x}] - \int (-)\sin(x)\cdot(-)e^{-x}\,dx$$

$$= [-\cos(x)\cdot e^{-x}] - \int \sin(x)\cdot e^{-x}\,dx$$

We have to integrate the new integral by using integration by parts again.

Let $f = \sin(x) \Longrightarrow f' = \cos(x)$ and $g' = e^{-x} \Longrightarrow g(x) = -e^{-x}$. Thus

$$I = [-\cos(x) \cdot e^{-x}] - [\sin(x) \cdot (-)e^{-x}] + \kappa + \int \cos(x) \cdot (-)e^{-x}\, dx$$

But the last term is the integral I we started with, which means

$$I = [-\cos(x) \cdot e^{-x}] + [\sin(x) \cdot e^{-x}] + \kappa - I$$

Thus we have $2I = [\sin(x) \cdot e^{-x}] - [\cos(x) \cdot e^{-x}] + \kappa$, which gives

$$I = \int \cos(x) \cdot e^{-x}\, dx = \frac{1}{2}[\sin(x) \cdot e^{-x} - \cos(x) \cdot e^{-x}] + \kappa$$

Example in Medicine: The area under the curve (AUC) of plasma drug concentrations over time reflects the actual exposure of the body to a drug after administration of a drug dose and is expressed in $\mathrm{mg\,h\,L^{-1}}$. This AUC is dependent on the rate of elimination of the drug from the body and also on the dose administered.

During clinical trials, the patient's plasma drug concentration-time profile was plotted by measuring the plasma concentration C in $\mathrm{mg\,L^{-1}}$ at several time points t in hours, h, and the resulting curve was modelled by the equation $C(t) = \dfrac{t}{\alpha} \cdot e^{2-\beta t}$, where α is a positive constant with units of $\mathrm{L\,mg^{-1}}$ and β is another positive constant with units of $\mathrm{h^{-1}}$.

Estimate the AUC between time $t = 0\,\mathrm{h}$ and $t = 8\,\mathrm{h}$ with $\alpha = \beta = 1$.

Solution: The AUC is the integral of the function of plasma concentration C over time $C(t) = \dfrac{t}{\alpha} \cdot e^{2-\beta t}$, and it can be calculated using integration by parts and putting $\alpha = \beta = 1$.

Remembering the acronyms LATE (for choosing f) and ETAL (for choosing g'), let

$$f = t \Longrightarrow f' = 1 \text{ and } g' = e^{2-t} \Longrightarrow g(x) = -e^{2-t}$$

(here $g(x)$ is obtained by using the substitution method).

Putting these into the integral, we then have

$$\begin{aligned}\text{AUC} &= \int_0^8 t \cdot e^{2-t}\,dt = [-te^{2-t}]_0^8 - \int_0^8 (-)e^{2-t}\,dt \\ &= [-te^{2-t}]_0^8 - [e^{2-t}]_0^8 = [-8e^{-6} + 0] - [e^{-6} - e^2] \\ &= -\frac{9}{e^6} + e^2 \approx 7.37\,\text{mg h L}^{-1}\end{aligned}$$

8.2.3 Integration by method of partial fractions

We now have two methods of integrating compound functions, but there is a third type of function for which we still have no integration method.

These functions are products of fractions and before we can integrate them, we need to split them up into so-called 'partial fractions'.

An example of such a function is $f(x) = \dfrac{x+3}{(x-2)(x+4)}$.

- **Partial fractions**

If we consider first a function such as $f(x) = \dfrac{2}{x+1} + \dfrac{x}{x^2+1}$, each of the fractions in this expression has two important properties:

(1) the numerator has degree smaller than the denominator;
(2) the denominator is a power of a polynomial that cannot be reduced any further.

Functions with these two properties are called **partial fractions**.

Whenever we encounter the sum of partial fractions, we can add them together to form a single, but more complicated, quotient of polynomials whose denominator will be the common denominator of the constituent fractions.

For the function above,

$$f(x) = \frac{2}{x+1} + \frac{x}{x^2+1} = \frac{2(x^2+1) + x(x+1)}{(x+1)(x^2+1)} = \frac{3x^2+x+2}{(x+1)(x^2+1)}$$

It turns out that partial fractions behave in simple and predictable ways.

Thus, when working with more complicated quotients of polynomials, it is often useful to be able to reverse the addition process and rewrite them as sums of partial fractions.

This process is called **expressing**, or **decomposing, $f(x)$ into partial fractions**. We do this using the **method of undetermined coefficients.**

For instance, if we take a function with a higher power of x in the denominator such as $f(x) = \dfrac{x+3}{(x-2)(x+4)}$ and express it as the sum of two (or in some cases more) separate fractions, we get

$$\frac{x+3}{(x-2)(x+4)} = \frac{A}{x-2} + \frac{B}{x+4}$$

where A and B are constants to be determined.

However, if we have a function with a higher power of x in the denominator such as $g(x) = \dfrac{x+7}{(x-5)(x^2+1)}$, it can be expressed as $\dfrac{A}{x-5} + \dfrac{Bx+C}{x^2+1}$, where A, B, and C are constants to be determined.

In general, decomposition of rational functions into the sum of partial fractions is possible by using the method of undetermined coefficients and the use of a few simple rules.

Given a rational function of the type $f(x) = \dfrac{N(x)}{D(x)}$, where the numerator's degree of the power of x is smaller than that of the denominator,

(i) ensure that the denominator $D(x)$ is decomposed into factors;

(ii) split the function $f(x)$ into the sum of partial fractions, each one having one single decomposing factor as denominator and an appropriate numerator with undetermined coefficients, depending on the form of the factor in the denominator:

(a) a linear factor (*i.e.* x not raised to a power higher than one) in the denominator gives a partial fraction of the form

$$\boxed{\frac{\mathbf{A}}{\mathbf{ax+b}}} : \text{e.g. } \frac{x-2}{(x+3)(x-4)} = \frac{A}{x+3} + \frac{B}{x-4}$$

(b) a quadratic factor in the denominator gives a Partial Fraction of the form

$$\boxed{\frac{\mathbf{Ax+B}}{\mathbf{x^2+bx+c}}} : \textit{e.g.}\ \frac{x^2+1}{(x^2+2)(x-1)} = \frac{Ax+B}{x^2+2} + \frac{C}{x-1}$$

(c) a repeated factor in the denominator gives two Partial Fractions of the form

$$\boxed{\frac{\mathbf{A}}{\mathbf{ax+b}} + \frac{\mathbf{B}}{(\mathbf{ax+b})^2}} : \textit{e.g.}\ \frac{2x-1}{(x-2)^2} = \frac{A}{x-2} + \frac{B}{(x-2)^2}$$

(iii) multiply both sides of the resulting equation by the denominator of the original fraction;

(iv) equate coefficients of x, including the units on each side (*i.e.* the x^0 terms);

(v) solve the simultaneous equations obtained to determine the coefficients (*e.g.* $A, B, C, \ldots$) in the numerator of the partial fractions.

Example: Decompose the function $f(x) = \dfrac{1}{(x^2-1)}$ into a sum of partial fractions.

Solution: We can follow the five steps outlined in Section 8.2.3 to decompose $f(x)$ into the sum of partial fractions:

(i) First we decompose the denominator into factors:

$$\frac{1}{(x^2-1)} = \frac{1}{(x-1)(x+1)}$$

(ii) We then split the function $f(x)$ into a sum of partial functions with unknown coefficients A and B:

$$\frac{1}{(x-1)(x+1)} = \frac{A}{(x-1)} + \frac{B}{(x+1)}$$

(iii) We now multiply both sides by the denominator, $(x-1)(x+1)$, to get $1 = A(x+1) + B(x-1) \Longleftrightarrow 1 = Ax + A + Bx - B$ $\Longleftrightarrow 1 = x(A+B) + A - B$.

(iv) We can now equate coefficients of x to get equation (i) and equate the units (x^0 components) to obtain equation (ii):

(a) $A + B = 0$,
(b) $A - B = 1$.

(v) Finally we solve these simultaneous equations to determine A and B. The sum (i)+(ii) gives $A = \frac{1}{2}$ and the difference (a)–(b) gives $B = -\frac{1}{2}$.

Thus, $f(x) = \dfrac{1}{(x^2 - 1)} = \dfrac{1}{2(x-1)} - \dfrac{1}{2(x+1)}$.

Example: Express the function $f(x) = \dfrac{x^2 - 3}{(x-1)(x^2+1)}$ as the sum of partial fractions.

Solution: Again we follow the five steps outlined in Section 8.2.3.

(i) The denominator of $f(x)$ is already decomposed into factors.

(ii) We then express $f(x)$ as the sum of partial fractions:

$$\frac{x^2 - 3}{(x-1)(x^2+1)} = \frac{A}{x-1} + \frac{Bx + C}{x^2 + 1}$$

(iii) We multiply both sides by $(x-1)(x^2+1)$ to get

$$\begin{aligned} x^2 - 3 &= A(x^2+1) + (Bx + C)(x-1) \\ \iff x^2 - 3 &= Ax^2 + A + Bx^2 + Cx - Bx - C \\ \iff x^2 - 3 &= x^2(A+B) + x(C-B) + A - C \end{aligned}$$

(iv) We equate the coefficients of x^2, x and units to get the simultaneous equations:

(a) $A + B = 1$
(b) $C - B = 0$
(c) $A - C = -3$

(v) Solving the simultaneous equations gives $A = -1$ (using (a) + (c) + (b)), $B = 2$, and $C = 2$.

Thus, $f(x) = \dfrac{x^2 - 3}{(x-1)(x^2+1)} = -\dfrac{1}{x-1} + \dfrac{2x+2}{x^2+1}$.

Example: Express the function $f(x) = \dfrac{(3x+1)}{(x+1)^2(x-2)}$ as the sum of partial fractions.

Solution: We use the five steps outlined in Section 8.2.3:

(i) The denominator of $f(x)$ is already decomposed into factors but in this case one of them is repeated $(x+1)^2$, so we need three undetermined coefficients.

(ii) We then express $f(x)$ as the sum of partial fractions:

$$\frac{(3x+1)}{(x+1)^2(x-2)} = \frac{A}{(x+1)^2} + \frac{B}{(x+1)} + \frac{C}{(x-2)}$$

(iii) We multiply both sides by $(x+1)^2(x-2)$ to get

$$3x+1 = A(x-2) + B(x+1)(x-2) + C(x+1)^2$$

(iv) Usually we equate the coefficients of x^2, x and units to get useful simultaneous equations.

(v) However, instead we can insert particular values of x to make finding A, B, and C easier.

By inspection, when $x = -1$, we get

$$-2 = A(-1-2) \Longleftrightarrow A = \frac{2}{3}$$

Likewise, when $x = 2$, we obtain $7 = 9C \Longleftrightarrow C = \dfrac{7}{9}$.

We cannot actually find B directly, but at $x = 0$, we get

$$1 = -2A - 2B + C = -\frac{4}{3} - 2B + \frac{7}{9} \Longleftrightarrow 1 + \frac{4}{3} - \frac{7}{9} = -2B = \frac{14}{9}$$

which gives $B = -\dfrac{7}{9}$.

Thus, $$f(x) = \frac{(3x+1)}{(x+1)^2(x-2)}$$

$$= \frac{2}{3(x+1)^2} - \frac{7}{9(x+1)} + \frac{7}{9(x-2)}.$$

• Integration of partial fractions

Decomposing a rational function into a sum of partial fractions is mainly useful to simplify integrating it, as illustrated in the following example.

Example: Calculate $\int \frac{1}{(2x+1)(x-5)}\,dx$.

Solution: In order to calculate this integral, we first follow the five steps outlined in Section 8.2.3 to decompose the rational function into a sum of partial fractions:

(i) The denominator of $f(x)$ is already decomposed into factors.
(ii) We then express $f(x)$ as the sum of partial fractions:

$$\frac{1}{(2x+1)(x-5)} = \frac{A}{(2x+1)} + \frac{B}{(x-5)}$$

(iii) We now multiply both sides by $(2x+1)(x-5)$ to get

$$1 = A(x-5) + B(2x+1) \Longleftrightarrow 1 = Ax - 5A + B2x + B$$
$$\Longleftrightarrow 1 = x(A+2B) - 5A + B$$

(iv) We then equate coefficients of x and the units to get the simultaneous equations:
(a) $A + 2B = 0$
(b) $-5A + B = 1$

(v) We solve the simultaneous equations to A and B:
$A = -\frac{2}{11}\left((i) - 2\cdot(ii)\right)$ and $B = \frac{1}{11}$ by substituting A into (i).
Thus:

$$\int \frac{1}{(2x+1)(x-5)}\,dx = -\int \frac{2}{11(2x+1)}\,dx + \int \frac{1}{11(x-5)}\,dx$$

Now we can use the method of substitution (see Section 8.2.1) to integrate this:

- $u = 2x+1$ and $du/dx = 2$ so $dx/du = 1/2$;
- $v = x-5$ so $dv/dx = 1$ and $dx/dv = 1$.

Therefore,

$$\int \frac{1}{(2x+1)(x-5)}\,dx = -\int \frac{2}{2\cdot 11\cdot u}\,du + \int \frac{1}{11\cdot v}\,dv$$
$$= -\frac{\ln|u|}{11} + \frac{\ln|v|}{11} + \kappa = -\frac{\ln|2x+1|}{11} + \frac{\ln|x-5|}{11} + \kappa$$

Example in Chemistry: HCl is introduced into a solution at a constant rate of 1 mole per minute for the first three minutes and then subsequently for $t \geq 3$ at a diminishing rate given by $r(t) = \dfrac{4}{(2t-4)(t-1)}$ in $\mathrm{mol\,min^{-1}}$. How many moles of HCl have been introduced into the solution after six minutes?

Solution: During the first three minutes, three moles were introduced into the solution because the rate was constant at 1 mole per minute.

To find the amount introduced during the next three minutes (three to six min), we can integrate the rate with respect to t:

$$\int_3^6 \frac{4}{(2t-4)(t-1)}dt$$

We follow the five steps outlined in Section 8.2.3 to decompose $r(t)$ into a sum of partial fractions:

(i) The denominator of $r(t)$ is already decomposed into factors.
(ii) We split the function $r(t)$ into the sum of partial functions:

$$\frac{4}{(2t-4)(t-1)} = \frac{A}{(2t-4)} + \frac{B}{(t-1)}$$

(iii) We now multiply through by the common denominator to get $4 = (t-1)A + (2t-4)B \iff 4 = t(A+2B) + (-A-4B)$ by collecting terms.

(iv) We then equate coefficients of x and the units to get the simultaneous equations:
   (a) $A + 2B = 0$
   (b) $-A - 4B = 4$

(v) Now we solve these simultaneous equations to determine the coefficients of partial fractions: $A = 4$ and $B = -2$. Thus,

$$r(t) = \frac{4}{(2t-4)} - \frac{2}{(t-1)}$$

These partial fractions are easy to integrate by using the substitution method:

$$\int_3^6 \frac{4}{(2t-4)(t-1)} dt = \int_3^6 \left(\frac{4}{(2t-4)} - \frac{2}{(t-1)}\right) dt$$

$$= [2\ln|2t-4| - 2\ln|t-1|]_3^6 = \left[2\ln\left|\frac{2t-4}{t-1}\right|\right]_3^6$$

$$= 2\left[\ln\left(\frac{8}{5}\right) - \ln\left(\frac{2}{2}\right)\right] = 2\left[\ln\left(\frac{8}{5}\right) - 0\right] \approx 0.94$$

This means that a total of 3.94 mol of HCl have been introduced into the solution during the first six minutes.

8.3 Improper Integrals

To compute the AUC using the fundamental theorem of calculus (see Section 8.1.1), the function $f(x)$ between the two limits x_1 and x_2 must satisfy two necessary conditions:

(1) $f(x)$ is a real-valued **continuous function** on the closed interval $[x_1, x_2]$;
(2) the **antiderivative** $F(x)$ of $f(x)$ exists in $[x_1, x_2]$.

Sometimes the interval between x_1 and x_2 where the area is required is **one-sided or even two-sided open** since the function $f(x)$ is not continuous at one or both of the limits.

Additionally, one or both limits of the interval might be **infinity** ($\pm\infty$).

In all these cases, the integral is called **improper**, and the fundamental theorem of calculus cannot be applied to it as it stands.

However, the calculation of an improper integral is possible if the following three conditions are met:

(1) $f(x)$ is a real-valued **continuous function** on the open interval (x_1, x_2);
(2) the **antiderivative** $F(x)$ of $f(x)$ exists in (x_1, x_2);
(3) both limits $\lim\limits_{x\to x_2} F(x)$ and $\lim\limits_{x\to x_1} F(x)$ are **finite real values** (both limits or one limit as required).

So

$$\int_{x_1}^{x_2} f(x)\,dx = [F(x)]_{x\to x_1}^{x\to x_2} = \lim_{x\to x_2} F(x) - \lim_{x\to x_1} F(x) \tag{8.11}$$

Example: Calculate the integral $\int_0^{+\infty} \frac{1}{(x+1)^3}\,dx$, if it exists.

Solution: The first two required conditions for us to calculate the improper integral are satisfied since $f(x) = \frac{1}{(x+1)^3}$ is a real-valued continuous function and its antiderivative is $F(x) = -\frac{1}{2(x+1)^2} + \kappa$.

For this example, only one limit is required:

$\lim\limits_{x\to+\infty}\left(-\frac{1}{2(x+1)^2}\right) = -\frac{1}{2(+\infty+1)^2} = 0$, which is a finite real value, and so it also meets the third condition.

Thus by using equation (8.11),

$$\int_0^{+\infty} \frac{1}{(x+1)^3}\,dx = \left[-\frac{1}{2(x+1)^2}\right]_0^{x\to+\infty} = 0 - \left(-\frac{1}{2}\right) = \frac{1}{2}$$

Graphically, we can colour the area given by the improper integral in a Cartesian plane:

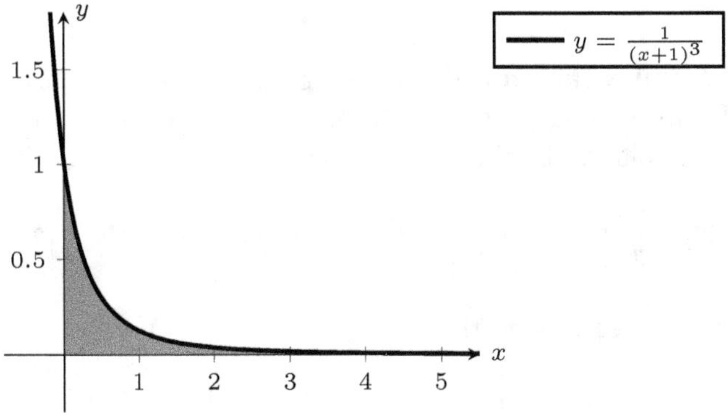

Special Example: Integrate $\int_0^{+\infty} x^3 e^{-x}\, dx.$

Solution: Since the first two conditions to calculate this improper integral are met, by applying integration by parts three times in succession where the f is the power function (initially $f = x^3$) and $g' = e^{-x}$, we obtain

$$\int_0^{+\infty} x^3 e^{-x}\, dx = \left[-e^{-x}x^3 - 3e^{-x}x^2 - 6e^{-x}x - 6e^{-x}\right]_0^{+\infty}$$

Then we calculate $\lim\limits_{x\to+\infty}(-e^{-x}x^3 - 3e^{-x}x^2 - 6e^{-x}x - 6e^{-x})$.

In general, since

$$\lim_{x\to+\infty} e^{-x} = 0 \text{ and } \lim_{x\to+\infty} x^n e^{-x} = \lim_{x\to+\infty}\frac{x^n}{e^x} = \frac{(\pm)\infty}{+\infty}$$

for all positive integer n values $\neq 0$, we can use L'Hôpital's rule as many times as necessary (usually $n-1$ times) to obtain $\lim\limits_{x\to+\infty}\frac{n!}{e^x} = 0$.

Thus, $\lim\limits_{x\to+\infty}(-e^{-x}x^3 - 3e^{-x}x^2 - 6e^{-x}x - 6e^{-x}) = 0$, which is a finite real value.

Finally, since $e^0 = 1$, we have:

$$[-e^{-x}x^3 - 3e^{-x}x^2 - 6e^{-x}x - 6e^{-x}]_0^{+\infty} = 0 - (-6) = 6$$

Important note: Sometimes a simpler version of the integral can be used to allow its easier reiteration for the calculation by the 'by parts' method. For instance for the integral $\int_0^{\infty} x^n e^{-x} dx$, we can let $I(n) = \int_0^{\infty} x^n e^{-x} dx$.

We have already seen that

$$I(0) = \int_0^{\infty} e^{-x} dx = [-e^{-x}]_0^{\infty} = 0 - (-1) = 1$$

Using integration by parts with $f = x^n$ and $g' = e^{-x}$ we obtain

$$\begin{aligned}\int_0^{\infty} x^n e^{-x} dx &= [x^n(-e^{-x})]_0^{\infty} - \int_0^{\infty} (-)e^{-x}\, n\, x^{n-1}\, dx \\ &= [x^n(-e^{-x})]_0^{\infty} + n \int_0^{\infty} x^{n-1} e^{-x} dx \\ &= 0 - (-0) + n \cdot I(n-1) = n \cdot I(n-1)\end{aligned}$$

Thus, $I(n) = n \cdot I(n-1)$ and iterating gives

$$\begin{aligned}I(n) &= n \cdot I(n-1) = n \cdot (n-1) \cdot I(n-2) \\ &= n \cdot (n-1) \cdot (n-2) \cdot I(n-3) = \cdots \\ &= n(n-1)(n-2)(n-3)\ldots 3 \cdot 2 \cdot 1 = n!\end{aligned}$$

Example in Biology: A piece of volcanic rock, when held close to a Geiger counter, emits radiation at an average rate which gives one 'click' per second.

For this situation, it can be shown that the probability P of the first click falling between times $t = a$ and $t = b$ is given by $P = \int_a^b e^{-t}\, dt$. What is the probability that we must wait at least three seconds for the first click?

Solution: The integral P can be evaluated using the substitution method:

$$\int_a^b e^{-t}\,dt = -\left[e^{-t}\right]_a^b = e^{-a} - e^{-b}$$

We can find the probability of the first click occurring at *some* time after a seconds (here $a = 3\text{s}$) by taking the limit of this expression as $b \to +\infty$:

$$\int_3^{+\infty} e^{-t}\,dt = \left[-e^{-t}\right]_3^{b\to+\infty} = e^{-3} - \lim_{b\to+\infty} e^{-b} = e^{-3} - 0 = e^{-3}$$

Example in Biochemistry (Part 3): In order to complete the calculation of the average speed $\overline{v}$ of a molecule m in a gas at temperature, T, as given by the Maxwell–Boltzmann equation, verify whether or not the initial improper integral $\overline{v} = 2Q\left(\frac{kT}{m}\right)^2 \int_0^{-\infty} ze^z\,dv$ gives a finite value.

Solution: The first two conditions required to enable us to calculate the improper integral are met since $f(z) = z\,e^z$ is a real-valued continuous function and its antiderivative is $F(z) = z\,e^z - e^z + \kappa$.

Next, we calculate $\lim_{z\to-\infty}(ze^z - e^z)$.

Since $\lim_{x\to+\infty} e^{-x} = 0$ and $\lim_{x\to+\infty} x^n e^{-x} = \lim_{x\to+\infty} \frac{x^n}{e^x} = \frac{(\pm)\infty}{+\infty}$ for all positive integer n values $\neq 0$, we can use L'Hôpital's rule as many times as necessary (usually $n-1$ times) to obtain $\lim_{x\to+\infty} \frac{n!}{e^x} = 0$. Calculation of these two last limits is required to verify that the condition (iii) is met for the whole improper integral.

Now we can calculate $\lim_{z\to-\infty}(z\,e^z - e^z) = 0 - e^{-\infty} = 0$.

Thus, by using equation (8.11),

$$\overline{v} = \int_0^{\infty} v\, f(v)\, dv = 2Q\left(\frac{kT}{m}\right)^2 \int_0^{-\infty} z\, e^z dz$$

$$= 2Q\left(\frac{kT}{m}\right)^2 \left\{ \left[z\, e^z\right]_0^{-\infty} - \left[e^z\right]_0^{-\infty} \right\}$$

$$= 2Q\left(\frac{kT}{m}\right)^2 \left\{ \left[\left(\lim_{z\to-\infty} z\cdot e^z\right) - 0\cdot e^0\right] - \left[\left(\lim_{z\to-\infty} e^z\right) - e^0\right] \right\}$$

$$= 2Q\left(\frac{kT}{m}\right)^2 \cdot \left\{ [0-0] - [0-1] \right\}$$

which gives $\overline{v} = 2Q\left(\frac{kT}{m}\right)^2 \cdot 1$.

Substituting Q back in and then collecting together all instances of each constant together in order to simplify the expression, we get:

$$\overline{v} = 2\cdot 4\pi \left(\frac{m}{2\pi kT}\right)^{\frac{3}{2}} \left(\frac{kT}{m}\right)^2 = 8\sqrt{\frac{kT}{\pi m}}$$

8.4 The Mean Value Theorem for Integrals

An expectation value is the average value of a distribution. It is obtained by summing the numbers obtained by multiplying the probability of a certain value by the value itself.

For instance, the probability of throwing any of the values on the die is the same and is one sixth. The average value of the throw of an unbiased six-sided die, $\overline{d}$, is

$$\overline{d} = \sum_{n=1}^{n=6} \frac{1}{6} n = \frac{1}{6}(1+2+3+4+5+6) = \frac{21}{6} = 3.5$$

This is an example which has discrete values, whereas for continuous distributions the summation sign can be replaced by an integral over lower and upper bounds of the variable being averaged.

The mean value theorem for integrals gives us a valid tool to find a way of calculating the mean value of a continuous function $f(x)$ over an interval $[a, b]$ rather than of a set of discrete (*i.e.* separate) numbers

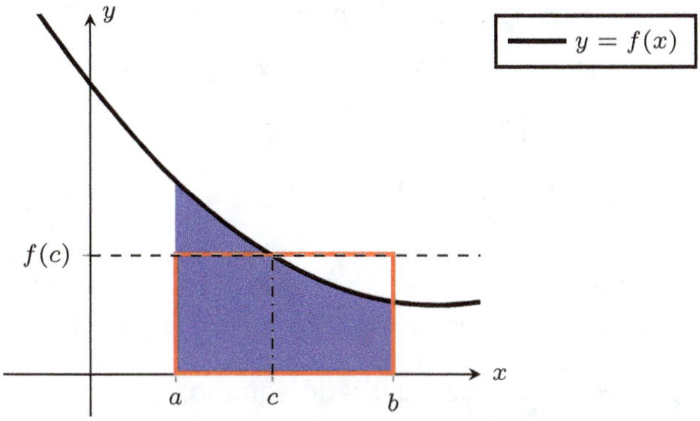

Figure 8.3: Mean value theorem for integrals

as we obtained for the arithmetic mean (or average value) discussed in Chapter 1 (see Section 1.4.1).

The **mean value theorem for integrals** states that, for a continuous function $f(x)$, there is at least one point c inside the interval $[a, b]$ at which

$$f(c) = \frac{\int_a^b f(x)\,dx}{b-a} \tag{8.12}$$

where $f(c)$ is the **mean value** of the function $f(x)$ in the interval $[a, b]$ and can also be written as $\overline{f}$ (Figure 8.3).

If we multiply both sides of equation (8.12) by $(b - a)$, we get

$$\int_a^b f(x)\,dx = f(c) \cdot (b-a)$$

This means that the area under the curve (blue area in Figure 8.3) of $f(x)$ in the interval $[a, b]$ equals the area of the rectangle $f(c) \cdot (b - a)$ (box outlined in red in Figure 8.3).

Example: Calculate the mean value $\overline{f}$ of $f(x) = 4x - x^2$ in the interval $[0, 4]$. Which x values within the interval $[0, 4]$ give $\overline{f}$?

Solution: In order to calculate the mean value of $f(x) = 4x - x^2$ in the interval $[0, 4]$, we use equation (8.12): $\overline{f} = f(c) = \dfrac{\int_0^4 (4x - x^2)\,dx}{4-0}$.

We start by calculating the definite integral:

$$\int_0^4 (4x - x^2)\, dx = \left[4 \cdot \frac{x^2}{2} - \frac{x^3}{3}\right]_0^4 = \left(4 \cdot \frac{4^2}{2} - \frac{4^3}{3}\right) - \left(4 \cdot \frac{0^2}{2} - \frac{0^3}{3}\right) = \frac{4^3}{6}$$

Thus, $\overline{f} = \dfrac{1}{4-0} \cdot \dfrac{4^3}{6} = \dfrac{4^2}{6} = \dfrac{8}{3}$.

Lastly, in order to determine the values of c that satisfy the mean value theorem for integrals of the function $f(x) = 4x - x^2$ on the interval $[0, 4]$, we solve the quadratic equation $4x - x^2 = \dfrac{8}{3}$, which gives two possible values:

$$c_1 = \frac{2\left(3 - \sqrt{3}\right)}{3} \approx 0.85 \text{ and } c_2 = \frac{2\left(3 + \sqrt{3}\right)}{3} \approx 3.15$$

We now plot the results. The region under the curve $f(x)$ over the interval $[0, 4]$ is coloured in blue, while the area of the rectangle $f(c_1) \cdot (4 - 0) = f(c_2) \cdot (4 - 0)$ is in the box outlined in red:

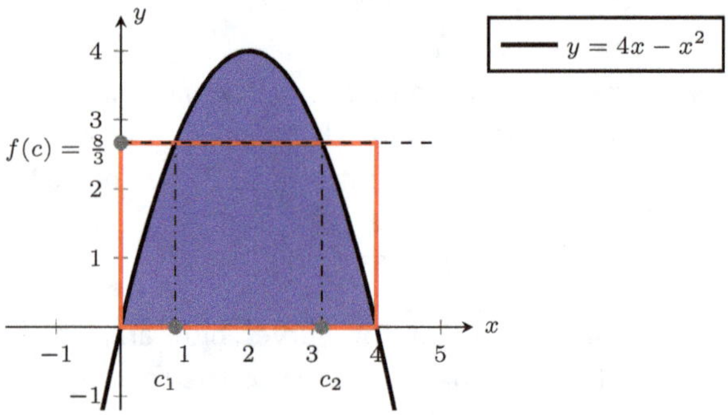

Example in Medicine: The time dependence of the action potential, ΔV, in an axon can be modelled by using the function

$$\Delta V(t) = (430t - 160t^2) \cdot e^{-t} - 70$$

with ΔV in mV and time, t, in ms. Find the mean of the action potential of an axon between 0 and $7\,\mathrm{ms}$.

Solution: In order to find the mean of the action potential ΔV of an axon over the time interval 0 and 7 ms, we use equation (8.12):

$$\overline{\Delta V} = \Delta V(t_c) = \frac{\int_0^7 \left(\left(430t - 160t^2\right) \cdot e^{-t} - 70\right)\, dt}{7 - 0}$$

Using Rule **B**, we can split the definite integral into the sum of two integrals:

$$\int_0^7 \left(\left(430t - 160t^2\right) \cdot e^{-t} - 70\right)\, dt = \int_0^7 \left(430t - 160t^2\right) \cdot e^{-t}\, dt - \int_0^7 70\, dt$$

We can straightforwardly calculate $\int_0^7 70\, dt = [70t]_0^7$, and we can solve the other integral $\int_0^7 (430t - 160t^2) \cdot e^{-t}\, dt$ by using integration by parts. Let $f = 430t - 160t^2 \Longrightarrow f' = 430 - 320t$ and $g' = e^{-t} \Longrightarrow g = -e^{-t}$.

Putting these into the integral, we have $\int_0^7 (430t - 160t^2)\, e^{-t}\, dt$ $= \left[-\left(430t - 160t^2\right) e^{-t}\right]_0^7 - (-1)\int_0^7 (430 - 320t)\, e^{-t}\, dt$, which needs another application of the by parts method.

Let $f = 430 - 320t \Longrightarrow f' = -320$ and $g' = e^{-t} \Longrightarrow g = -e^{-t}$. Putting these into the integral, we have

$$\begin{aligned}\int_0^7 (430 - 320t)\, e^{-t}\, dt &= \left[-(430 - 320t)\, e^{-t}\right]_0^7 - \int_0^7 -(-320)\, e^{-t}\, dt \\ &= \left[-(430 - 320t)\, e^{-t}\right]_0^7 - \left[-(320)\, e^{-t}\right]_0^7\end{aligned}$$

Summing all the results obtained so far and paying very careful attention to the signs, we get

$$\int_0^7 ((430t - 160t^2)\cdot e^{-t} - 70)\,dt$$
$$= [-(430t - 160t^2)\,e^{-t}]_0^7 + [-(430 - 320t)\,e^{-t}]_0^7 + [320e^{-t}]_0^7 - [70t]_0^7$$
$$= \left[-\left(430\cdot 7 - 160\cdot 7^2\right)e^{-7} - 0\right] + \left[-(430 - 320\cdot 7)\,e^{-7} + 430\right] + \left[320e^{-7} - 320\right] - [490 - 0]$$
$$= \frac{4830}{e^7} - 0 + \frac{1810}{e^7} + 430 + \frac{320}{e^7} - 320 - 490$$
$$= \frac{6960}{e^7} - 380 - 373.65\,\text{mV ms}$$

Thus, $\overline{\Delta V} \approx \dfrac{-373.65}{7} \approx -53.38\,\text{mV}$.

Let us plot the results obtained.

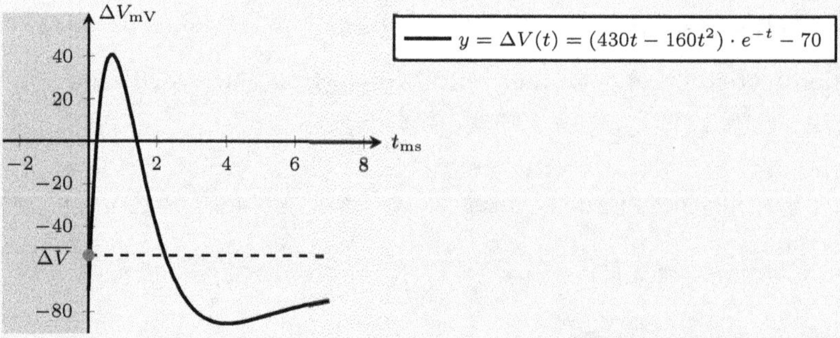

8.5 Exercises

(1) For each of the following functions, draw a graph and calculate the area under the curve between the specified points by dividing it into the given number of strips. Then find the definite integral of each function and compare the two answers.

(a) $y = x^2$ between $x = 2$ and $x = 4$, use four strips
(b) $y = x^4 + x^2$ between $x = -2$ and $x = 0$, use four strips
(c) $y = \sqrt{x}$ between $x = 5$ and $x = 6$, use two strips

(2) Find the following indefinite integrals, checking all your answers by differentiating them.

(a) $\displaystyle\int \left(7x^3 - x^2\right)dx$ (b) $\displaystyle\int \left(x^4 + 2x\right)dx$ (c) $\displaystyle\int \left(\frac{1}{2x^{\frac{1}{2}}}\right)dx$

(d) $\displaystyle\int \left(\frac{1}{x^2} + \frac{1}{x^3}\right)dx$ (e) $\displaystyle\int \left(8x^8 - 3x^{-2}\right)dx$ (f) $\displaystyle\int \left(4 - 5x^2\right)dx$

(3) A biochemical process produces an unwanted by-product X at the rate of $\frac{1}{3}\sqrt[4]{t}$ grams per minute.

(a) What is the rate of production of X six minutes into the process?

(b) How much by-product has unfortunately been produced by that time?

(4) The rate of change of the population of a colony of fungi $r(t)$ in a laboratory experiment can be expressed by the function $r(t) = 13t^2 - \frac{1}{2}t^3$, with t in days.

(a) When is the maximum population observed?

(b) What function, $P(t)$, describes the population size, if at $t = 0$ days there are 12 fungi?

(c) What is the total number of fungi after 20 days? How many are there after 26 days? And after 30 days?

(5) Find the following indefinite and definite integrals.

(a) $\int (5x + 2)\, dx$ (b) $\int \left(4x^3 - \frac{3}{x^3}\right) dx$ (c) $\int \left(3e^x - \frac{2}{x}\right) dx$

(d) $\int_1^2 \sqrt{x^3}\, dx$ (e) $\int_2^4 \left(x^3 - 4x^2\right) dx$ (f) $\int_1^4 \frac{7}{x^3}\, dx$

(g) $\int_0^{\frac{\pi}{2}} 3\sin(x)\, dx$ (h) $\int_1^e \frac{1}{3x}\, dx$ (i) $\int -\frac{\cos(x)}{3}\, dx$

(6) Integrate the following functions using the substitution method, and check your answers by differentiating them.

(a) $\int (3x + 4)^{100}\, dx$ (b) $\int \sin(2x - 3)\, dx$ (c) $\int \frac{1}{1 - 2x}\, dx$

(d) $\int (9 + x)^2\, dx$ (e) $\int x(45x^2 - 4)^2\, dx$ (f) $\int \frac{4}{1 + (3x)^2}\, dx$

(g) $\int_3^5 \frac{1}{\sqrt{(x-1)^3}}\, dx$ (h) $\int_1^2 \frac{1}{(9x - 2)^2}\, dx$ (i) $\int_0^{\frac{\pi}{2}} \frac{1}{\cos^2\left(\frac{x}{2}\right)}\, dx$

(7) Find the following integrals using the substitutions suggested.

(a) $\int x^2 e^{-4x^3}\, dx$ by substituting $g = -4x^3$

(b) $\int \frac{x^2}{7 - x^3}\, dx$ by substituting $g = 7 - x^3$

(c) $\int 2x\sqrt{1 + x^2}\, dx$ by substituting $g = 1 + x^2$

(d) $\int x^4(1+x^5)^3\,dx$ by substituting $g = 1 + x^5$

(e) $\int \frac{1}{1+\sqrt{x}}\,dx$* by substituting $g = \sqrt{x}$

* Hint: $\frac{t}{1+t} = 1 - \frac{1}{1+t}$

(8) Find the following definite and indefinite integrals, perhaps by using the two sub-rules derived from the substitution method. For indefinite integrals express your answer in terms of the original variable.

(a) $\int_3^6 \frac{1}{\sqrt{x-2}}\,dx$ (b) $\int_1^5 \frac{2x}{x^2+5}\,dx$ (c) $\int_0^{\sqrt{\frac{\pi}{2}}} x\cos(x^2)\,dx$

(d) $\int_2^4 2xe^{x^2}\,dx$ (e) $\int e^{ax}\,dx$ (f) $\int \frac{1}{1-5x}\,dx$

(g) $\int \sqrt{1+2x}\,dx$ (h) $\int \sin 3x\,dx$ (i) $\int \frac{e^{\sqrt{x}}}{\sqrt{x}}\,dx$

(j) $\int \frac{\cos(3x)}{2\sin 3x}\,dx$ (k) $\int x^2 e^{(3x^3-1)}\,dx$ (l) $\int \frac{\sin(x)}{2\cos x + 1}\,dx$

(m) $\int_{0.5}^1 \frac{2x+3}{x^2+3x+1}\,dx$ (n) $\int_{1.5}^2 \frac{x^4}{5x^5-8}\,dx$ (o) $\int_6^7 \frac{14x}{x^2+1}\,dx$

(p) $\int \frac{4x^3-3}{x^4-3x}\,dx$ (q) $\int \frac{4x+8}{x^2+4x-25}\,dx$

(9) The rate at which the world's oil is being consumed is continuously increasing. Suppose the rate, r (in billions of barrels per year), is given by the function $r = f(t)$, where t is measured in years and $t = 0$ is the start of 1995.

(a) Write down a definite integral which represents the total quantity of oil used between the start of 1995 and the beginning of 2001.

(b) Calculate this integral using the function $f(t) = 32e^{0.05t}$ to find the total number of billions of barrels of oil used during this time period.

(10) The flow of water (in litres per day) pumped upwards through the xylem of a tree, F, is given by the relationship $F = W\left(1 + \frac{q}{p}\cdot t\right)^{\frac{3}{4}}$ where t is the age of the tree in days, p and q are positive constants, and $W = M_0 p^{\frac{3}{4}}$ is the flow of pumped water in the tree at the time of planting, when $t = 0$.

If $p = 12$, $q = 0.012\,\text{day}^{-1}$, and $M_0 = 0.98\,\text{l}\,\text{day}^{-1}$, determine the total volume of water pumped up the tree in the tenth year.

(11) Use the method of integration by parts to calculate the following integrals.

(a) $\int xe^{3x}\,dx$ (b) $\int x\sin(5x)\,dx$

(c) $\int x^2\ln x\,dx$ (d) $\int (x^2+2x+1)e^{7x}\,dx$

(e) $\int x\cos 3x\,dx$ (f) $\int \frac{\ln x}{x^5}\,dx$

(g) $\int \cos(2x)\cdot 3x\,dx$ (h) $\int \ln(x)(3-8x^2)\,dx$

(i) $\int \frac{\ln x}{\sqrt{x}}$ (j) $\int_0^{\pi/2} \sin xe^{-2x}\,dx$

(k) $\int \cos(x)\sin(x)\,dx$ (l) $\int_{1/e}^{e} \ln(x^4)\,dx$

(m) $\int \sin^3 x\,dx$ * (n) $\int_0^{\frac{\pi}{2}} \cos^3 x\,dx$

(o) $\int_0^{\frac{\pi}{2}} \sin(2x)\,e^{-x}\,dx$ **

* Hint: $\sin^3(x)=\sin^2(x)\cdot\sin(x)$ and $\sin^2(x)+\cos^2(x)=1$.
** Hint: What is the sign of the integrand function $\sin(2x)\,e^{-x}$ in the interval $\left[-\frac{\pi}{2},\frac{\pi}{2}\right]$?

(12) During clinical trials, the patient's plasma drug concentration profile was plotted against time by measuring the plasma concentration C in mg L^{-1} at several time points t in h. The resulting curve was modelled by the equation $C(t)=\alpha t^2\cdot e^{3-\beta t}$, where α is a positive constant with units of $\text{mg L}^{-1}\,\text{h}^{-2}$ and β is another positive constant with units of h^{-1}.

(a) Estimate the area under the curve (AUC) expressed in mg h L^{-1} between time $t=0\,\text{h}$ and $t=12\,\text{h}$ with $\alpha=\beta=1$ in order to evaluate the actual exposure of the body to a drug after administration of a drug dose.
(b) At what time, namely (t_{max}), was the maximum plasma drug concentration observed?
(c) What is the value of AUC between $t=0\,\text{h}$ and twice the t_{max}? What about between twice the t_{max} and $t=12\,\text{h}$? What conclusion can be drawn by comparing these AUC values?

(13) Decompose these functions into partial fractions (Basic).

(a) $\frac{4}{(x+1)(x-3)}$ (b) $\frac{2x+12}{(x+4)(x+8)}$ (c) $\frac{-7x-49}{(x+6)(x-1)}$

(d) $\frac{1-6x}{(5-2x)(x+2)}$ (e) $\frac{1}{x^2+5x+4}$ (f) $\frac{2x-5}{x^2-3x-4}$

(14) Decompose these functions into partial fractions (Advanced).

(a) $\dfrac{x^2-2}{(x-1)(x^2+2)}$ (b) $\dfrac{x+1}{(x-4)^2}$ (c) $\dfrac{6x^2+7x+2}{(x^2-1)(x+2)}$

(d) $\dfrac{x+1}{(x-3)(x+5)}$ (e) $\dfrac{3x^2+7x-10}{(x+2)^2(x+8)}$ (f) $\dfrac{7x^2+31x-35}{(x-1)^2(x+8)}$

(g) $\dfrac{3x^2+3}{(x^2-1)^2}$ (h) $\dfrac{x^2+8x-14}{(2-2x^2)(x+4)}$ (i) $\dfrac{2x+1}{(x-1)^3}$

(15) Find these integrals by first using partial fractions to simplify them.

(a) $\displaystyle\int \frac{1}{(1+x)(2+3x)}\,dx$ (b) $\displaystyle\int \frac{6}{x^2-1}\,dx$

(c) $\displaystyle\int \frac{7x-3}{(3x+1)(x-5)}\,dx$ (d) $\displaystyle\int \frac{2x+3}{x^2-9}\,dx$

(e) $\displaystyle\int \frac{2-x}{x^2+5x}\,dx$ (f) $\displaystyle\int \frac{x^2-1}{x^2-16}\,dx$ *

(g) $\displaystyle\int \frac{x^2+x-1}{x(x^2-1)}\,dx$ (h) $\displaystyle\int \frac{x+7}{x^2(x+2)}\,dx$

(i) $\displaystyle\int \frac{x^2-x+1}{(x+1)^3}\,dx$ (j) $\displaystyle\int \frac{x^2+x+1}{x^3+x}\,dx$

(k) $\displaystyle\int \frac{x^3+4}{(x^2-1)(x^2+3x+2)}\,dx$

* Hint: $\dfrac{x^2-1}{x^2-16} = 1 + \dfrac{15}{x^2-16}$

(16) A reagent X is introduced into a solution at a constant rate of 0.5 moles per minute for the first four minutes, and then subsequently at a diminishing rate given by $r(t) = \dfrac{15t}{t^2+4t+3}$ in $\mathrm{mol\,min^{-1}}$. How many moles of X have been introduced into the solution after 10 minutes?

(17) Find these integrals by using the appropriate method.

(a) $\displaystyle\int_1^4 \frac{2x}{x^2+1}\,dx$ (b) $\displaystyle\int \frac{3x^2-1}{x^3-x+5}\,dx$

(c) $\displaystyle\int_0^{\pi} \cos^4 x\,dx$ (d) $\displaystyle\int_2^3 5e^{5x}\,dx$

(e) $\displaystyle\int (x+4)\ln x\,dx$ (f) $\displaystyle\int xe^{x^2}\,dx$

(g) $\displaystyle\int \sqrt{2x-3}\,dx$ (h) $\displaystyle\int (1+3e^{-3x})e^{-3x}\,dx$

(i) $\displaystyle\int \frac{\sqrt{x}+1}{x-1}\,dx$

(18) Calculate the following improper integrals.

(a) $\displaystyle\int_2^{+\infty} \left(\frac{1}{x^2} - \frac{1}{x^3}\right) dx$ (b) $\displaystyle\int_{-\infty}^{-4} \frac{1}{(x-2)^2}\, dx$

(c) $\displaystyle\int_0^{+\infty} e^{-3x}\, dx$ (d) $\displaystyle\int_{-\infty}^{0} -xe^{2x}\, dx$

(e) $\displaystyle\int_0^{\infty} x^3 e^{-x}\, dx$ (f) $\displaystyle\int_0^{\infty} x^{12} e^{-x}\, dx$

(19) Baranov developed expressions for commercial yields of caught fish in terms of lengths, L, of the fish. His formula gave the total number of fish of length L as $k \cdot e^{-cL}$, where c (units L^{-1}) and k are constants (k is positive).

(a) Make a sketch of the graph $f(L) = k \cdot e^{-cL}$. On your sketch, introduce marks on the horizontal axis that represent lengths $L = 1, L = 2, L = 3, L = 4$, and $L = 5$. Now shade the region on your sketch that represents the number of fish whose lengths are between $L = 3$ and $L = 4$.

(b) Explain how we can represent the total number of fish N as an area. Show that this number N equals k/c.

(c) Only fish longer than L_0 count as commercial. Hence, assuming that the fish are all similar in shape (*i.e.* their width and breadth scale with their length) and of equal density ρ, show that the mass, M, of the commercial fish population is $M = \displaystyle\int_{L_0}^{+\infty} a\, k\rho\, L^3\, e^{-cL}\, dL$, where a is a constant.

(d) Calculate the integral.

(20) Calculate the mean value of the following functions in the specified intervals.

(a) $f(x) = x\, e^{1-x}$, $[0, 6]$ (b) $f(x) = \dfrac{x^2}{x^3+1}$, $[0, 10]$

(c) $f(x) = \dfrac{1}{4-x^2}$, $[-1, 1]$

(21) The rate of growth per day at time t of an insect population in a laboratory experiment is given by $\dfrac{dp}{dt} = \alpha t \cdot e^{-\frac{t}{\beta}}$ with t in days, where α is a positive constant with units of day^{-2} and β is another positive constant with units of day.

(a) When (which day) is the maximum growth rate observed, if $\alpha = 25\ \text{day}^{-2}$ and $\beta = 5$ day?

(b) Determine the size of the population P_{10} after 10 days, if the initial population was 65 insects.

(c) Estimate the mean population $\bar{P}$ added per day to the initial population between the $t = 0\,\text{day}$ and the $t = 10\,\text{day}$.

Answers

(1) In order: area using $f(x)$ value of side of strip with lower x value (underestimate), area using $f(x)$ value of side of strip with upper x value (overestimate), and AUC from integration:
(a) $15.75, 21.70, 18.7$ (b) $4.81, 14.81, 9$ (c) $2.3, 2.4, 2.35$

(2) (a) $y = \frac{7x^4}{4} - \frac{x^3}{3} + \kappa$ (b) $y = \frac{x^5}{5} + x^2 + \kappa$ (c) $y = x^{\frac{1}{2}} + \kappa$
(d) $y = -\frac{1}{x} - \frac{1}{2x^2} + \kappa$ (e) $y = \frac{8x^9}{9} + \frac{3}{x} + \kappa$ (f) $y = 4x - \frac{5x^3}{3} + \kappa$

(3) (a) $\frac{1}{3}\sqrt[4]{6} = 0.52\,\mathrm{g\,min^{-1}}$ (b) $\int_0^6 \frac{1}{3}\sqrt[4]{x}dx = \frac{1}{3}\left[\frac{4}{5}x^{\frac{5}{4}}\right]_0^6 = \frac{8\sqrt[4]{6}}{5} = 2.5\,\mathrm{g\,min^{-1}}$

(4) (a) $r(t) = 0 \Longleftrightarrow t = 26$ (b) $P(t) = \frac{13t^3}{3} - \frac{t^4}{8} + 12$
(c) $P(20) = 14{,}678,\ P(26) = 19{,}053,\ P(30) = 15{,}762$

(5) (a) $y = \frac{5x^2}{2} + 2x + \kappa$ (b) $y = x^4 + \frac{3}{2x^2} + \kappa$ (c) $y = 3e^x - 2\ln|x| + \kappa$
(d) $\left[\frac{2}{5}x^{\frac{5}{2}}\right]_1^2 = \frac{8\sqrt{2}-2}{5}$ (e) $\left[\frac{x^4}{4} - \frac{4x^3}{3}\right]_2^4 = -\frac{44}{3}$ (f) $7\left[-\frac{1}{2x^2}\right]_1^4 = \frac{105}{32}$
(g) $3\left[-\cos(x)\right]_0^{\frac{\pi}{2}} = -3$ (h) $-\frac{1}{3}\left[\ln|x|\right]_1^e = -\frac{1}{3}$ (i) $y = -\frac{1}{3}\sin(x) + \kappa$

(6) (a) $y = \frac{1}{303}(3x+4)^{101} + \kappa$ (b) $y = -\frac{1}{2}\cos(2x-3) + \kappa$
(c) $y = -\frac{1}{2}\ln|1-2x| + \kappa$ (d) $y = \frac{(9+x)^3}{3} + \kappa$
(e) $\frac{(45x^2-4)^3}{270} + \kappa$ (f) $\frac{4}{3}\arctan(3x) + \kappa$
(g) $\left[-\frac{2}{\sqrt{x}}\right]_3^5 = -\frac{2}{\sqrt{5}} + \frac{2}{\sqrt{3}}$ (h) $\frac{1}{9}\left[-\frac{1}{9x-2}\right]_1^2 = \frac{1}{112}$
(i) $2\left[\tan\left(\frac{x}{2}\right)\right]_0^{\frac{\pi}{2}} = 2$

(7) (a) $-\frac{1}{12}e^{-4x^3} + \kappa$ (b) $-\frac{1}{3}\ln\left|7 - x^3\right| + \kappa$ (c) $\frac{2}{3}\left(1+x^2\right)^{\frac{3}{2}} + \kappa$
(d) $\frac{1}{20}\left(1+x^5\right)^4 + \kappa$ (e) $2\left(1 + \sqrt{x} - \ln\left|1+\sqrt{x}\right|\right) + \kappa$

(8) (a) $\left[2\sqrt{x-2}\right]_3^6 = 2$ (b) $\left[\ln\left|x^2+5\right|\right]_1^5 = \ln(5)$
(c) $\frac{1}{2}\left[\sin\left(x^2\right)\right]_0^{\sqrt{\frac{\pi}{2}}} = \frac{1}{2}$ (d) $\left[e^{x^2}\right]_2^4 = e^{16} - e^4$
(e) $\frac{1}{a}e^{ax} + \kappa$ (f) $-\frac{1}{5}\ln|1-5x| + \kappa$
(g) $\frac{1}{3}(1+2x)^{\frac{3}{2}} + \kappa$ (h) $-\frac{1}{3}\cos(3x) + \kappa$
(i) $2e^{\sqrt{x}} + \kappa$ (j) $\frac{1}{6}\ln|\sin(3x)| + \kappa$
(k) $\frac{1}{9}e^{3x^3-1} + \kappa$ (l) $-\frac{1}{2}\ln|2\cos(x)+1| + \kappa$
(m) $\left[\ln|2x+3|\right]_{0.5}^1 = 0.598$ (n) $\frac{1}{25}\left[\ln\left|5x^5-8\right|\right]_{1.5}^2 = 0.065$
(o) $7\left[\ln\left|x^2+1\right|\right]_6^7 = 7(\ln(50) - \ln(37))$ (p) $\ln\left|x^4 - 3x\right| + \kappa$
(q) $2\ln\left|x^2 + 4x - 25\right| + \kappa$

(9) (a) $\int_0^6 f(t)\,dt$ (b) $\int_0^6 32e^{0.05}tdt = \frac{32}{0.05}(e^{0.3} - 1) = 224$ billions of barrels

(10) $\int_{9\cdot365}^{10\cdot365} M_0\, p^{\frac{3}{4}} \left(1+\frac{q}{p}t\right)^{\frac{3}{4}} dt = M_0\, p^{\frac{3}{4}} \left[\frac{4\,(qt+p)^{\frac{7}{4}}}{7qp^{\frac{3}{4}}}\right]_{9\cdot365}^{10\cdot365} = 7100$ litres in tenth year of the tree's life

(11) (a) $\frac{1}{9}\left(e^{3x}\cdot 3x - e^{3x}\right)+\kappa$ (b) $-\frac{1}{5}x\cos(5x)+\frac{1}{25}\sin(5x)+\kappa$
(c) $\frac{1}{3}x^3\ln(x)-\frac{x^3}{9}+\kappa$ (d) $\frac{1}{343}\left(49e^{7x}x^2+84e^{7x}x+37e^{7x}\right)+\kappa$
(e) $\frac{1}{9}(3x\sin(3x)+\cos(3x))+\kappa$ (f) $-\frac{\ln(x)}{4x^4}-\frac{1}{16x^4}+\kappa$
(g) $\frac{3}{4}(2x\sin(2x)+\cos(2x))+\kappa$
(h) $3(x\ln(x)-x)-8\left(\frac{1}{3}x^3\ln(x)-\frac{x^3}{9}\right)+\kappa$
(i) $2\sqrt{x}\ln(x)-4\sqrt{x}+\kappa$
(j) $\left[-\frac{e^{-2x}\cos(x)}{5}-\frac{2e^{-2x}\sin(x)}{5}\right]_0^{\pi/2} = -\frac{2}{5e^{\pi}}+\frac{1}{5}$
(k) $\frac{\sin^2(x)}{2}+\kappa$ (l) $\left[x\ln\left(x^4\right)-4x\right]_{\frac{1}{e}}^{e}=\frac{8}{e}$
(m) $-\cos(x)+\frac{\cos^3(x)}{3}+\kappa$ (n) $\left[\sin(x)-\frac{\sin^3(x)}{3}\right]_0^{\frac{\pi}{2}}=\frac{2}{3}$
(o) $\left[-\frac{2e^{-x}\cos(2x)}{5}-\frac{e^{-x}\sin(2x)}{5}\right]_0^{\frac{\pi}{2}} = \frac{2}{5e^{\frac{\pi}{2}}}+\frac{2}{5}$

(12) (a) $\int_0^{12} t^2e^{3-t}dt = \left[-e^{3-t}t^2-2\left(e^{3-t}t+e^{3-t}\right)\right]_0^{12} = -\frac{170}{e^9}+2e^3$
$= 40.15\,\mathrm{mg\,h\,L^{-1}}$
(b) $t_{max}=2\,\mathrm{h}$
(c) $\int_0^4 t^2e^{3-t}dt = -\frac{26}{e}+2e^3 = 30.6\,\mathrm{mg\,h\,L^{-1}}$,
$\int_4^{12} t^2e^{3-t}dt = -\frac{170}{e^9}+\frac{26}{e} = 9.5\,\mathrm{mg\,h\,L^{-1}}$; thus the bioavailability of the drug in the plasma is much greater in the first four hours following drug administration than it is after that

(13) (a) $-\frac{1}{x+1}+\frac{1}{x-3}$ (b) $\frac{1}{x+4}+\frac{1}{x+8}$ (c) $\frac{1}{x+6}-\frac{8}{x-1}$
(d) $\frac{28}{9(2x-5)}+\frac{13}{9(x+2)}$ (e) $\frac{1}{3(x+1)}-\frac{1}{3(x+4)}$ (f) $\frac{7}{5(x+1)}+\frac{3}{5(x-4)}$

(14) (a) $-\frac{1}{3(x-1)}+\frac{4x+4}{3(x^2+2)}$ (b) $\frac{1}{x-4}+\frac{5}{(x-4)^2}$
(c) $-\frac{1}{2(x+1)}+\frac{5}{2(x-1)}+\frac{4}{x+2}$ (d) $\frac{1}{2(x-3)}+\frac{1}{2(x+5)}$
(e) $-\frac{1}{2(x+2)}-\frac{2}{(x+2)^2}+\frac{7}{2(x+8)}$ (f) $\frac{134}{27(x-1)}+\frac{1}{3(x-1)^2}+\frac{55}{27(x+8)}$
(g) $\frac{3}{2(x+1)^2}+\frac{3}{2(x-1)^2}$ (h) $-\frac{7}{4(x+1)}+\frac{1}{4(x-1)}+\frac{1}{x+4}$
(i) $\frac{2}{(x-1)^2}+\frac{3}{(x-1)^3}$

(15) (a) $-\ln|x+1|+\ln|3x+2|+\kappa$ (b) $-6\left(\frac{1}{2}\ln|x+1|-\frac{1}{2}\ln|x-1|\right)+\kappa$
(c) $\frac{1}{3}\ln|3x+1|+2\ln|5-x|+\kappa$ (d) $\frac{1}{2}\ln|x+3|+\frac{3}{2}\ln|x-3|+\kappa$
(e) $\frac{2}{5}\ln|x|-\frac{7}{5}\ln|x+5|+\kappa$ (f) $-\frac{15}{8}\left(\ln\left|\frac{x}{4}+1\right|-\ln\left|\frac{x}{4}-1\right|\right)+x+\kappa$
(g) $\frac{1}{2}\ln\left|x^2-1\right|-\ln|x+1|+\ln|x|+\kappa$

(h) $-\frac{5}{4}\ln|x| - \frac{7}{2x} + \frac{5}{4}\ln|x+2| + \kappa$
(i) $\ln|x+1| + \frac{3}{x+1} - \frac{3}{2(x+1)^2} + \kappa$
(j) $\arctan(x) + \ln(x) + \kappa$
(k) $-\frac{3}{4}\ln|x+1| + \frac{3}{2(x+1)} + \frac{5}{12}\ln|x-1| + \frac{4}{3}\ln|x+2| + \kappa$

(16) $\displaystyle\int_4^{10} \frac{15t}{t^2+4t+3}dt = 15\left[\frac{3}{2}\ln(t+3) - \frac{1}{2}\ln(t+1)\right]_4^{10} = 8.0\,\text{mol}$ of X between minutes 4 and 10; total moles of X introduced after 10 minutes is $8{+}0.5{\cdot}4 = 10$

(17) (a) $\left[\ln\left|x^2+1\right|\right]_1^4 = \ln\left(\frac{17}{2}\right)$ (b) $\ln\left|x^3-x+5\right| + \kappa$
(c) $\left[\frac{1}{32}\left(32\cos^3(x)\sin(x) + 3(4x - \sin(4x))\right)\right]_0^{\pi} = \frac{3\pi}{8}$
(d) $\frac{1}{5}\left[e^{5x}\right]_2^3 = \frac{e^{15}-e^{10}}{5}$
(e) $\ln(x)\left(\frac{x^2}{2}+4x\right) - \frac{1}{2}\left(\frac{x^2}{2}+8x\right) + \kappa$ (f) $\frac{1}{2}e^{x^2} + \kappa$
(g) $\frac{1}{3}(2x-3)^{\frac{3}{2}} + \kappa$ (h) $-\frac{1}{3}e^{-3x} - \frac{1}{2}e^{-6x} + \kappa$
(i) $2\left(\sqrt{x} + \ln\left|1-\sqrt{x}\right|\right) + \kappa$

(18) (a) $\left[-\frac{1}{x} + \frac{1}{2x^2}\right]_2^{+\infty} = \frac{3}{8}$ (b) $\left[-\frac{1}{x-2}\right]_{-\infty}^{-4} = \frac{1}{6}$
(c) $\left[-\frac{1}{3}e^{-3x}\right]_0^{+\infty} = \frac{1}{3}$ (d) $\left[-\frac{1}{4}\left(e^{2x}\cdot 2x - e^{2x}\right)\right]_{-\infty}^{0} = \frac{1}{4}$
(e) $\left[-e^{-x}x^3 - 3\left(e^{-x}x^2 - 2\left(-e^{-x}x - e^{-x}\right)\right)\right]_0^{\infty} = 6 = 3!$ (f) $12!$

(19) (a)

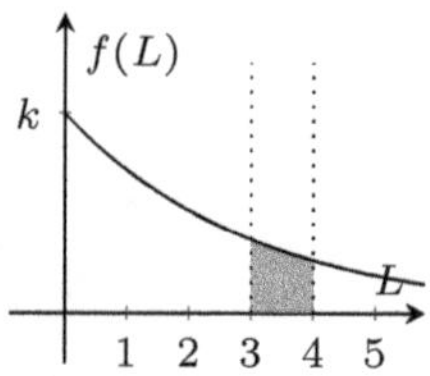

(b) $N = \displaystyle\int_0^{+\infty} k\,e^{-cL}\,dL = \left[-\frac{k\,e^{-cL}}{c}\right]_0^{+\infty} = \frac{k}{c}$
(c) Given $M = \rho{\cdot}V$ where M is mass of one fish, ρ the density of fish, $V = \alpha\,L^3$ the volume of fish, the total mass $M = \displaystyle\int_{L_0}^{+\infty} \alpha\,k\,\rho\,L^3\,e^{-cL}\,dL$
(d) $M = \frac{\alpha N\rho}{c^3}e^{-cL}\left[(cL_0)^3 + 3(cL_0)^2 + 6(cL_0) + 6\right]$ where $N = \frac{k}{c}$

(20) (a) $\frac{1}{6}\cdot\left[-e^{1-x}x - e^{1-x}\right]_0^6 = \frac{1}{6}\cdot\left(-\frac{7}{e^5} + e\right)$
(b) $\frac{1}{10}\cdot\frac{1}{3}\left[\ln\left|x^3\right|\right]_0^{10} = \frac{1}{30}\ln(1001)$
(c) $\frac{1}{2}\cdot\frac{1}{4}\left[\ln\left(\frac{x+2}{2-x}\right)\right]_{-1}^{1} = \frac{\ln(3)}{4}$

(21) (a) $\frac{d}{dt}\left(\frac{dp}{dt}\right) = 25e^{-\frac{t}{5}} - 5e^{-\frac{t}{5}}t = 0$ at $t = 5\,\text{days}$
(b) If $P_{10} = \displaystyle\int_0^{10} 25te^{-\frac{t}{5}}dt = 25\left[-5e^{-\frac{t}{5}}t - 25e^{-\frac{t}{5}}\right]_0^{10} = -\frac{1875}{e^2} + 625$ ≈ 371, then population after 10 days is $371 + 65 = 436$ insects
(c) $\bar{P} = \frac{P_{10}}{10} \approx 37$ insects are added per day to the original population

Chapter 9

Differential Equations

Preamble

Most of the mathematical formulae we have met so far have involved the determination of physical quantities that can be measured.

However, there are problems where we do not need to find a numerical value, but we want to define a suitable functional form to describe a particular physical process. Usually, we know several properties of the process: its derivatives, the link between the derivatives, or some other definite information, such as that the graph of the function passes through a specific point with coordinates (x, y).

Since derivatives are involved, the equations that describe these situations are called **differential** equations.

In many fields, including Biosciences, there is often a known or assumed relationship between some unknown quantity and its rate of change over time. These relationships involve the first derivative but sometimes also a derivative of higher order. The independent variable, t, representing time, is used, as many applications are based on rates of change with time.

In this chapter we focus on studying only these so-called 'first-order differential equations'.

9.1 Differential Equations

A **differential equation** is an equation which contains an unknown function, y, of a real variable x and that establishes a mathematical relationship between x, y and at least one of its derivatives with respect to x:

$$F(x, y, y', y'', \ldots, y^{(n)}) = 0$$

for instance, $-3y''' + 2y'' - \frac{x}{y} + e^x = 0$.

The **order** of a differential equation is the **maximum order** n **of the derivative** defining the unknown function: *i.e.* 3 in the example above.

A **first-order differential equation** has the form

$$F(x, y, y') = 0$$

where F is a function of three variables which are labelled x, y, and y', for example, $y' + \frac{y}{x} - x^2 = 0$, where y and y' are functions of x.

A **solution** of a first-order differential equation is a function $y = f(x)$ that satisfies the condition $F(x, f(x), f'(x)) = 0$ for every value of x.

The differential term y' will explicitly appear in the differential equation, although x and y may not. The variable y itself is dependent on x, so y' must be the derivative of y with respect to x.

Since only the first derivative of y appears, but no higher-order derivative, this is a **first-order differential equation**.

Often the y' can be expressed in terms of y and x. In this case, if y is a variable that depends on the variable x, then a **first-order differential equation in** x **and** y is one that relates $\frac{dy}{dx}$ to x, to y, or to both, and can also be written as

$$\frac{dy}{dx} = f(x, y) \tag{9.1}$$

If we are trying to solve the differential equation $\frac{dy}{dx} = f(x, y)$, we have found a **solution** y if its derivative is the same expression as in our original equation for $f(x, y)$.

There are usually many different solutions for the same differential equation.

The solution that models the situation of interest will be one whose graph passes through one or more points already known from theory or experiment. These specific requirements are called the **initial conditions**.

The **initial condition(s)** are a set of points that have to be satisfied by the solution (or by its derivatives). For a first-order differential equation involving a function $f(x)$, initial conditions can be of the form: $f(x_0) = f_0$ and/or $f'(x_0) = f_1$, for instance, if we are given that at $x = 0$, then $f(x) = 5$.

The first-order equation of the form given in equation (9.1) is too general, and there is no single method with which we can solve them all, or even the majority of them.

However, there are techniques for making progress on specific kinds of first-order differential equations which are relevant and useful in the Biosciences.

9.2 Separation of Variables Method

In order to find the **solutions** to a first-order differential equation, a versatile technique is the **separation of variables method**.

All the equations presented in the first part of this chapter can be solved by this method.

It can be summarised in the following four steps using the acronym **SISI**:

(a) **S**-separate;
(b) **I**-integrate;
(c) **S**-simplify;
(d) **I**-initial conditions (if they are specified).

Practically, our solution depends on rewriting the equation so that all instances of y are on one side of the equation and all instances of x are on the other, *i.e.* by 'separating the variables.'

9.2.1 Elementary differential equations

An elementary differential equation is of the form

$$\frac{dy}{dx} = f(x) \tag{9.2}$$

where $f(x)$ is a function of the independent variable x.

Note that there is nothing new here as this is a standard integration problem that you have already studied in Chapter 8. We can now use the separation of variables method:

(a) Separate the variables by integrating both sides of the equation over dx, so on the **left hand side (LHS)** we have simultaneously differentiated with respect to x and integrated over dx resulting in an integral over dy.

We can then rearrange the left-hand side so that all the terms in y are on one side and all the terms in x are on the other side of the equation:

$$\int \frac{dy}{dx}\, dx = \int f(x)\, dx \iff \int dy = \int f(x)\, dx$$

(b) Integrate both sides of the equation:

$$\int dy = \int f(x)\, dx \iff y + \kappa_1 = F(x) + \kappa_2$$

where $F(x)$ is the antiderivative of $f(x)$.

(c) Simplify the resulting antiderivatives and make y the subject of the equation:

$$y + \kappa_1 = F(x) + \kappa_2 \iff y = F(x) + \kappa$$

where $\kappa = \kappa_2 - \kappa_1$.

(d) Initial conditions (if they are specified) are used to determine κ and thus to find the particular solution.

Example: Solve $\dfrac{dy}{dx} = x^2 + 4$ and find the unique solution $y = f(x)$ which satisfies the boundary condition $f(3) = 0$.

Solution: This differential equation is of the elementary form (see equation (9.2)), so to solve it we can follow the four steps outlined above:

(a) **Separate:** $\displaystyle\int \frac{dy}{dx}\, dx = \int (x^2 + 4)\, dx \iff \int dy = \int (x^2 + 4)\, dx;$

(b) **Integrate:** $\displaystyle\int dy = \int (x^2 + 4)\, dx \iff y + \kappa_1 = \frac{x^3}{3} + 4x + \kappa_2;$

(c) **S**implify: $y + \kappa_1 = \dfrac{x^3}{3} + 4x + \kappa_2 \Longleftrightarrow y = \dfrac{x^3}{3} + 4x + \kappa$, where $\kappa = \kappa_2 - \kappa_1$;

(d) **I**nitial conditions: for $f(3) = 0$, we have:

$$0 = \frac{3^3}{3} + 4 \cdot 3 + \kappa \Longleftrightarrow \kappa = -21$$

Thus, the solution of the equation is $y = \dfrac{x^3}{3} + 4x - 21$.

Example in Biology: Most human skills, such as fitness and strength training, weight gain or loss, but also literacy and writing, grow quickly initially and then increase more slowly with time.

A function which is commonly used to model this rate of change is $\dfrac{dS}{dt} = \dfrac{A}{t+1}$, where S quantifies the level of skill, t is the time, and $A > 0$ relates to the strength of the dependence of the model on S.

Find an equation modelling skill acquisition over time, if at the beginning the skills are assessed as $S(0) = B$ which is > 0.

Lastly, sketch the function $S(t)$ that you obtain.

Solution: This differential equation is of the elementary form (see equation (9.2)), so we can solve it by following the four steps outlined above:

(a) **S**eparate: $\displaystyle\int \frac{dS}{dt}\,dt = \int \frac{A}{t+1}\,dt \Longleftrightarrow \int dS = \int \frac{A}{t+1}\,dt$;

(b) **I**ntegrate: $\displaystyle\int dS = \int \frac{A}{t+1}\,dt \Longleftrightarrow S + \kappa_1 = A \cdot \ln(t+1)^* + \kappa_2$

(*since t is always positive, we do not the modulus sign here);

(c) **S**implify: $S + \kappa_1 = A \cdot \ln(t+1) + \kappa_2 \Longleftrightarrow S = A \cdot \ln(t+1) + \kappa$ where $\kappa = \kappa_2 - \kappa_1$;

(d) **I**nitial conditions: at $t=0$, $S(0) = B$, therefore we have $B = A \cdot \ln(0+1) + \kappa \Longleftrightarrow B = 0 + \kappa \Longleftrightarrow \kappa = B$.

Thus, the solution of the equation is $S = A \cdot \ln(t+1) + B$.

We can now sketch the function $S(t)$ in a Cartesian plane in order to visualise the trend of S over time for different values of A.

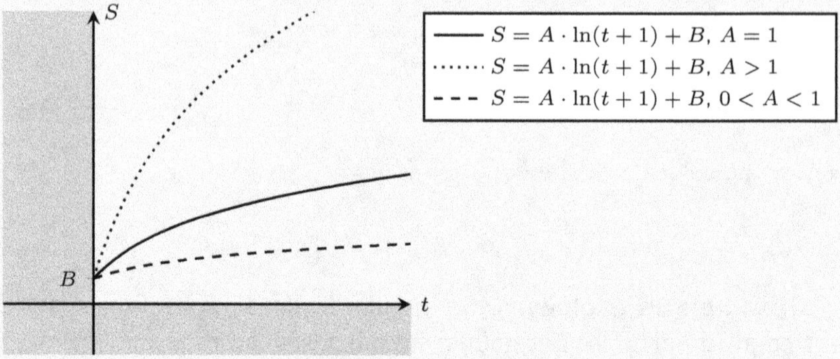

Special Example: Show that the complex number $z = \cos\theta + i\sin\theta$ can also be expressed as $z = e^{i\theta}$.

Solution: If we first treat $z = \cos\theta + i\sin\theta$ as being a function of θ, z can be differentiated with respect to θ:

$$(1) \quad \frac{dz}{d\theta} = \frac{d}{d\theta}\Big(\cos\theta + i\sin\theta\Big) = -\sin\theta + i\cos\theta$$

Now we can see that the right-hand side of this equation is just the product $i \cdot z$:

$$(2) \quad i \cdot z = i(\cos\theta + i\sin\theta) = i\cos\theta + i^2\sin\theta = -\sin\theta + i\cos\theta$$

Thus from (1) and (2), we have $\dfrac{dz}{d\theta} = iz$.

This implies that $z = \cos\theta + i\sin\theta$ satisfies the differential equation $\dfrac{dz}{d\theta} = iz$ which is separable, so

$$\int \frac{1}{iz}\frac{dz}{d\theta}d\theta = \int d\theta \Longleftrightarrow \int \frac{dz}{iz} = \int d\theta$$
$$\Longrightarrow \frac{1}{i}\ \ln z = \theta + c \Longleftrightarrow \ln z = i\theta + ic$$

Writing this in exponential form we get

$$z = e^{i\theta + ic} = e^{i\theta}\, e^{ic} = Ae^{i\theta}$$

where $A = e^{ic}$.

When $\theta = 0, z = 1$, which gives $A = 1$ so that:

$$\boxed{z = \cos\theta + i\sin\theta = e^{i\theta}}$$

9.2.2 Linear homogeneous differential equations

A linear homogeneous differential equation is of the form

$$\frac{dy}{dx} = f(x) \cdot y \tag{9.3}$$

where $f(x)$ is a function of the independent variable x and the exponent of y has to be equal to 1.

We can also use the separation of variables method on this form:

(a) **S**eparate the variables by integrating both sides over dx, so on the LHS we have simultaneously differentiated with respect to x and integrated over dx resulting in an integral over dy.

We can then rearrange it so that all the terms in y are on one side and all the terms in x are on the other side:

$$\int \frac{dy}{dx}\frac{1}{y}\,dx = \int f(x)\,dx \Longleftrightarrow \int \frac{dy}{y} = \int f(x)\,dx$$

(b) **I**ntegrate both sides of the equation:

$$\int \frac{dy}{y} = \int f(x)\,dx \Longleftrightarrow \ln|y| + \kappa_1 = F(x) + \kappa_2$$

where $F(x)$ is the antiderivative of $f(x)$.

(c) **S**implify the resulting antiderivatives and make y the subject of the equation:

$$\ln|y| + \kappa_1 = F(x) + \kappa_2 \Longleftrightarrow \ln|y| = F(x) + \kappa \Longleftrightarrow y = \pm e^{F(x)+\kappa}$$

where $\kappa = \kappa_2 - \kappa_1$.

(d) **I**nitial conditions (if they are specified) are now used to find the particular solution required.

Example: Solve $\dfrac{dy}{dx} = 5x^{\frac{3}{2}}y$ and find the unique solution $y = f(x)$ which satisfies the boundary condition $f(0) = 1$.

Solution: This differential equation is linear homogeneous, since the exponent of y is equal to 1 (see equation (9.3)), so we can solve it by following the four steps outlined above:

(a) **S**eparate: $\displaystyle\int \frac{dy}{dx}\frac{1}{y}\,dx = \int \left(5x^{\frac{3}{2}}\right)dx \iff \int \frac{dy}{y} = \int \left(5x^{\frac{3}{2}}\right)dx$;

(b) **I**ntegrate: $\displaystyle\int \frac{dy}{y} = \int \left(5x^{\frac{3}{2}}\right)dx \iff \ln|y| + \kappa_1 = \frac{5x^{\frac{5}{2}}}{\frac{5}{2}} + \kappa_2$;

(c) **S**implify: $\ln|y| = 2x^{\frac{5}{2}} + \kappa \iff y = \pm e^{2x^{\frac{5}{2}}+\kappa}$, where $\kappa = \kappa_2 - \kappa_1$, but the negative solution is forbidden by the boundary condition $f(0) = 1$;

(d) **I**nitial conditions: for $f(0) = 1$, we have $1 = e^{2\cdot 0^{\frac{5}{2}}+\kappa}$ $\iff 1 = e^{0+\kappa} \iff \kappa = 0$.

Thus, the solution of the linear homogeneous differential equation is $y = e^{2\cdot x^{\frac{5}{2}}}$.

Example in Medicine: When a drug is administered into body fluids, such as blood, its concentration diminishes over time due to elimination, destruction, or inactivation. Usually the rate of reduction of the concentration is found to be proportional to the concentration: $\dfrac{dC}{dt} = -\dfrac{C}{\beta}$, where β is a positive constant which determines the rate at which the concentration falls.

Solve the differential equation by assuming that at $t = 0$, $C = C_0$ and find out what happens when $t = \beta$.

Lastly, sketch the function $C(t)$ obtained.

Solution: This differential equation is linear homogeneous, since the exponent of y is equal to 1 (see equation (9.3)), so we can solve it by following the four steps outlined above:

(a) **S**eparate: $\displaystyle\int \frac{dC}{dt}\frac{1}{C}\,dt = -\int \frac{dt}{\beta} \Longleftrightarrow \int \frac{dC}{C} = -\int \frac{dt}{\beta}$;

(b) **I**ntegrate: $\displaystyle\int \frac{dC}{C} = -\int \frac{dt}{\beta} \Longleftrightarrow \ln|C| + \kappa_1 = -\frac{t}{\beta} + \kappa_2$;

(c) **S**implify: $\ln|C| = -\dfrac{t}{\beta} + \kappa \Longleftrightarrow C = +e^{-\frac{t}{\beta}+\kappa}$, where $\kappa = \kappa_2 - \kappa_1$, since the concentration, C, is a physical quantity that can only take positive values, so we reject the negative solution;

(d) **I**nitial conditions: for $t = 0$, we have $C = C_0$, so

$$C_0 = e^{0+\kappa} \Longleftrightarrow \ln(C_0) = \ln\left(e^{\kappa}\right) \Longleftrightarrow \kappa = \ln(C_0)$$

Thus, the solution of this linear homogeneous differential equation is

$$C = e^{-\frac{t}{\beta}+\ln(C_0)} \Longleftrightarrow C = e^{\ln(C_0)} \cdot e^{-\frac{t}{\beta}} \Longleftrightarrow C = C_0 \cdot e^{-\frac{t}{\beta}}$$

.

When $t = \beta$, the concentration has dropped to $\dfrac{C_0}{e}$ so that in time β the concentration reduces to $\dfrac{1}{e}$ of its initial value.

The time, β, is usually known as the 'lifetime' of the the process and is a significant parameter: the larger it is, the more slowly the drug disperses.

We can now sketch the function $C(t)$ in a Cartesian plane in order to visualise the trend of C over time for different values of β.

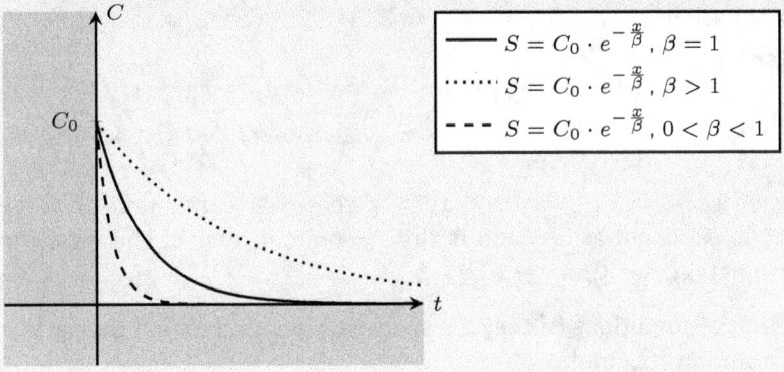

9.2.3 Non-linear differential equations

One common type of nonlinear differential equation is of the form

$$\frac{dy}{dx} = f(x) \cdot g(y) \tag{9.4}$$

where $f(x)$ is a function of the independent variable x, while $g(y)$ is a function of the dependent variable y.

We can again use the separation of variables method:

(a) Separate the variables by integrating both sides of the equation over dx, so on the LHS we have simultaneously differentiated with respect to x and integrated over dx resulting in an integral over dy.

We also rearrange it so that all the terms in y are on one side and all the terms in x are on the other side:

$$\int \frac{dy}{dx}\frac{1}{g(y)}\,dx = \int f(x)\,dx \Longleftrightarrow \int \frac{dy}{g(y)} = \int f(x)\,dx$$

(b) Integrate both sides of the equation:

$$\int \frac{dy}{g(y)} = \int f(x)\,dx \Longleftrightarrow G(y) + \kappa_1 = F(x) + \kappa_2$$

where $F(x)$ is the antiderivative of $f(x)$ and $G(x)$ is the antiderivative of the expression $\dfrac{1}{g(y)}$.

(c) Simplify the resulting antiderivatives and make y the subject of the equation:

$$\begin{aligned} G(y) + \kappa_1 = F(x) + \kappa_2 &\Longleftrightarrow G(y) = F(x) + \kappa \\ &\Longleftrightarrow y = G^{-1}\left(F(x) + \kappa\right) \end{aligned}$$

where $\kappa = \kappa_2 - \kappa_1$ and G^{-1} is the inverse function of G (e.g. the exponential function is the algebraic inverse of the logarithmic function: see Chapter 6).

(d) Initial conditions (if they are specified) are used to find the particular solution required.

Example: Solve $e^x \dfrac{dy}{dx} = \dfrac{x}{y}$ and find the unique solution $y = f(x)$ which satisfies the boundary condition $f(0) = 0$.

Solution: If we rearrange the equation $e^x \dfrac{dy}{dx} = \dfrac{x}{y} \iff \dfrac{dy}{dx} = \dfrac{x}{e^x \cdot y}$, we can recognise that it is a nonlinear differential equation (see equation (9.4)), so we can again follow the four steps outlined above:

(a) **S**eparate: $\displaystyle\int y \frac{dy}{dx}\, dx = \int \frac{x}{e^x}\, dx \iff \int y\, dy = \int x \cdot e^{-x}\, dx;$

(b) **I**ntegrate (by parts):

$$\int y\, dy = \int (x \cdot e^{-x})\, dx \iff \frac{y^2}{2} + \kappa_1 = -xe^{-x} + \int e^{-x}\, dx$$
$$\iff \frac{y^2}{2} + \kappa_1 = -xe^{-x} - e^{-x} + \kappa_2;$$

(c) **S**implify: $y^2 = -2e^{-x}(1 + x) + \kappa \iff y = \pm\sqrt{-2e^{-x}(1 + x) + \kappa}$ where $\kappa = \kappa_2 - \kappa_1$;

(d) **I**nitial conditions: for $f(0) = 0$, we have $0 = \pm\sqrt{-2e^{-0}(1 + 0) + \kappa}$ $\iff 0 = \pm\sqrt{-2 + \kappa} \iff \kappa = 2$.

Thus, the solution of this non-linear differential equation is

$$y = \pm\sqrt{-2e^{-x}(1 + x) + 2}$$

Example in Chemistry: According to Newton's law of cooling, the temperature T of a body changes at a rate proportional to the difference between the temperature of the body and the temperature of the surrounding medium T_m.

Determine the differential equation describing this law and find the equation for $T(t)$, with T_0 being the initial temperature of the body at $t = 0$. Then, sketch the function $T(t)$ obtained.

If a body is taken out of an oven at $T = 150\,^\circ\mathrm{C}$ and then left at a room temperature of $20\,^\circ\mathrm{C}$, it reaches $80\,^\circ\mathrm{C}$ after $10\,\mathrm{min}$. When will the body reach $30\,^\circ\mathrm{C}$?

Solution: Based on the definition above, the differential equation describing Newton's law is

$$\frac{dT}{dt} = -k\,[T - T_m]$$

where k is a positive constant and the minus sign indicates that the temperature of the body decreases with time if it is greater than the temperature of the surroundings, or increases if it is lower.

The differential equation can be recognised as a nonlinear differential equation (see equation (9.4)), so we can follow the four steps outlined above:

(a) **S**eparate: $\displaystyle\int \frac{dT}{dt}\frac{1}{[T - T_m]}\,dt = \int -k\,dt$

$$\iff \int \frac{dT}{[T - T_m]} = -k\int dt;$$

(b) **I**ntegrate (by substitution):

$$\int \frac{dT}{[T - T_m]} = -k\int dt \iff \ln\,[T - T_m]^* + \kappa_1 = -k\cdot t + \kappa_2$$

(*since $T > T_m$, there is no need to use the modulus signs here);

(c) **S**implify: $\ln\,[T - T_m] = -k\cdot t + \kappa \iff T - T_m = e^{-k\cdot t+\kappa}$

$\iff T = T_m + e^{-k\cdot t+\kappa}$, where $\kappa = \kappa_2 - \kappa_1$;

(d) **I**nitial conditions: at $t = 0$ we know that $T = T_0$, so

$$T_0 = T_m + e^{-k\cdot 0+\kappa} \iff T_0 - T_m = e^{\kappa} \iff \kappa = \ln\,(T_0 - T_m)$$

Thus, the solution of the equation is:

$$T = T_m + e^{-k\cdot t+\ln(T_0 - T_m)} \iff T = T_m + e^{\ln(T_0 - T_m)}\cdot e^{-k\cdot t}$$

$$\iff T = T_m + (T_0 - T_m)\,e^{-k\cdot t}$$

We can now sketch the function $T(t)$ in a Cartesian plane in order to visualise the trend of T over time.

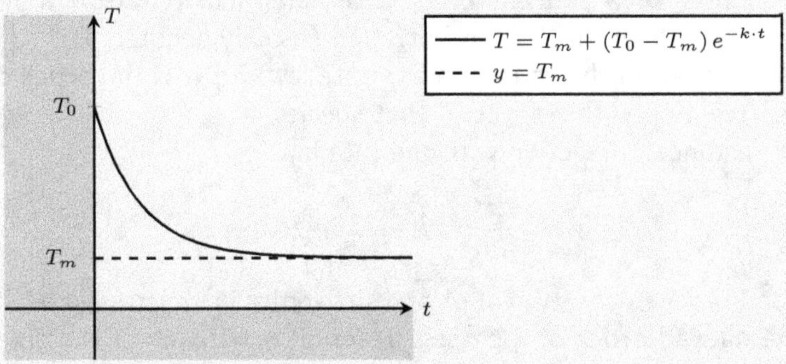

For the body taken out of the oven, Newton's law becomes

$$T = 20 + (150 - 20)\, e^{-k \cdot t}$$

The body reaches $80\,°\mathrm{C}$ after $10\,\mathrm{min}$, so we can use this information to calculate the constant k:

$$80 = 20 + (150 - 20)\, e^{-k \cdot 10} \iff 80 - 20 = (150 - 20)\, e^{-k \cdot 10}$$
$$\iff \frac{80 - 20}{150 - 20} = e^{-k \cdot 10} \iff \frac{6}{13} = e^{-k \cdot 10} \iff \ln\left(\frac{6}{13}\right) = -k \cdot 10$$
$$\iff k = -\frac{\ln\left(\frac{6}{13}\right)}{10} \approx 0.077\,\mathrm{min}^{-1} (*)$$

(*Note that the units of k must **always** be stated.)

To estimate when the body will reach $30\,°\mathrm{C}$, we can use our solution to Newton's law from above:

$$30 = 20 + (80 - 20)\, e^{-0.077 \cdot t} \iff \frac{30 - 20}{80 - 20} = e^{-0.077 \cdot t}$$
$$\iff \ln\left(\frac{1}{6}\right) = -0.077 \cdot t \iff t = \frac{\ln\left(\frac{1}{6}\right)}{-0.077}\,\mathrm{min} = 23.27\,\mathrm{min}$$

9.3 Biochemical Processes

In Biosciences, Medicine, and Chemistry we frequently need to describe the characteristics of a process or a reaction between two or more components as it evolves with time. We can categorise the various types of processes using the behaviour of the mathematical functions that describe them. A very useful concept that we will now introduce is the **order** of such processes.

The **order of a process or of a biochemical reaction** is defined as the power to which the concentration (usually denoted as [...]) of a species is raised in the corresponding rate law, *i.e.* it is the **order** of the process/reaction with respect to that species.

For instance, a reaction with the rate law

$$\frac{dP}{dt} = k \cdot [A]^2 \cdot [B]$$

is second-order in constituent A and first-order in constituent B.

The **overall order** of a process/reaction is the sum of the orders of all the components.

The rate law of the example above is thus third order overall.

9.3.1 Zeroth-order process

There are several processes that occur in nature for which the **rate of change (growth or decay)** of a species is constant.

The equation of such a rate of change can be written as

$$\frac{d[A]}{dt} = \pm k \qquad (9.5)$$

where $\boxed{+\mathbf{k}}$ indicates a constant rate of **growth**, while $\boxed{-\mathbf{k}}$ indicates a constant rate of **decay**.

This equation represents the change with time for a **zeroth-order process**.

Important note: In the following equations if the symbols $\pm$ are used, the *upper symbol* always relates to the *growth* process, while the *lower symbol* always relates to the *decay* process.

Since equation (9.5) is an elementary differential equation (see equation (9.2)), we can use the separation of variables method:

(a) **S**eparate: $\displaystyle\int \frac{d[A]}{dt}\,dt = \int \pm k\,dt \Longleftrightarrow \int d[A] = \pm k \int dt;$

(b) **I**ntegrate: $\displaystyle\int d[A] = \pm k \int dt \Longleftrightarrow [A_t] + c_1 = \pm kt + c_2;$

(c) **S**implify: $[A_t] = \pm kt + c$ where $c = c_2 - c_1$;

(d) **I**nitial conditions: A 'boundary condition' is that at

$$t = 0,\ [A_t] = [A_0]\text{: } [A_0] = \pm k \cdot 0 + c \Longleftrightarrow c = [A_0]$$

The general solution for a **zeroth-order process** is thus

$$[A_t] = \pm kt + [A_0] \tag{9.6}$$

and its plot in a Cartesian plane is shown in Figure 9.1.

If $t_0 \neq 0$, we obtain a more general expression (which reduces to equation (9.6) when $t = 0$):

$$[A_t] = \pm k(t - t_0) + [A_0]$$

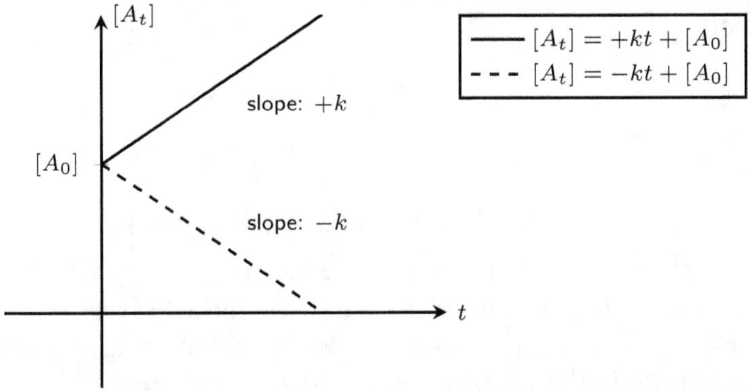

Figure 9.1: Zeroth-order process

Example in Biochemistry (Part 1): The Michaelis–Menten equation relates the rate of an enzyme reaction (substrate consumption) to its substrate concentration and has the form (if there is a large quantity of enzyme):

$$-\frac{ds}{dt} = \frac{V.s}{K + s}$$

where $s = [S]$ is the substrate concentration, and V and K are constants (for simplicity s is used as a short hand for $[S]$ in the text and formulae).

By integration, derive the mathematical expression relating the substrate concentration to the time when $s \gg K$ and sketch a graph of s *versus* t.

Lastly, estimate the time taken for half of the substrate to disappear if $V = 10\,\mu\text{mol dm}^{-3}\,\text{min}^{-1}$, $K = 100\,\mu\text{mol dm}^{-3}$, and the initial substrate concentration is s_0 (μmol dm^{-3}).

Solution: When $s \gg K$, the K term in the denominator of the differential equation can be neglected: thus, $-\dfrac{ds}{dt} = \dfrac{V.s}{s} = V$.

The rate of change in concentration is a constant, so this is a *zeroth-order process.*

The Michaelis–Menten equation then becomes an elementary differential equation and we can use the separation of variables method:

(a) **S**eparate: $\dfrac{ds}{dt} = -V \Longleftrightarrow \displaystyle\int \frac{ds}{dt}\,dt = \int -V\,dt$
$\Longleftrightarrow \displaystyle\int ds = -V \int dt;$

(b) **I**ntegrate: $\displaystyle\int ds = -V \int dt \Longleftrightarrow s + c_1 = -Vt + c_2;$

(c) **S**implify: $s = -Vt + c$ where $c = c_2 - c_1$;

(d) **I**nitial conditions: a usual 'boundary condition' is that at $t = 0$, $s = s_0$ (μmol dm^{-3}) where s_0 is the initial concentration of substrate in the assay: $s_0 = -V \cdot 0 + c \Longleftrightarrow c = s_0$.

The general solution for the case $s \gg K$ is thus approximately $s = -Vt + s_0$.

This equation is of the form $y = mx + c$ and so is a straight line with a slope of $-V$ and s crosses the y-axis at $y = s_0$, as shown in the following figure.

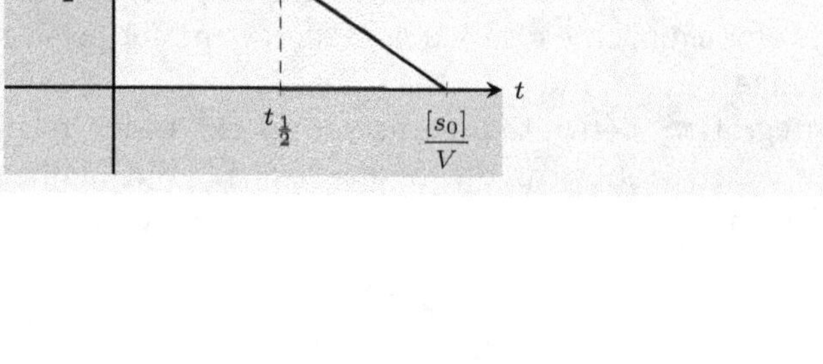

We can now estimate the time taken for half of the substrate to disappear: $\int_{s_0}^{\frac{s_0}{2}} ds = -\int_0^{t_{\frac{1}{2}}} V\,dt \Longleftrightarrow \Big[s\Big]_{s_0}^{\frac{s_0}{2}} = -\Big[Vt\Big]_0^{t_{\frac{1}{2}}}$
$\Longleftrightarrow \frac{s_0}{2} - s_0 = -Vt_{\frac{1}{2}} \Longleftrightarrow \frac{s_0}{2V} = t_{\frac{1}{2}}$.
We know that $V = 10\,\mu\text{mol}\,\text{dm}^{-3}\,\text{min}^{-1}$, so $t_{\frac{1}{2}} = \frac{s_0}{20}\,\text{min}$.

Important note: This time is **dependent** on the **initial concentration** of the substrate s_0.

9.3.2 First-order process

In other natural processes, the **rate of change (growth or decay)** of the concentration of a compound A increases or decreases as the concentration of A increases or decreases.

The equation of such a change can be written as

$$\frac{d[A]}{dt} = \pm k \cdot [A] \tag{9.7}$$

where $\boxed{+\mathbf{k}}$ indicates **growth**, while $\boxed{-\mathbf{k}}$ indicates **decay**.

This equation represents the change for a **first-order process**.

Important note: In the following equations if the symbols $\pm$ are used, the *upper symbol* always relates to the *growth* process, while the *lower symbol* always to the *decay* process.

Since equation (9.7) is an elementary differential equation (see equation (9.2)), we can use the separation of variables method:

(a) **S**eparate: $\int \frac{d[A]}{dt}\frac{1}{[A]}\,dt = \int \pm k\,dt \Longleftrightarrow \int \frac{d[A]}{[A]} = \pm k \int dt$;

(b) **I**ntegrate: $\int \frac{d[A]}{[A]} = \pm k \int dt \Longleftrightarrow \ln[A_t] + c_1 = \pm kt + c_2$;

(c) **S**implify: $\ln \mid [A_t] \mid = \pm kt + c \Longleftrightarrow [A_t] = e^{\pm kt+c}$ where $c = c_2 - c_1$, where the modulus is not necessary for $[A_t]$, since the variable $[A]$ is a physical entity which is always positive;

(d) **I**nitial conditions: A usual 'boundary condition' is that at

$$t = 0,\ [A_t] = [A_0]\colon\ [A_0] = e^{\pm k\cdot 0+c} \Longleftrightarrow c = \ln[A_0]$$

We can rearrange the expression:

$$[A_t] = e^{\pm kt+\ln[A_0]} = e^{\ln[A_0]} \cdot e^{\pm kt} = [A_0] \cdot e^{\pm kt}$$

Thus, the general solution for a **first-order process** is

$$[A_t] = [A_0] \cdot e^{\pm kt} \tag{9.8}$$

and its plot in a Cartesian plane is shown in Figure 9.2. If $t_0 \neq 0$, we obtain a more general expression of equation (9.8) (which reduces to 9.8. when $t_0 = 0$):

$$[A_t] = [A_0] \cdot e^{\pm k(t-t_0)}$$

If we take the natural logarithm of both sides of equation (9.8), we have

$$\ln[A_t] = \ln[A_0] \pm k(t)$$

which is of the form $y = mx + c$ and so is a straight line with a slope of $\pm k$ and crosses the y-axis at $y = \ln[A_0]$, as shown in Figure 9.2.

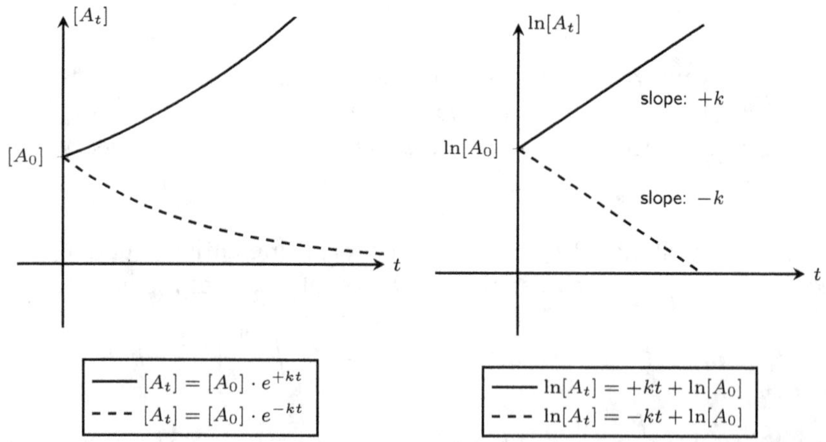

Figure 9.2: First-order process (left) and its natural logarithm transposition (right)

Example in Biochemistry (Part 2): By integrating the Michaelis–Menten equation $-\frac{ds}{dt} = \frac{V.s}{K+s}$, where $s = [S]$ is the substrate concentration, and V and K are constants, derive an expression relating substrate concentration to time when $s \ll K$ and sketch a graph of s *versus* t.

If $V = 10\,\mu\text{mol}\,\text{dm}^{-3}\,\text{min}^{-1}$ and $K = 100\,\mu\text{mol}\,\text{dm}^{-3}$, estimate the time taken for half of the substrate to disappear, if the initial substrate concentration is s_0 ($\mu\text{mol}\,\text{dm}^{-3}$).

Solution: When $s \ll K$, the s term in the denominator of the differential equation can be neglected, thus $-\frac{ds}{dt} = \frac{V.s}{K+s} = \frac{V \cdot s}{K}$. The rate of change in concentration depends on the concentration itself, so this is a *first-order process.*

Thus, the Michaelis–Menten equation becomes a linear homogeneous differential equation and we can use the separation of variables method:

(a) **S**eparate:

$$-\frac{ds}{dt} = \frac{V \cdot s}{K} \Longleftrightarrow -\frac{K}{V}\int \frac{ds}{dt}\frac{1}{s}\,dt = \int dt$$
$$\Longleftrightarrow -\frac{K}{V}\int \frac{ds}{s} = \int dt$$

(b) **I**ntegrate: $-\frac{K}{V}\int \frac{ds}{s} = \int dt \Longleftrightarrow -\frac{K}{V}\ln(s) + c_1 = t + c_2$, where the modulus is not necessary for the logarithm term, $\ln$, since the variable s is a physical entity which is always positive;

(c) **S**implify: $-\frac{K}{V}\ln(s) + c_1 = t + c_2 \Longleftrightarrow \ln(s) = -\frac{V \cdot t}{K} + c$, where $c = (c_2 - c_1)\cdot\left(-\frac{V}{K}\right)$; then make s the subject of the equation: $s = e^{-\frac{V \cdot t}{K}+c}$;

(d) **I**nitial conditions: a usual 'boundary condition' is that at $t = 0$, $s = s_0$ ($\mu\text{mol}\,\text{dm}^{-3}$), where s_0 is the initial concentration of substrate in the assay:

$$s_0 = e^{-\frac{V \cdot 0}{K}+c} \Longleftrightarrow s_0 = e^c \Longleftrightarrow c = \ln(s_0)$$

Thus, the general solution is $s = e^{-\frac{V \cdot t}{K} + \ln(s_0)}$, which can be arranged as follows: $s = e^{-\frac{V \cdot t}{K}} \cdot e^{\ln(s_0)} \iff s = s_0 \cdot e^{-\frac{V \cdot t}{K}}$. So, s crosses the y-axis at $y = s_0$ and decreases exponentially with time, as shown in the following figure.

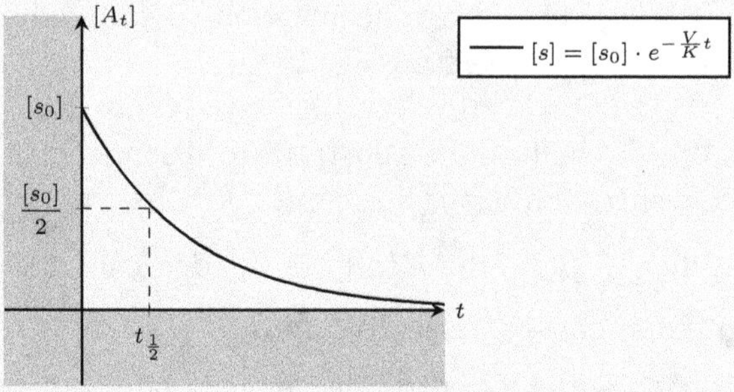

We can now estimate the time taken for half of the substrate to disappear if $V = 10\,\mu\text{mol}\,\text{dm}^{-3}\,\text{min}^{-1}$, $K = 100\,\mu\text{mol}\,\text{dm}^{-3}$, and the initial substrate concentration is s_0 (μmol dm^{-3}):

$$\int_{s_0}^{\frac{s_0}{2}} -\frac{K\,ds}{V \cdot s} = \int_0^{t_{\frac{1}{2}}} dt \iff -\left[\frac{K}{V}\ln s\right]_{s_0}^{\frac{s_0}{2}} = t_{\frac{1}{2}}$$

$$\iff -\frac{K}{V}\ln\left(\frac{s_0}{2}\right) + \frac{K}{V}\ln(s_0) = \frac{K}{V}\ln\left(\frac{s_0}{\frac{s_0}{2}}\right) = \frac{K}{V}\ln 2 = t_{\frac{1}{2}}$$

$$\iff t_{\frac{1}{2}} = \frac{K}{V}\ln(2)$$

Now, $V = 10\,\mu\text{mol}\,\text{dm}^{-3}\,\text{min}^{-1}$ and $K = 100\,\mu\text{mol}\,\text{dm}^{-3}$, so $t_{1/2} = 10 \cdot \ln(2) = 10 \cdot 0.693 = 6.93\,\text{min}$.

Important note: This time is **independent** of the **initial concentration** of the substrate s_0. Such behaviour is **different** from that of a zeroth-order process.

Example in Biology: In a first-order growth process of bacterial growth, if there are N_0 bacteria at $t = 0\,\text{h}$ and $3N_0$ at $t = 10\,\text{h}$, when will there be $11N_0$?

Solution: For this problem, we can use equation (9.7):

$$[A_t] = [A_0] \cdot e^{+kt} \Longleftrightarrow 3N_0 = N_0 \cdot e^{+k\cdot 10} \Longleftrightarrow 3 = e^{+k\cdot 10}$$

We take the natural logarithm of both sides of the equation:

$$\ln(3) = k \cdot 10 \Longleftrightarrow k = \frac{\ln(3)}{10} \approx 0.11\,\mathrm{h}^{-1}$$

So using equation (9.7), $11N_0 = N_0 \cdot e^{+0.11\,t} \Longleftrightarrow 11 = e^{+0.11t}$ $\Longleftrightarrow \ln(11) = 0.11 \cdot t \Longleftrightarrow t = \dfrac{\ln(11)}{0.11} = 21.8\,\mathrm{h}$ by when there will be $11N_0$.

Example in Chemistry: For the radioactive decay of the nucleus of an atom, the rate of change of the number of radioactive atoms over time is proportional to the number of atoms left. The decay is thus a first-order process.

Find the differential equation describing this phenomenon and then obtain the general equation for the radioactive decay process, with N_t being the number of atoms at time t, N_0 the number of atoms at $t = t_0$, and λ the 'decay constant'.

Estimate the time taken for $[N]$ to fall to $\dfrac{[N]}{2}$ (*i.e.* the number of radioactive atoms is halved).

The element radium (mass = 226) has a decay constant (λ) of $13.6 \cdot 10^{-12}\,\mathrm{s}^{-1}$. How many disintegrations per second will $1\,\mathrm{g}$ of radium produce?

Solution: Since the rate of change is proportional to the amount left, we can use equations (9.7) and (9.8) to define the differential equation for the process, and use the general equation for the radioactive decay:

$$\frac{d[N]}{dt} = -\lambda[N] \Longleftrightarrow N_t = N_0\; e^{-\lambda(t-t_0)}$$

with the terms defined as above.

By taking the natural logarithm of the general decay process equation and using the properties of logarithms, we get $\ln[N] = -\lambda t + \ln[N_0]$.

By comparing this equation with the equation of a straight line $y = mx + c$, we get a straight line plot of $\ln[N]$ *versus* t, with slope $-\lambda$ and intercept with the y-axis of $\ln[N_0]$.

In order to calculate the time taken for $[N]$ to fall to $\frac{[N]}{2}$, we put $[N] = \frac{[N_0]}{2}$ and $t = t_0 + t_{\frac{1}{2}}$ into the general expression derived from equation (9.8).

Thus, $\frac{[N_0]}{2} = [N_0]\, e^{-\lambda(t_0 + t_{\frac{1}{2}} - t_0)} \iff \frac{1}{2} = e^{-\lambda t_{\frac{1}{2}}}$

$$\iff \ln\left(\frac{1}{2}\right) = -\lambda t_{\frac{1}{2}}.$$

Since $-\ln\left(\frac{1}{2}\right) = -[\ln 1 - \ln 2] = \ln 2$, we have

$$\ln 2 = \lambda t_{\frac{1}{2}} \iff \boxed{t_{\frac{1}{2}} = \frac{\ln 2}{\lambda} = \frac{0.693}{\lambda}}$$

Important note: This time is **independent** of the **concentration** of $[N]$. This is a fundamental property of a **first-order process**.

Finally, in order to calculate how many disintegrations per second 1 g of radium will produce, we should first find the number of atoms in a gram of radium:

1 g of radium contains $\frac{6.02 \cdot 10^{23}}{226} = 2.66 \cdot 10^{21}$ atoms

We can now calculate the number of disintegrations per second:

$$\frac{d[N]}{dt} = -\lambda[N] = 13.6 \cdot 10^{-12} \cdot 2.66 \cdot 10^{21} = 3.6 \cdot 10^{10} \text{ disintegrations } s^{-1}$$

9.3.3 Second-order process

In **second-order processes**, the rate of change depends on the concentrations of two components.

The reaction $A + B \longrightarrow X$ follows the rate law:

$$\frac{dX}{dt} = k_2 \cdot [A] \cdot [B] \tag{9.9}$$

This is first-order in A and first-order in B, but since the overall order of a reaction is the sum of the orders of all the components, it is **second-order** overall.

If the initial concentrations are $[A_0]$ and $[B_0]$, and after time t there are X moles per litre of product X formed, then

$$\frac{dX}{dt} = k_2 \cdot [A] \cdot [B] = k_2 \cdot (\,[A_0] - X) \cdot (\,[B_0] - X)$$

We can then rearrange and integrate this by using the separation of variables method:

$$\int \frac{dX}{dt} \frac{1}{(\,[A_0] - X) \cdot (\,[B_0] - X)}\, dt = \int k_2\, dt \tag{9.10}$$

To decompose the rational function into a sum of partial fractions, we first use the five steps described in Section 8.2.3:

(i) The denominator is composed of two factors;

(ii) Thus we can split the function into a sum of partial fractions with unknown coefficients D and E:

$$\frac{1}{(\,[A_0] - X) \cdot (\,[B_0] - X)} = \frac{D}{(\,[A_0] - X)} + \frac{E}{(\,[B_0] - X)}$$

(iii) We multiply both sides of the resulting equation by the denominator $(\,[A_0] - X) \cdot (\,[B_0] - X)$ and rearrange the equation:

$$1 = D([B_0]-X)+E([A_0]-X) \Longleftrightarrow 1 = D[B_0]+E[A_0]-X(D+E)$$

(iv) We now equate coefficients of x and of the units to get the simultaneous equations:

(a) $-D - E = 0$
(b) $D[B_0] + E[A_0] = 1$

(v) We then solve the simultaneous equations to determine the partial fraction coefficients: since $E = -D$ from (a), in (b)

$$D[B_0] - D[A_0] = 1 \Longleftrightarrow D = \frac{1}{[B_0] - [A_0]}$$

Thus, we obtain

$$\frac{1}{([A_0] - X)([B_0] - X)} = \frac{1}{([B_0] - [A_0])}\left(\frac{1}{[A_0] - X} - \frac{1}{[B_0] - X}\right) = \frac{1}{[Q_0]}\left(\frac{1}{[A_0] - X} - \frac{1}{[B_0] - X}\right)$$

where, for simplicity, $[Q_0] = [B_0] - [A_0]$.

The integral to be solved is thus:

$$\frac{1}{[Q_0]}\int \frac{dX}{([A_0] - X)} - \frac{1}{[Q_0]}\int \frac{dX}{([B_0] - X)} = \int k_2\, dt$$

and by using some of the properties of integrals (see Chapter 8), we get

$$\frac{1}{[Q_0]}[-\ln([A_0] - X) + \ln([B_0] - X)] = k_2 t + b \tag{9.11}$$

where b is an unknown constant of integration and moduli in the natural logarithms are not needed as their arguments are always positive. Since we have the boundary condition (or initial condition) that at $t = 0$, $X = 0$, we can determine b:

$$b = \frac{1}{[Q_0]}[-\ln[A_0] + \ln[B_0]] = \frac{1}{[Q_0]}\ln\left[\frac{[B_0]}{[A_0]}\right]$$

Putting this value of b into equation (9.11),

$$\begin{aligned} k_2 t &= \frac{1}{[Q_0]}\ln\left[\frac{([B_0] - X)}{([A_0] - X)}\right] - \frac{1}{[Q_0]}\ln\left[\frac{[B_0]}{[A_0]}\right] \\ &= \frac{1}{[Q_0]}\ln\left[\frac{[A_0]([B_0] - X)}{[B_0]([A_0] - X)}\right] \end{aligned}$$

This last equation gives us the **time dependence of a second-order process**.

Note that it depends on the initial concentrations of A and B.

Comparing this with the familiar equation $y = mx + c$, we can see that a plot of $\ln\left[\dfrac{[A_0]([B_0]-X)}{[B_0]([A_0]-X)}\right]$ *versus* t will be a straight line (Figure 9.3), with slope $k_2[Q_0]$, where $[Q_0] = [B_0] - [A_0]$, and intercept 0.

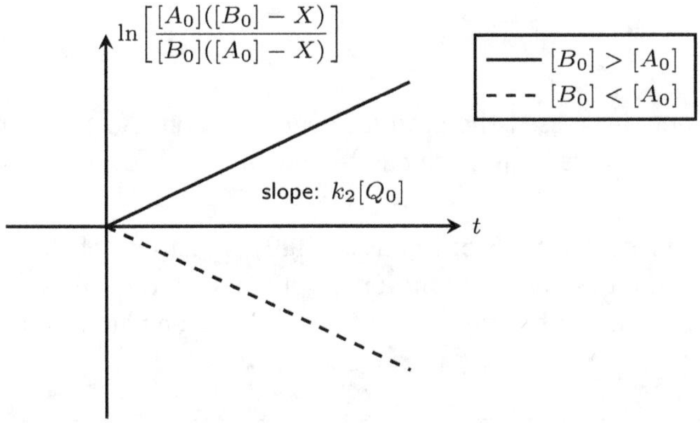

Figure 9.3: Second-order process

In an experiment, X and t would be measured, and then the above function plotted against t to test whether or not it was a straight line. If it was a straight line, the data points could be fitted to determine the reaction constant, k_2.

However, this equation does not yet give us X as a function of time t and we may want to make X the subject of the equation instead of t. We can rearrange it algebraically. We start with

$$k_2 t = \frac{1}{[Q_0]} \ln\left[\frac{[A_0]([B_0]-X)}{[B_0]([A_0]-X)}\right]$$

We multiply both sides by $[Q_0]$, take the exponential of both sides, and then collect all the terms in X together:

$$\begin{aligned} e^{k_2[Q_0]t} &= \frac{[A_0]([B_0]-X)}{[B_0]([A_0]-X)} \iff \frac{[B_0]}{[A_0]}e^{k_2[Q_0]t} = \frac{[B_0]-X}{[A_0]-X} \\ &\iff \frac{[B_0]}{[A_0]}e^{k_2[Q_0]t}([A_0]-X) = [B_0]-X \end{aligned}$$

$$\iff [B_0]e^{k_2[Q_0]t} - \frac{[B_0]}{[A_0]} X\, e^{k_2[Q_0]t} - [B_0] = -X$$

$$\iff [B_0](e^{k_2[Q_0]t} - 1) = X\left(\frac{[B_0]}{[A_0]}e^{k_2[Q_0]t} - 1\right)$$

$$X = \frac{[B_0](e^{k_2[Q_0]t} - 1)}{\left(\frac{[B_0]}{[A_0]}e^{k_2[Q_0]t} - 1\right)} \tag{9.12}$$

We have now expressed the product concentration, X, as a function of time, and we can use this to calculate how much of X there is at time t.

Important note: Equation (9.9) is just one of the possibilities for a second-order process: if the initial concentrations are the same *i.e.* $[A_0] = [B_0]$, then equation (9.9) becomes $\frac{dX}{dt} = k_2\left([A_0] - X\right)^2$.

Example in Biology: In ecology, HO_2 is one of the highly reactive species which forms in the atmosphere and which is removed through the following reaction:

$$2\,HO_2(g) \longrightarrow H_2O_2(g) + O_2(g)$$

The disappearance of HO_2 is known to follow a second-order process over time.

Form a differential equation describing the disappearance of HO_2 with time, and solve it to obtain an equation giving the concentration of HO_2 with respect to time t.

Next find an equation such that plotting the dependent variable *versus* time gives a straight line, and finally estimate the time taken for half of the HO_2 to disappear.

Solution: For this second-order process, the rate equation to describe this reaction is $-\frac{d[HO_2]}{dt} = k_2 \cdot [HO_2]^2$

$\iff -\frac{d[A]}{dt} = k_2 \cdot [A]^2$, where $A = HO_2$.

If the initial concentration of A is $[A_0]$, we can integrate the differential equation by using the separation of variables method:

$$-\int \frac{d[A]}{dt}\frac{1}{[A]^2}\,dt = \int k_2\,dt \Longleftrightarrow -\int \frac{d[A]}{[A]^2} = k_2\int dt$$

We can then carry out the integration:

$$-\int \frac{d[A]}{[A]^2} = k_2\int dt \Longleftrightarrow \frac{1}{[A]} = k_2 t + c$$

Since at $t = 0$, $[A] = [A_0]$, we have $c = \dfrac{1}{[A_0]}$.

By comparing $\dfrac{1}{[A]} = k_2 t + \dfrac{1}{[A_0]}$ with the equation of a straight line $y = mx + c$, we can see that a plot of $\dfrac{1}{[A]}$ *versus* t would give a straight line, with slope k_2 and intercept with the y-axis of $\dfrac{1}{[A_0]}$.

We can now rearrange the equation algebraically to make $[A]$ its subject:

$$\frac{1}{[A]} = k_2 t + \frac{1}{[A_0]} \Longleftrightarrow \frac{1}{[A]} = \frac{k_2[A_0]t + 1}{[A_0]} \Longleftrightarrow [A] = \frac{[A_0]}{k_2[A_0]t + 1}$$

In order to calculate the time taken for $[A_0]$ to fall to $\dfrac{[A_0]}{2}$, we put:

$$-\int_{[A_0]}^{\frac{[A_0]}{2}} \frac{d[A]}{[A]^2}\,dt = \int_0^{t_{\frac{1}{2}}} k_2\,dt \Longleftrightarrow \left[\frac{1}{[A]}\right]_{[A_0]}^{\frac{[A_0]}{2}} = k_2 t_{\frac{1}{2}}$$

$$\Longleftrightarrow \left[\frac{2}{[A_0]} - \frac{1}{[A_0]}\right] = k_2 t_{\frac{1}{2}} \Longleftrightarrow t_{\frac{1}{2}} = \frac{1}{k_2[A_0]}$$

Important note: This time is **dependent** on the **initial concentration** of the substrate $[A_0]$ and the kinetic constant k_2. Such behaviour is **different** from that of both zeroth- and first-order processes.

Table 9.1 summarises the key properties of zeroth-, first-, and second-order processes, where

- $[A]$ is the concentration of the **product appearing** or **reagent disappearing**;
- $[A_0]$ is its corresponding initial concentration;
- k_n is the reaction constant;
- $t_{\frac{1}{2}}$ is the time taken for half of reagent A to disappear *i.e.* for $\frac{[A_0]}{2}$ to have disappeared.

Table 9.1: Properties of common biochemical processes

Parameter	Zeroth-Order	First-Order	Second-Order
Differential rate law	$\pm\frac{d[A]}{dt} = k_0$	$\pm\frac{d[A]}{dt} = k_1 \cdot [A]$	$\pm\frac{d[A]}{dt} = k_2 \cdot [A]^2$
Integrated rate law	$[A] = [A_0] \pm k_0 t$	$[A] = [A_0]e^{\pm k_1 t}$	$[A] = \frac{[A_0]}{k_2[A_0]t \mp 1}$
Equation for straight line	$[A] = [A_0] \pm k_0 t$	$\ln[A] = \ln[A_0] \pm k_1 t$	$\frac{1}{[A]} = \frac{1}{[A_0]} \mp k_2 t$
Half-life	$t_{\frac{1}{2}} = \frac{[A_0]}{2k_0}$	$t_{\frac{1}{2}} = \frac{\ln(2)}{k_1}$	$t_{\frac{1}{2}} = \frac{1}{k_2[A_0]}$
Units of rate constant k	$\mathrm{M\,s^{-1}}$	$\mathrm{s^{-1}}$	$\mathrm{M^{-1}\,s^{-1}}$

Note: when the symbols $\pm$ or $\mp$ are used, the **upper symbol** always relates to the process of *product appearing*, while the **lower symbol** always relates to the process of *reagent disappearing*.

9.4 Advanced Differential Equations

Other physical and natural processes can be described by various types of first-order differential equations that cannot be solved through using the separation of variables method.

Amongst the wide range of possible cases, in this section we focus only on linear non-homogeneous differential equations.

9.4.1 Linear non-homogeneous differential equations

A linear non-homogeneous differential equation is of the form

$$\frac{dy}{dx} = f(x) \cdot y + g(x) \tag{9.13}$$

where $f(x)$ and $g(x)$ are functions of the independent variable x and the exponent of y must be equal to 1.

Both functions $f(x)$ and $g(x)$ can also just be real constants.

For linear non-homogeneous differential equations, it is unlikely that the separation of variables method will be appropriate.

However, a common way used to solve them is the **integrating factor method**, where an **integrating factor** is introduced into the equation to transform the non-separable original equation into a separable one.

A few steps demonstrate how to obtain a general solution y of such differential equations.

If we subtract $f(x) \cdot y$ from both sides of equation (9.13), we obtain

$$\frac{dy}{dx} - f(x) \cdot y = g(x)$$

If we then multiply both sides of the new equation by $e^{-F(x)}$ (which is called the **integrating factor**), where $F(x) = \int f(x)\,dx$, *i.e.* $F(x)$ is the antiderivative of $f(x)$, we get

$$\frac{dy}{dx} \cdot e^{-F(x)} - f(x) \cdot y \cdot e^{-F(x)} = g(x) \cdot e^{-F(x)}$$

The left side of this equation is just the derivative of $y \cdot e^{-F(x)}$ according to the Product Rule (see Section 5.3.2):

$$\frac{d\left[y \cdot e^{-F(x)}\right]}{dx} = g(x) \cdot e^{-F(x)}$$

We can integrate both sides of the equation over dx by using the separation of variables method:

$$\int \frac{d\left[y \cdot e^{-F(x)}\right]}{dx}\,dx = \int \left(g(x) \cdot e^{-F(x)}\right) dx \Longleftrightarrow y \cdot e^{-F(x)} = H(x) + c$$

where $H(x)$ – is the antiderivative of $\left(g(x) \cdot e^{-F(x)}\right)$ with respect to x.

Finally, the general solution of a linear non-homogeneous differential equation is of the form

$$y = e^{F(x)} \cdot [H(x) + c] \tag{9.14}$$

where,

- $F(x) = \int f(x)\,dx$,
- $H(x) = \int \left(g(x) \cdot e^{-F(x)}\right)\,dx$,
- c is the **integration constant**, which includes all the combined integration constants generated through the intermediate integrals.

If $f(x) = A$ and $g(x) = B$ where A and B are constants (real numbers), the general solution of the corresponding linear non-homogeneous differential equations becomes simpler:

$$y = e^{Ax} \cdot \left[\int \left(B \cdot e^{-Ax}\right)\,dx + c\right] \tag{9.15}$$

In general, given a linear non-homogeneous differential equation $\frac{dy}{dx} = f(x) \cdot y + g(x)$, the **integrating factor method** can be summarised in the following four steps using the acronym CIWI:

(a) **C**-calculate $F(x) = \int f(x)\,dx$; *
(b) **I**-integrate $H(x) = \int \left(g(x) \cdot e^{-F(x)}\right)\,dx$; *
(c) **W**-write and simplify the general solution as $y = e^{F(x)} \cdot [H(x) + c]$;
(d) **I**-initial conditions (if they exist) should be used to find the particular solution required by them (*i.e.* they allow the constant c to be determined).

* In steps (a) and (b) the constants of integration κ can be omitted, since they will be included in the integration constant c in step (c).

Example: Solve $y' = 2xy - 2x$ and find the unique solution $y = f(x)$ which satisfies the boundary condition $f(0) = 0$.

Solution: Since $y' = 2xy - 2x$ is a linear non-homogeneous differential equation $\frac{dy}{dx} = f(x) \cdot y + g(x)$ with $f(x) = 2x$ and $g(x) = -2x$, the **integrating factor method** can be used to solve it:

(a) **Calculate:** $F(x) = \int f(x)\,dx = \int 2x\,dx = x^2$;
(b) **Integrate:** $H(x) = \int \left(g(x) \cdot e^{-F(x)}\right)\,dx = \int \left(-2x \cdot e^{-x^2}\right)\,dx = e^{-x^2}$ by using the substitution method (see Section 8.2.1);

(c) **S**implify the general solution:
$y = e^{F(x)} \cdot [H(x) + c] = e^{x^2} \cdot \left[e^{-x^2} + c\right] = e^{x^2 - x^2} + c \cdot e^{x^2} = 1 + c \cdot e^{x^2}$
(check solution: $y' = c \cdot 2xe^{x^2}$, so putting our solution for y back into $y' = -2xy - 2x$ gives us $y' = 2x(1 + ce^{x^2}) - 2x = c \cdot 2xe^{x^2}$, so this agrees!);

(d) **I**nitial conditions: $0 = 1 + c \cdot e^{0^2} \Longleftrightarrow 0 = 1 + c \Longleftrightarrow c = -1$.

Thus, the solution of the linear non-homogeneous differential equation is $y = 1 - e^{x^2}$.

Example in Medicine: If during physical exercises under pressure an athlete's body consumes more oxygen than the oxygen breathed in, they will accumulate an increasing oxygen debt (D) and their breathing will become more wheezy.

However, the rate at which oxygen is breathed into the lungs is proportional to the accumulated oxygen debt (with the proportionality constant $k > 0$) and the amount of oxygen required for any exercise under pressure is constant per exercise (β).

Based on these observations, form a differential equation modelling the oxygen debt $D(t)$ of the athlete over time and find the unique solution satisfying the initial condition $D(0) = 0$.

Then, sketch the function $D(t)$ obtained.

Solution: Based on the information given above, the differential equation modelling the oxygen debt $D(t)$ by the athlete is

$$\frac{dD}{dt} = -k \cdot D + \beta$$

which is a linear non-homogeneous differential equation (equation (9.13)), with $f(t) = -k$ and $g(t) = \beta$.

Therefore, we can use the **integrating factor method** to solve it:

(a) **C**alculate: $F(t) = \int f(t)\,dx = \int -k\,dt = -kt$;

(b) **I**ntegrate: $H(t) = \int \left(g(t) \cdot e^{-F(t)}\right) dt = \int \left(\beta \cdot e^{kt}\right) dt = \frac{\beta}{k} e^{kt}$ by using the substitution method (see Section 8.2.1);

(c) **S**implify the general solution: $D(t) = e^{F(t)} \cdot [H(t) + c]$
$= e^{-kt} \cdot \left[\frac{\beta}{k} e^{kt} + c\right] = \frac{\beta}{k} e^{-kt+kt} + c \cdot e^{-kt} = \frac{\beta}{k} + c \cdot e^{-kt}$;

(d) **I**nitial conditions: $0 = \frac{\beta}{k} + c \cdot e^{-k \cdot 0} \Longleftrightarrow 0 = \frac{\beta}{k} + c \Longleftrightarrow c = -\frac{\beta}{k}$.

Thus, the solution of the linear non-homogeneous differential equation is

$$D(t) = \frac{\beta}{k} - \frac{\beta}{k} e^{-kt} = \frac{\beta}{k}\left(1 - e^{-kt}\right)$$

We can now sketch the function $D(t)$ in a Cartesian plane in order to visualise the trend of the oxygen debt, D, over time.

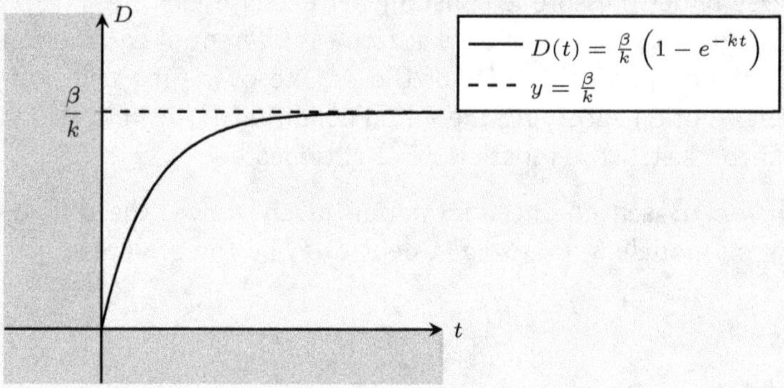

9.5 Exercises

(1) Solve the following elementary differential equations.

(a) $\frac{dy}{dx} = x - \cos(x)$ (b) $\frac{dy}{dx} = -4x + \frac{3}{2x}$ (c) $\frac{dy}{dx} = \sqrt{5 - 4x}$

(d) $\frac{dy}{dx} = x \cdot e^{-x^2}$ (e) $\frac{dy}{dx} = \frac{1}{x^2 - 3x}$ (f) $\frac{dy}{dx} = x \cdot \ln(x)$

(2) The differential equation $-\dfrac{dW}{dt} = \dfrac{\alpha}{22t+1}$ is used to model the rate of change of weight according to a specific training programme, where W is the person's weight, t is the time in months, and $\alpha = 44\,\text{kg}$.

(a) Find the equation modelling weight loss over time, if at the beginning of the program the person's weight is $100\,\text{kg}$.

(b) Plot the resulting function $W(t)$.

(c) What is the current weight of this person if they started this training programme one year ago?

(3) Normally the climate has a dynamic structure and varies continuously. Basic climatological features, such as precipitation in a region, depend on many meteorological variables. The changes in precipitation in a Mediterranean region have revealed a sinusoidal form of yearly precipitation rate: $\dfrac{dP}{dt} = 60 + 50\cos\left(\dfrac{2\pi t}{12}\right)$, where precipitation P is measured in mm and time t in months.

(a) When are the maximum and minimum precipitation rates observed?

(b) Solve the differential equation.

(c) Estimate the amount of precipitation from April to September, and then also in the rest of the year.

(4) Solve the following linear homogeneous differential equations.

(a) $\dfrac{dy}{dx} = y\sin x$ (b) $\dfrac{dy}{dx}\ln x = \dfrac{y}{x}$ (c) $\dfrac{dy}{dx}\cdot\dfrac{1}{\cos(2x)} = 3y$

(d) $2x\dfrac{dy}{dx} = \dfrac{y}{2}$ (e) $\dfrac{dy}{dx}\cdot e^{-x} = y^2$ (f) $-\dfrac{dy}{dx} + 4\sqrt[3]{x^2}y = 0$

(5) The initial number of nuclei in a radioactive sample is N_0 and they decay at a rate proportional to the number present at time t.

(a) Form and solve a differential equation describing this process.

(b) Plot the resulting function, $N(t)$.

(c) If after three hours the sample has decayed to half its original activity, how long would it take for $N = \dfrac{N_0}{5}$?

(6) The rate of growth of bacteria in a Petri dish is dependent on the number of bacteria already present.

(a) Form and solve a differential equation describing this process.

(b) If initially there were 1000 bacteria and two hours later there were 3000, how long will it take for the number to reach 1 million?

(7) For a chemical reaction $A \longrightarrow P$, the mass of A at any given time t is m_t and the rate of the reaction is proportional to m_t.

(a) If the original mass of A is $10\,\mathrm{mg}$ and after one minute $25\,\mu\mathrm{g}$ were transformed to substance P, form and solve a differential equation describing this reaction.

(b) Find how much of A has been transformed after 10 minutes.

(8) The amount of ^{14}C (radioactive carbon-14) in a sample is measured using a Geiger counter, which records the disintegration of each atom. The rate at which ^{14}C decays is proportional through a constant k to the amount present. The half-life of ^{14}C is about 5730 years, so the half of the ^{14}C present will have disintegrated after 5730 years. In living tissue, ^{14}C disintegrates at a rate of about 13.5 atoms per minute per gram of all the carbon present. Because living tissue is constantly exchanging carbon with its environment, the proportion of ^{14}C compared with the number of ^{12}C atoms remains constant over time. Once the tissue is no longer living, this constant exchange of carbon ceases and the ratio of ^{14}C to ^{12}C begins to decrease. Consequently, the disintegration rate drops.

Note that measurement of the ^{14}C to ^{12}C ratio can thus provide a way to determine the date of death of a living organism: this is the method of carbon 14 dating.

In 1977, a charcoal fragment found at Stonehenge on the Salisbury Plain was recorded to give 8.2 disintegrations per minute per gram of carbon – about 60% of that for living tissue.

(a) Let $y(t)$ denote the amount of ^{14}C in the fragment at time t. Let the units for y be measured in number of disintegrations of ^{14}C atoms per minute per gram and those for t be measured in half-lives of ^{14}C. Assuming that the charcoal was formed during the building of the site, form the differential equation modelling the ^{14}C decay.

(b) Find the equation $y = f(t)$ modelling the ^{14}C decay.

(c) Let N_0 denote the number of ^{14}C atoms present in the fragment when it was produced, which means that $y(0) = N_0$. After $t = 1$ half-life (*i.e.* 5730 years), we know that $y(1)$ will be half the original amount, so $y(1) = \frac{1}{2}N_0$. Use these facts to find the particular solution that models this situation.

(d) Now use the information to estimate the date at which Stonehenge was built in BCE (Before Common Era) years, which is a secular version of BC (Before Christ) but has the same definition as BC.

(9) Solve the following non-linear differential equations.

(a) $\dfrac{dy}{dx} = \dfrac{1+x^2}{y}$ (b) $\dfrac{dy}{dx} = -9x^2y^2$ (c) $\dfrac{dy}{dx} = \dfrac{x+2}{xy}$

(d) $\dfrac{dy}{dx} = \sin(2x) \cdot \dfrac{1}{y}$ (e) $x\dfrac{dy}{dx} = 2e^{-y}$ (f) $\dfrac{dy}{dx} = e^{3x} \cdot \dfrac{1}{y}$

(g) $y^2 \dfrac{dy}{dx} = \sin(x^5) \cdot x^4$ (h) $\dfrac{dy}{dx} = 3e^y x^2$ (i) $(1-x^2)\dfrac{dy}{dx} = xe^y$

(j) $y^4 \dfrac{dy}{dx} = e^x x^2$ (k) $\dfrac{dy}{dx} = \dfrac{y^3 x}{\sqrt{1+4x^2}}$ (l) $\dfrac{dy}{dx} = \dfrac{\sin x}{y \cos y}$

(m) $\dfrac{1}{x}\dfrac{dy}{dx} = \dfrac{\sin 5x}{y}$ (n) $\dfrac{dy}{dx} = \cos(5x)\, e^y$ (o) $\dfrac{dy}{dx} = xe^{-3x^2} y^{\frac{1}{3}}$

(10) Newton's Law of Cooling states that the rate of change in temperature of an object is proportional to the extent to which its temperature differs from that of the surroundings.

(a) If T_0 is the initial temperature of the object, T is its temperature after time t, and T_s is the temperature of the surroundings, form and solve a differential equation describing this process.

(b) A biological sample at $25\,^\circ\mathrm{C}$ was put on ice ($0\,^\circ\mathrm{C}$), and it cooled down to $20\,^\circ\mathrm{C}$ in four minutes. Assuming that there was an excess of ice and so the temperature of the ice did not change, calculate how long it took for the biological sample to cool to $10\,^\circ\mathrm{C}$.

(11) The increase in radius, r, of a sore on an infected patient is inversely proportional through a constant A to the radius of the sore at time t.

(a) Form and solve the differential equation describing the radius of the sore with time.

(b) Plot the function $r(t)$ obtained.

(c) If the sore has a radius of $0.5\,\mathrm{cm}$ at $t = 0$ days and a radius of $3\,\mathrm{cm}$ after 2 days, determine the constants A and the integration constant κ for the function $r(t)$.

(d) How big would the sore be after four days if left untreated?

(12) In the second-order rate equation $\dfrac{dX}{dt} = k_2\,[A]\,[B]$ (see Section 9.3.3), consider the case where the initial concentrations are $[A_0] = [B_0]$, and after time t there are X moles of product formed.

(a) Form the differential equation describing this process.

(b) Solve the differential equation, by making X the subject of the final equation.

(13) The process of cell division depends not only on the number of cells but also on some additional factors, such as the quantity of nutrients present, the oxygen availability, and having an adequate cell pH or concentrations of enhancing-growth factors. Sometimes cell multiplication is restricted if some of these factors are at a low level or are missing, thereby causing crowding effects during the process.

The differential equation $\dfrac{dN}{dt} = \alpha N - \beta N^2$ (also known as a *logistic equation*) is often used to model the process of cell division, with αN accounting for the cell number increase due to cell division and $-\beta N^2$ for the cell-division inhibitory effects caused by crowding.

(a) Find the corresponding equation for $N(t)$, commonly known as a *logistic law of growth*, which models the cell division process. N_0 is the initial number of cells and $\beta N_0 < \alpha$, which is consistent with the fact that multiplication factors can overcome inhibitory factors.

(b) Differentiate $N(t)$ with respect to t and show that $N(t)$ always increases if $t > 0$.

(c) To what maximum value, N_{max}, could the function $N(t)$ tend as $t \to +\infty$?

(d) When is there half of N_{max} left (*i.e.* what is $t_{\frac{1}{2}}$)?

(e) Show also that at $t_{\frac{1}{2}}$ the function $N(t)$ has a critical point. What kind of critical point is it? What is its physical meaning?

(f) Plot the function $N(t)$, and highlight its key features to show its sigmoidal trend.

(14) One model for the spread of the Covid-19 epidemic assumed that the number of people infected changes over time at a rate proportional *via* a constant $\alpha > 0$ to the product of the number of people already infected and the number of people yet to be infected.

(a) If $I(t)$ denotes the function giving the number of people infected at time t, and P the total population, form the differential equation modelling this assumed form of the spread of the pandemic.

(b) Show that $I(t) = \dfrac{P \cdot I_0}{I_0 + (P - I_0)e^{-\alpha P t}}$, assuming that the initial number of people infected was I_0.

(c) What happens if the pandemic becomes endemic worldwide, i.e. what happens at $\lim_{t \to +\infty} I(t)$?

(15) In a model of bacterial population growth, the growth rate is proportional to the size of the population B through a constant α. However, some environmental factors also impose limitations on population growth and, in particular, the growth rate decreases by $\dfrac{\alpha B^2}{C}$, where C is the maximum carrying capacity of the environment.

(a) Form and solve the differential equation modelling the bacterial population growth.

(b) Assuming that at time $t = 0$ the number of bacteria is B_0, show that the solution can be expressed as $B(t) = \dfrac{C}{1 + ke^{-\alpha t}}$, where $k = \dfrac{C - B_0}{B_0}$, with $C >> B_0$.

(c) What happens to $B(t)$ if $t \to +\infty$?

(d) Sketch the function $B(t)$ in a Cartesian plane.

(16) Measuring the optical density (OD) at $600\,\mathrm{nm}$, usually denoted OD600, is the easiest way to gauge the growth stage of a bacterial culture. OD measures the amount of light scattered by the bacteria within a culture; the more bacteria there are, the more light is scattered.
The OD600 of $1\,\mathrm{mL}$ of a sample of bacterial culture after $1\,\mathrm{h}$ of incubation at $30\,^\circ\mathrm{C}$ is 0.1, while after $3\,\mathrm{h}$ its OD600 reaches 0.5.

(a) Assuming exponential bacterial growth according to the equation $B(t) = B_0 \cdot e^{\alpha t}$, predict the OD600 value (B_0) at time $t = 0\,\mathrm{h}$ and the growth constant α by using the OD600 values at times $1\,\mathrm{h}$ and $3\,\mathrm{h}$. (* Hint: the simultaneous equations obtained can be solved using the substitution method, see Section 2.4.2.)
(b) According to the exponential model, what is the OD600 value after $6\,\mathrm{h}$ of incubation?
(c) In fact, the value of OD600 observed after $6\,\mathrm{h}$ is 2.7. Using the equation obtained from a logistic model: $B(t) = \dfrac{C}{1 + ke^{-\alpha t}}$, where $k = \dfrac{C - B_0}{B_0}$ (see Exercise 15), and the B_0 and α values are obtained in part (a), calculate C, which is the OD600 value of the maximum carrying capacity of the environment.

(17) The Gompertz equation $\dfrac{dP}{dt} = \alpha \cdot e^{-\beta t} P$, where α and β are positive constants, is used as a self-limiting growth model for animals and tumours.

(a) Solve the differential equation.
(b) Show that the unique solution for the boundary condition $P(0) = \lambda$ can be expressed as $P(t) = \lambda e^{\left[\frac{\alpha}{\beta}\left(1-e^{-\beta t}\right)\right]}$.
(c) What happens to $P(t)$ when $t \to +\infty$?

(18) Solve the following linear non-homogeneous differential equations.

(a) $\dfrac{dy}{dx} = 2y - 4$ (b) $\dfrac{dy}{dx} + 5y - 2 = 0$ (c) $\dfrac{dy}{dx} - 7y = 4x$
(d) $\dfrac{dy}{dx} = xy - 2x$ (e) $\dfrac{dy}{dx} - 2y = e^{3x}$ (f) $\dfrac{dy}{dx} = \dfrac{y}{x} + \dfrac{\ln(x)}{x}$

(19) Let us consider the widely held theory of learning: 'The more an employee knows of a particular task the slower they learn the rest of the task'. One of the reasons for this might be the growing complexity involved in finally achieving complete mastery. The theory can be also reformulated this way: 'As y increases, its derivative $\dfrac{dy}{dt}$ decreases'.

(a) By assuming that the rate at which a person learns is equal to the percentage of the task not yet learned, express the above theory mathematically by letting the variable y represent the portion of the task already mastered by week t.
(b) Find the unique solution for y by using the boundary condition $y(0) = 0$.

(20) Many analyses in Pharmacology lead to equations arising in so-called turnover models. The fundamental equation of these models is $\frac{dR}{dt} = K_{in} - K_{out}R$, where R denotes the response of the system to the drug and K_{in} and K_{out} are positive rate constants.

(a) Solve the differential equation with the boundary condition $R(0) = 0$.

(b) Plot the resulting function, $R(t)$.

(21) The glucose in the bloodstream is usually converted into other substances useful to cells and is removed from the bloodstream at a constant rate.

(a) If k denotes the positive constant of proportionality for glucose removal, $G(t)$ the function quantifying the amount of glucose in the bloodstream over time, and G_0 the amount of glucose in the bloodstream at time $t = 0$, form the differential equation modelling glucose consumption and solve it to find the equation for $G(t)$.

(b) If a glucose solution is administrated intravenously into the bloodstream at a constant rate R per unit of time, reformulate the differential equation describing the glucose variation in the body and find the corresponding equation for $G(t)$, with $G(0) = G_0$.

(22) Solve the following differential equations using the specified boundary conditions.

(a) $\frac{dy}{dx} = -7x$ if $y = 3$ at $x = 0$

(b) $\frac{dy}{dx} = \frac{x}{2}y$ if $y = 2$ at $x = 1$

(c) $\frac{dy}{dx} = -e^{-x}y$ if $y = 1$ at $x = 0$

(d) $\frac{dy}{dx} = \frac{y \ln y}{x}$ if $y = 3$ at $x = 2$

(e) $\frac{dy}{dx} = (1 + x)(1 - y)$ if $y = 0$ at $x = 2$

(f) $\frac{dy}{dx} + 2y = 6$ if $y = 1$ at $x = 0$

(23) Solve the following differential equations from Pharmacology by using the appropriate method and the specified boundary conditions. The function $C(t)$ obtained will give the drug concentrations in plasma at a given time t.

(a) $\frac{dC}{dt} = (A - Bt)\,e^{-\lambda t}$, if $C(0) = 0$, and A, B and λ are positive constants.

(b) $\frac{dC}{dt} = -\frac{A}{V}C$, if $C(0) = C_0$, A and V are positive constants.

(c) $\frac{dC}{dt} = \frac{A}{C(t + B)^2}$, if $C(0) = C_0$, A and B are positive constants.

(d) $\dfrac{dC}{dt} = \dfrac{1}{V}(R - A \cdot C)$ if $C(0) = 0$, A, V, and R are positive constants.

(e) $\dfrac{dC}{dt} = -k \cdot C + \dfrac{V}{A}$ if $C(0) = C_0$, A, V and k are positive constants.

(24) The equation for the variation of the equilibrium constant, K, for a reaction with temperature, T, is given by $\ln(K) = -\dfrac{\Delta H^\circ}{RT} + \dfrac{\Delta S^\circ}{R}$, where H is the enthalpy, S is the entropy, and R is the universal gas constant.

(a) Assuming that ΔH° and ΔS° are independent of temperature, differentiate the equation with respect to T to obtain the Van't Hoff isochore equation, *i.e.* an expression for $\dfrac{d(\ln(K))}{dT}$.

(b) For an initial temperature of T_1 and a final temperature of T_2 corresponding to equilibrium constants of K_1 and K_2, respectively, reintegrate the Isochore equation to obtain an expression for K_2.

(c) When analysing the results of an experiment to measure K as a function of temperature, what would you choose as your x and y axes so that you obtain a graph of the form $y = mx + c$? What would be the slope and the y-intercept?

Answers

(1) (a) $y = \frac{x^2}{2} - \sin(x) + \kappa$ (b) $y = -2x^2 + \frac{3}{2}\ln|x| + \kappa$

(c) $y = -\frac{1}{6}(5 - 4x)^{\frac{3}{2}} + \kappa$ (d) $y = -\frac{1}{2}e^{-x^2} + \kappa$

(e) $y = -\frac{1}{3}\ln(x) + \frac{1}{3}\ln(x-3) + \kappa$ (f) $y = \frac{1}{2}x^2\ln(x) - \frac{x^2}{4} + \kappa$

(2) (a) $W = 100 - 2\ln(22t + 1)$

(b)

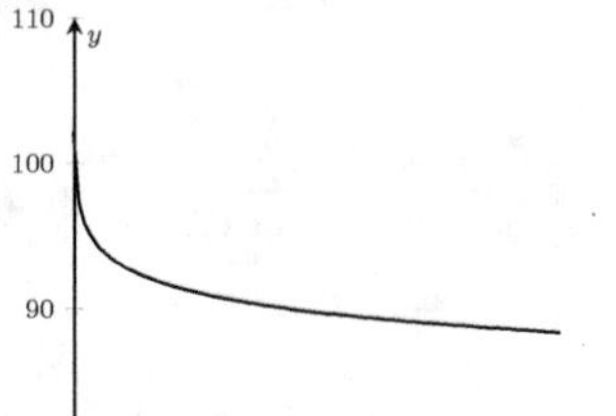

(c) 89 kg

(3) (a) $\frac{d^2P}{dt^2} = -\frac{25\pi}{3}\sin\left(\frac{\pi t}{6}\right) = 0$ gives max at the end of December when $t = 12$, $P = 110\,\mathrm{mm}$, min at the end of June when $t = 6$, $P = 10\,\mathrm{mm}$
(b) $P(t) = 60t + \frac{300}{\pi}\sin\left(\frac{\pi t}{6}\right) + \kappa$
(c) In the interval $[3-9]$ $P = 360 + \frac{600}{\pi} = 551\,\mathrm{mm}$, in the interval $[0-3]$ and $[9-12]$, $P = 360 - \frac{600}{\pi} = 169\,\mathrm{mm}$

(4) (a) $y = e^{-\cos(x)+\kappa}$ (b) $y = \kappa\ln(x)$ (c) $y = e^{\frac{3}{2}\sin(2x)+\kappa}$
(d) $y = \kappa\sqrt[4]{x}$ (e) $y = -\frac{1}{e^x+\kappa}$ (f) $y = e^{\frac{12}{5}x^{\frac{5}{3}}+\kappa}$

(5) (a) $\frac{dN}{dt} = -\lambda N$, $N = N_0 e^{-\lambda t}$
(b)

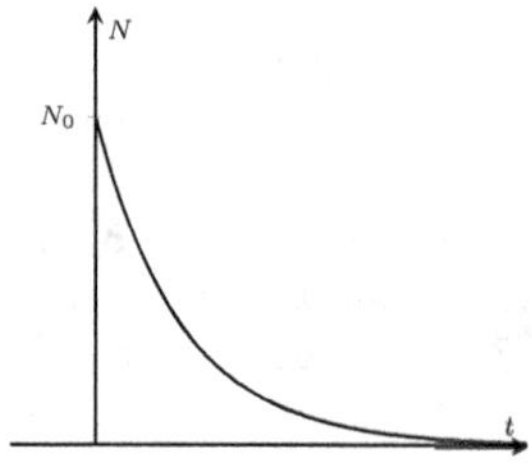

(c) $\lambda = 0.23\,\mathrm{h}^{-1}$, $t = 7\,\mathrm{h}$

(6) (a) $\frac{dN}{dt} = kN$, $N = N_0 e^{kt}$ (b) $k = 0.55\,\mathrm{h}^{-1}$, $t = 12.56\,\mathrm{h}$

(7) (a) $\frac{dm}{dt} = -k\cdot m_t$, $m_t = m_0 e^{-kt}$
(b) $k = 2.3\,\mathrm{min}^{-1}$, $m_t = 1.03\cdot 10^{-12}$ g, *i.e.* most of the substrate is transformed

(8) (a) $\frac{dy}{dt} = -ky$ (b) $y = e^{-kt+c}$ (c) $y = N_0\left(\frac{1}{2}\right)^t$ (d) 2144 BCE

(9) (a) $y = \pm\sqrt{2x + \frac{2x^3}{3} + \kappa}$ (b) $y = -\frac{1}{-3x^3+\kappa}$
(c) $y = \pm\sqrt{2x + 4\ln(x) + \kappa}$ (d) $y = \pm\sqrt{-\cos(2x) + \kappa}$
(e) $y = \ln(2\ln(x) + \kappa)$ (f) $y = \pm\sqrt{\frac{2e^{3x}+\kappa}{3}}$
(g) $y = \sqrt[3]{\frac{3\cos(x^5)+\kappa}{5}}$ (h) $y = -\ln\left(-x^3 - \kappa\right)$
(i) $y = -\ln\left(\frac{1}{2}\ln\left(1-x^2\right) - \kappa\right)$ (j) $y = \sqrt[5]{5x^2e^x - 10\left(xe^x - e^x\right) + \kappa}$
(k) $y = \pm\frac{\sqrt{2}}{\sqrt{\kappa - \sqrt{4x^2+1}}}$ (l) $y = \pm\sqrt{-2\cos(x) + \kappa}$
(m) $y = \pm\frac{\sqrt{-10x\cos(5x)+2\sin(5x)+\kappa}}{5}$ (n) $y = -\ln\left(-\frac{1}{5}\sin(5x) - \kappa\right)$
(o) $y = \frac{1}{27}\left(\kappa - e^{-3x^2}\right)^{\frac{3}{2}}$

(10) (a) $\frac{dT}{dt} = -k(T-R)$, $y = (T_0 - R)e^{-kx} + R$
(b) $k = 0.056\,\mathrm{min}^{-1}$, $T = 10\,°\mathrm{C}$ in $16\,\mathrm{min}\,21\,\mathrm{s}$

(11) (a) $\frac{dr}{dt} = \frac{A}{r}$, $r = \sqrt{2At + 2\kappa}$

(b)

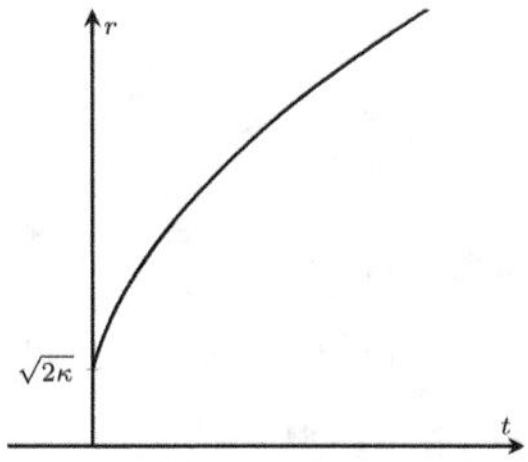

(c) $A = 2.1875\,\text{cm}^2\,\text{days}^{-1}$, $\kappa = 0.125\,\text{cm}^2$

(d) $r = 4.2\,\text{cm}$

(12) (a) $\frac{dX}{dt} = k_2\left([A_0] - X\right)^2$

(b) $k_2 t = \frac{1}{([A_0]-X)} - \frac{1}{[A_0]} \Longleftrightarrow X = \frac{k_2 t [A_0]^2}{(1+k_2 t[A_0])}$

(13) (a) $N(t) = \frac{\alpha N_0}{\beta N_0 + (\alpha - \beta N_0)e^{-\alpha t}}$ (b) $\frac{dN}{dt} = \frac{N_0 e^{-\alpha t}\alpha^2(\alpha - N_0\beta)}{\left(N_0\beta + e^{-\alpha t}(\alpha - N_0\beta)\right)^2} > 0$

(c) $N_{max} = \frac{\alpha}{\beta}$ (d) $t_{\frac{1}{2}} = \frac{1}{\alpha}\ln\left(\frac{\alpha}{\beta N_0} - 1\right)$

(e) At $t_{\frac{1}{2}}$ there is an inflection point, where $\frac{d^2N}{dt^2} = \frac{N_0 e^{-2\alpha t}\alpha^3\left(-N_0\beta - N_0 e^{\alpha t}\beta + \alpha\right)(\alpha - N_0\beta)}{\left(N_0\beta + e^{-\alpha t}(\alpha - N_0\beta)\right)^3} = 0$, before $t_{\frac{1}{2}}$, the cell division occurs faster than after $t_{\frac{1}{2}}$ has elapsed

(f)

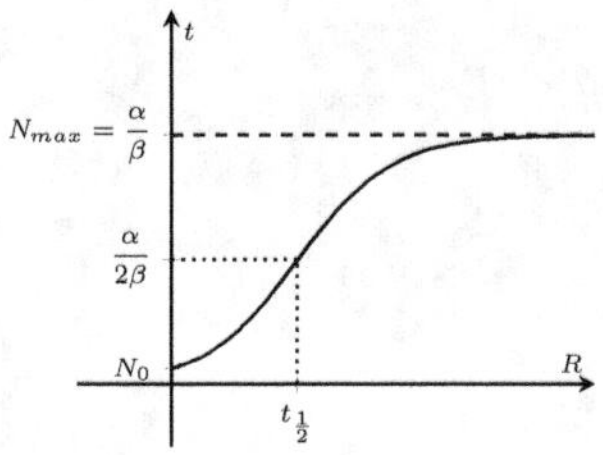

(14) (a) $\frac{dI}{dt} = \alpha I\left(P - I\right)$ (b) $y = \frac{P}{e^{-\alpha P t + \kappa} + 1}$

(c) $\lim_{t\to+\infty} I(t) = P$, *i.e.* the entire population will be infected

(15) (a) $\frac{dB}{dt} = \alpha B\left(1 - \frac{B}{C}\right)$; $B = \frac{C}{1+e^{-\alpha t + \kappa}}$

(b) if $B = B_0$ at $t = 0$, rearrange the resulting equation

(c) $\lim_{t\to+\infty} B(t) = C$

(d)

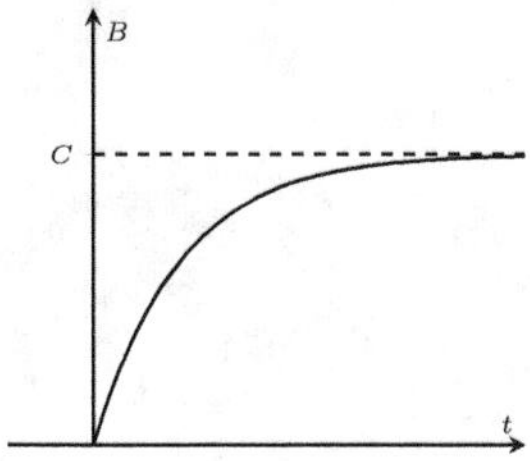

(16) (a) $B(t) = 0.045 \cdot e^{\frac{\ln(5)}{2}t}$ (b) at $t = 6\,\mathrm{h}$ OD600 $= 5.6$
(c) $C = 5.2$ (OD600)

(17) (a) $P = e^{-\frac{\alpha e^{-\beta t}}{a}+\kappa}$ (b) substitute appropriately
(c) $\lim_{t\to+\infty} P(t) = \lambda \cdot e^{\frac{\alpha}{\beta}}$

(18) (a) $y = e^{2x+\kappa} + 2$ (b) $y = -\frac{e^{-5x-\kappa}}{5} + \frac{2}{5}$ (c) $y = -\frac{4(7x+1)}{49} + \kappa e^{7x}$
(d) $y = e^{\frac{x^2}{2}+\kappa} + 2$ (e) $y = e^{3x} + \kappa e^{2x}$ (f) $y = -\ln(x) - 1 + \kappa x$

(19) (a) $\frac{dy}{dt} = 100 - y$, $y = 100 + \kappa \cdot e^{-t}$ (b) $y = 100 + 100 \cdot e^{-t}$

(20) (a) $R(t) = -\frac{e^{(-K_{out}t-\kappa \cdot K_{out})}}{K_{out}} + \frac{K_{in}}{K_{out}}$; $R(t) = -\frac{K_{in} \cdot e^{(-K_{out}t)}}{K_{out}} + \frac{K_{in}}{K_{out}}$
(b)

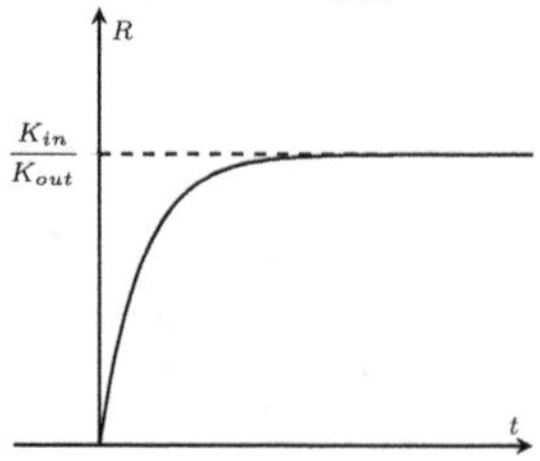

(21) (a) $\frac{dG}{dt} = -kG$; $G(t) = G_0 e^{-kt}$ (b) $\frac{dG}{dt} = R - kG$; $G(t) = \frac{R}{k} + \left(G_0 - \frac{R}{k}\right) e^{-kt}$

(22) (a) $y = -\frac{7x^2}{2} + 3$ (b) $y = 2e^{\frac{x^2-1}{4}}$ (c) $y = e^{e^{-x}-1}$
(d) $y = e^{\frac{\ln(3)x}{2}}$ (e) $y = -e^{-x-\frac{x^2}{2}+4} + 1$ (f) $y = -2e^{-2x} + 3$

(23) (a) $C(t) = -\frac{Ae^{-\lambda t}}{\lambda} - \frac{B\left(-\lambda e^{-\lambda t}t - e^{-\lambda t}\right)}{\lambda^2} - \frac{B}{\lambda^2} + \frac{A}{\lambda}$
(b) $C(t) = C_0 e^{-\frac{At}{v}}$
(c) $C(t) = \sqrt{\frac{2At}{B(t+B)}}$
(d) $C(t) = -\frac{Re^{-\frac{At}{V}}}{A} + \frac{R}{A}$
(e) $C(t) = C_0 e^{-kt} - \frac{Ve^{-kt}}{kA} + \frac{V}{kA}$

(24) (a) $\frac{d(\ln(K))}{dT} = \frac{\Delta H^\circ}{RT^2}$
(b) $\int_{K_1}^{K_2} d(\ln(K)) = \int_{T_1}^{T_2} \left(\frac{\Delta H^\circ}{R} \cdot \frac{1}{T^2}\right) dT \Rightarrow K_2 = K_1 \cdot e^{-\frac{\Delta H^\circ}{R}\left(\frac{1}{T_2} - \frac{1}{T_1}\right)}$
(c) plot $\ln(K)$ (y-axis) *versus* $\frac{1}{T}$ (x-axis): $\ln(K) = -\frac{\Delta H^\circ}{RT} + \frac{\Delta S^\circ}{R}$

Index

E

F

G

S

T

U

V

W

X

Z

www.ingramcontent.com/pod-product-compliance
Lightning Source LLC
LaVergne TN
LVHW010329220926
848484LV00004BB/861
* 9 7 8 1 8 0 0 6 1 6 0 3 5 *